Casimir S. Arpentigny

The Science of the Hand

The art of recognising the tendencies of the human mind by the observation of the

formations of the hands - La science de la main

Casimir S. Arpentigny

The Science of the Hand
*The art of recognising the tendencies of the human mind by the observation of the
formations of the hands - La science de la main*

ISBN/EAN: 9783337367497

Printed in Europe, USA, Canada, Australia, Japan

Cover: Foto ©berggeist007 / pixelio.de

More available books at **www.hansebooks.com**

The Science of the Hand

آنانکه محیط فضل و آداب شدند

از جمع کمالِ شمعِ اصحاب شدند

ره زین شبِ تاریک نبردند برون

گفتند فسانه‌ای و در خواب شدند

Dedication.

Te . . . canam. Quamquam me cognita virtus
Terret ut infirmae valeant subsistere vires,
Incipiam tamen. At meritas si carmina laudes
Deficiant, tantis humilis si conditor actis,
Nec tua, te praeter, chartis interere quisquam
Facta queat, dictis ut non majora supersint:
Est nobis voluisse satis. Nec numera parva
Respueris

       *   *   *   *

Parvaque coelestes placabit mica; nec illis
Semper inaurato taurus cadit hostia cornu.
Hic quoque sit gratus parvus labor, ut tibi possim
Inde alios aliosque memor componere versus.

       *   *   *   *

At quodcumque meae poterunt audere Camoenae
Seu tibi par poterunt, seu quod spes abnuit ultro,
Sive minus, certeque canent minus, omne vovemus
Hoc tibi; nec tanto careat mihi carmine charta.

       *   *   *   *

Sum quodcumque tuum est; nostri si parvula cura
Sit tibi quanta libet: si sit modo: non mihi regna
Lydia, non magni potior sit fama Gylippi;
Posse Meletaeas nec mallem mittere chartas.
Quod tibi si versus noster totusve, minusve,
Vel bene sit notus, summo vel inerret in ore:
Nulla mihi statuent finem te fata canendi.
Quin etiam mea cum tumulus contexerit ossa;
Seu matura dies celerem properat mihi mortem,
Quandocumque hominem me longa receperit aetas
Inceptis de te subtexam carmina chartis.

[Tib., "Poema," i.]

PREFACE.

CHEIROGNOMY, the science of declaring the characters, aptitudes, and mental conditions of men by a glance at the formations of their hands, came suddenly into existence, without any known precedent. Adrien Desbarrolles, the cheiromant, whose death has created only recently so much stir among cheirosophists, not only in Paris, but all over the world, tells us in his book " *Les Mystères de la Main* " [p. 107], that one day he asked d'Arpentigny how he had discovered his system. " By a Divine inspiration," quoth he.

However this may be, we know how our author tabulated his system, and from what materials. M. Gourdon de Genouillac has told us how his attention was first drawn thereto in Spain. M. Desbarrolles has told us how he completed his investigations, and how he formulated his observations so as to be able to act upon his experience. It appears that when he was still quite young and living in the country, he was in the habit of going to the parties of a rich land-owner who lived near him. This gentleman being imbued with a strong taste for exact sciences in general and mechanics in particular, his assemblies were mainly composed of geometricians and mechanicians. His wife, on the other hand, was passionately fond of art, and received none but artists ; the natural result was, that each of them received on different days, and d'Arpentigny, being neither a mechanician nor an artist, went to the receptions of both. Having himself hands which were very beautiful [*vide* Appendix A., p. 417], he continually contrasted the hands of the people by whom he was surrounded with his own, and it gradually occurred to him that the fingers of the arithmeticians and ironworkers were knotty, whilst those of the artists were smooth [*vide* ¶ 110]. Having made this observation, he commenced to' develop it, examining as many hands' as he came across, with the invariable result that he found the hands of manual labourers to be prominent-jointed, whilst those of artists were smooth. He studied and observed for thirty years before he considered that he had sufficiently established his system to render it capable of reliable practice. He did not try to account for his science as I have done in the Introductory Arguments to this volume, and to "*A Manual of Cheirosophy;*" all he said was, "Here are facts ; here are indications which are *invariably* connected with certain characteristics ; these things speak for themselves."

There is a great tendency in the present day to dress up old dogmata in new forms and cry "Ecco ! a new science !" a constantly recurring state of things that reminds one of the grim, gigantic helmet in Dryden's "*Battle of the Books,*" in the farthest corner of which was found a tiny head the size of a walnut ; but in this science I do not think there are to be found—as there are in the newer cheiromancy,—traces of a long-disused cultus, "an old idea," as Longfellow says in "*Hyperion*," "folded in a new garment, which looks you in the face and pretends not to know you, though you have been familiar friends from childhood." In vain has Desbarrolles endeavoured to surround this branch of the science with a protoplasm of gnostic mysticism by devoting several pages [275-282] of his book to a process of what he calls "enriching the system of M. d'Arpentigny by means of the Kabbala." When, however, he tabulates the cheirognomical indications of the seven cardinal sins, he becomes interesting, and when in his larger and more recent volume, "*Suite et Fin*" (Paris: 1879), he gives a few pages on how to put the science rapidly into practice [p. 90], he becomes most valuable, and I can warmly recommend a perusal of those pages to all who are interested in the subject-matter of this volume.

It will be noticed, as M. Desbarrolles has remarked [*vide* p. 420], that our author, being of Bacon's opinion that "it is good to vary and intermingle speech of the present occasion with arguments, tales with reasons, asking of questions with telling of opinions, and jest with earnest," betrays a continual tendency to fly off at a tangent and talk delightfully and interestingly about something else concerning which he has somewhat to say, with the natural result that he frequently loses and frequently repeats himself. I have endeavoured, by appending an extremely full index, to reduce as far as possible the irritation resulting from this state of things, which the philosophically-minded reader will naturally evince, rather than to say with Omar Khayyam—

> " They who by genius and by power of brain
> The rank of man's enlighteners attain,
>   Not even they emerge from this dark night,
> But tell their dreams, and fall asleep again ! "
>
> *[Whinfield]—*

a verse whose original terminates this work on p. 439.

Again, in alluding to living and dead celebrities as illustrations of certain types of hands, he has often thought it sufficient merely to mention a name, or a place, or an event, assuming that it is already familiar to his readers. All of us, however, not being on a par with M. d'Arpentigny in the matter of erudition or of scholarship, I have deemed it expedient to verify all, and in some cases to correct, his allusions, and to give for the benefit of my readers references to works where they may, if they choose, pursue the suggestions of our author. The result of this has been an enormous mass of notes, carefully selected from a much larger quantity which I have collected during the past five years with a view to their ultimate utilisation in this form. It may strike the reader at the first glance that I have somewhat overdone this matter, but a few moments' perusal of these pages will, I think, convince him [or her] that in amplifying my completion of the labours of M. d'Arpentigny I have added interest as well as biographical and bibliographical value to his work.

" There be some," said Don Quijote to Sancho, " who weary themselves in knowing and verifying things which, after knowing and verifying, are not

worth a farthing to the mind or the memory." I trust
I have avoided such a reproach as this, and I hope
that I am not like Orbaneja, the painter of Ubeda,
who, when he had painted a cock, found it
necessary to write underneath it, "*This is a cock.*"
I had rather be considered [to follow up Don
Quijote], "some sage enchanter, from which kind
of people nothing is hid on which they wish to
write." To such gibes I would answer as the
Knight of the Rueful Visage did to Sancho, when
the latter said to him, "I say of a verity that your
worship is the devil himself, for there is nothing
you do not know;"—"'Tis necessary to know
everything," answered Don Quijote, "in the office
which I profess." Seriously, I have been obliged
in my notes to follow the style of M. d'Arpentigny,
who seems to have studied like Imlac in Johnson's
"*Rasselas,*" considering that "he who knows
most will have most power of diversifying his
scenes and of gratifying his reader with remote
allusions and unexpected instruction."

This volume is, as it were, illustrative to "*A
Manual of Cheirosophy.*" I have established
throughout a series of cross references to that
volume, so that those of my readers who already
possess it will find as it were a still more amplified
commentary to this text therein. It is for this
reason that I have carefully avoided in any way
repeating any of the indications which are laid
down in that volume, excepting so far as the first
section of the "*Manual*" owes its origin to the
work of M. d' Arpentigny.

This book might, therefore, be entitled "A
Manual of Biographical Cheirosophy," or, "The
Natural History of Hands." Professor Drum-

mond, in the work which I have had occasion several times to quote during the composition of the pages which follow, has very justly remarked that biography is practically a branch of natural history; it is in this light that I have treated it in this new volume of the great book of Nature, and I hope that I might, like Lavater, inscribe upon the title-page of my book that it is "destined to promote the knowledge and the love of mankind."

In conclusion, I wish to record my thanks to the very large number of the readers of "*A Manual of Cheirosophy*" who have written me letters expressing their interest in, and sympathy with, my work. I should like to have printed some extracts from some of the letters I have received from some of the most eminent thinkers and scientists of the day, but I feel that they were not sent to me with this object, and I merely consign this new volume to the tender hands which received my last so kindly, saying to it in the words of Herrick :—

> " Make haste away, and let one be
> A friendly patron unto thee,
> Lest rapt from hence, I see thee lie
> Torn for the use of pastery ;
> Or see thy injured leaves strive well
> To make loose gowns for mackerel ;
> Or see the grocers in a trice,
> Make hoods of thee to serve out spice !"

ED. HERON-ALLEN.

St. John's, Putney Hill, S.W.,
22nd April, 1886.

PREFACE TO THE EDITION OF 1865.

THE accomplished author whose name we find inscribed upon the title-page of this book was a man of a strongly-marked individuality.

A man of refinement in every sense of the expression, he became, almost unconsciously, a man of science ; gifted with an ardent desire for knowledge, and singularly adapted by the nature of his highly-impressionable organisation for the rapid assimilation and comprehension of things, he was readily attracted by the revelations of a Science, the dicta of which struck his mind forcibly, and concerning which he determined, with the singleness of purpose which was with him a leading characteristic, to become the leading authority.

Before proceeding further, however, let us briefly review the life of this attractive man, who, as the poet Barthélemy has said, "was equally expert with the pen and with the sword." By this means will be demonstrated to the reader the chain of circumstances which led him to become the high priest of the Science of CHEIROGNOMY.

Born on March 13th, 1798, at Yvetot, Casimir Stanislas d'Arpentigny, destined from his earliest years for a military career, entered the military college of St. Cyr, where he began immediately to attract attention to himself, not only on account of his rapid progress in mastering the curriculum, but by reason, particularly, of the malicious humour which inspired his biting epigrams, against the stinging sarcasms of which no one in the institution was proof, not even the commander-in-chief, who woke one day to the fact that he had been lampooned after a fashion so drastic and effectual, that he felt himself bound to punish the offender with a severity which sufficiently revealed the depth of the wound which had been inflicted upon his self-esteem.

D'Arpentigny, who was on the point of being gazetted to a sub-lieutenancy, was ignominiously expelled. He had, however, the consolation of knowing that he would be amply revenged by his epigram, which became popular in every barrack throughout France. Still, baulked of the epaulette which he was, as it were, in the act of grasping, it was clear to him that he must use every endeavour to regain his right thereto, and nerved by this consideration he enlisted in the 29th regiment of the line. Three years later he obtained his commission, and, having been taken prisoner at Dantzic, he returned to France in 1814, and was placed upon the retired list.

Gazetted to the 66th regiment in 1815, he was again disbanded in the ordinary course of events, and re-entered the service in 1818. In 1820 he served as lieutenant in Spain, and on his return was made a member of the royal body-guard [Compagnie de Croy]. With the revolution of 1830, the throne being overturned, its supporters were scattered, and D'Arpentigny entered the

40th regiment with the rank of captain; he was decorated with the Cross of the Legion of Honour in 1833, and served until his retirement in 1844, after an honourable service of thirty years' duration.

At the end of his military career his literary one commenced.

We saw him taking part in the Peninsular War; it was during this campaign that his cheirognomical studies had their origin. One day whilst the young officer was walking along one of the high roads of Andalucía he was accosted by a *gilaña*, who offered to read for him his fortune by the inspection of his hands. This girl, a perfect specimen of the pure Moorish type, was extremely beautiful, and D'Arpentigny, willingly submitting to her powers of persuasion, laughingly extended his hand and listened to the string of experimental ratiocinations which the gipsy poured forth with a complimentary eloquence and euphemism, which was in a direct ratio to the liberality of the lieutenant, who, as he pursued his walk, began to find much food for thought in this form of divination by the lines of the hand, and in some of the quaintly-suggestive terms of which the fortune-teller had made use, which had forcibly attracted his attention.[1] He reflected that allowing Cheiro-

---

[1] Our author might have compared himself to the chevalier Duguesclin, to whom a nun [a converted Jewess] predicted that he would be honoured above all men in the kingdom of France. "We are told that astronomy confirmed the predictions of cheiromancy, and that Duguesclin, in consequence, always kept a wise divining-sibyl at his side through all his enterprises" [*vide* H. Martin's "*Histoire de France*" (Paris: 1878), vol. v., p. 244; and G. de Berville, "*Histoire de Bertrand du Guesclin*" (Paris: 1767)].

mancy, as practised by gipsies and ignorant mountebanks, to be merely an innocent fraud cultivated and perpetrated with a view to the extraction of coppers from the wayfarer, it was none the less significant that in practising their pretended "science," these people merely repeated phrases which they had learnt from their fathers, who, in turn, had learnt them from their forefathers.

Running through the chaotic nonsense which the gipsy recited, the listener had been struck by the recurrence of certain expressions which seemed to him to be echoes of a forgotten language,—of a language whereof the essential character retained much of its ancient force. Reflecting thus, D'Arpentigny set himself thoroughly to sift the matter to the bottom, and he left no stone unturned to discover the truth, which he wished to prove to be clearly and palpably evident. For twenty years he devoted himself with enthusiasm and energy to this arduous task, for, as is usually the case under similar circumstances, in proportion as his ideas advanced like pioneers before his studies, losing themselves in the profound obscurities of the road, so did the horizon enlarge itself, and so did new openings and obstacles in the way present themselves. He examined the writings of Avicenna and of Frætichius, and by their means he corroborated the opinions of Antiochus, Tibertus, and Taisnier; he dived into Plato and Aristotle, he interrogated Ptolemy, and sought inspiration from Averroës;[2] in short, he mastered the literature of the subject, learnt all that

[2] *Vide* notes [100–101], p. 169-71.

was to be learnt from others, and then, having stored his mind with the observations of his predecessors, he came to the conclusion that nothing but doubt could result from his studies until he had certified his knowledge by actual experience.

It was then that he commenced to compare the hands of all those in whose company he was thrown, and that he commenced to note the most infinitesimal details of their conformations, that he analysed their aspects, and that he, for the first time, formulated an exact system based upon logic and upon reason.[3] And after having minutely examined the obscurest arcana of the Mysteries of the Hand, from which, day by day, his intelligent perspicacity rent the veil of doubt, he finally resolved to publish his book ;—a book which is clear and precise, and which possesses the great advantage of being easily obtainable by all classes of readers ; which calls things clearly by their right names, and which does not aim at the marvellous.

There is no childish and pretentious *mise-en-scène;* it is simply and neatly expressed, concise, and—a quality which in no wise detracts from its value as a philosophical work—it is written with that fascinating charm, which carries away the reader by the spontaneity of its treatment, the power of its expression, and the wealth of its ideas. If further recommendation were desired let us see what the most eminent literary men of the day have thought of this ingenious

---

[3] *Vide* on this point p. 7 in the *Preface* to this volume.

and unique volume ; here, for instance, we have a letter which the author of " Elvire "[4] addressed to the author of "La Science de la Main," when the work first made its appearance :—

"I have read your work with great interest, for your style would recommend to my consideration the most hypothetical science. Even if there be not complete theoretical certainty in the system, there is a singular charm in your exposition thereof. I have delayed writing this to you, because I did not know your address, but many mutual friends will have conveyed to you these compliments and my regrets that you should have had to wait so long for them. If there is, indeed, revelation in the hand, believe, I pray you, in mine when it acknowledges all the pleasure which you have given me. We were comrades in the Royal Body-guard, and I congratulate myself that we

---

[4] I presume that it was not without reason that M. de Genouillac names "*Elvire*" in this place—"*Elvire*" being one of the obscurest of De Lamartine's unfinished works. It may be found in vol. 1. of the " *Œuvres de Lamartine* " [" Méditations poétiques avec Commentaires "] (Paris : 1849, *F. Didot*, 14 vols.), on p. 109 of which the author says :—" This *Méditation* is merely a fragment of a much larger piece which I composed a long while before I wrote the real ' *Médita-tions.*' It consisted of some love verses addressed to a young Neapolitan girl whose death I have related in ' *Les Confidences.*' Her name was Graziella. These verses formed part of a collection in two volumes of the poems of my earliest youth, which I burnt in 1821. My friends had preserved a few of them, and restored to me this one, when I printed my ' *Méditations.*' I separated these verses and wrote the name ' *Elvire* ' above them instead of that of ' *Graziella.*' It is obvious that they are not born of the same inspiration." A most interesting commentary to my mind.

continue to be so in the study of natural philosophy.

"*Recevez, Monsieur, etc., etc.,* LAMARTINE.

"*29th June,* 1857."   .

Jules Janin[5] also pays a tribute of respect to the author in a few amiable lines as follows :—

"Take care ! You give *me* a book ! if I were not of a considerate turn of mind I could overwhelm you with it ; especially as you have constructed a theory which is charming, ingenious, probable, well expressed, and curious. I shall profit by it on the spot. When I have read your book with the care that it deserves, I hope you will allow me to talk to my readers about it in the *Journal des Debats.* It will give me great pleasure to testify publicly my appreciation of your civilities to me ; meanwhile, under all circumstances, and in all places, believe me to be very absolutely, Yours,

"J. JANIN."

I do not propose in this place to publish all the sympathetic testimonials which were sent to M. d'Arpentigny on the publication of this volume.[6]

---

[5] Member of the *Académie Française* and editor of the "*Journal des Debats,*" author of "*Barnave*" (Paris: 1860), "*La Bretagne*" (1844), "*Les Catacombes*" (1839), "*Chefs d'Œuvres Dramatiques du xviiie Siècle*" (1879), "*Le Livre*" (1870), and of a vast quantity of novels and novelettes.

[6] I omit as superfluous a long letter in the same style, written under date 23rd March, 1858, by J. M. Dargaud, the historian, author of "*Histoire d'Elizabeth d'Angleterre*" (Paris: 1866) ; "*Histoire de Lady Jane Grey*" (1863) ; "*Histoire de Marie Stuart*" (1850) ; "*Histoire d'Oliver Cromwell*" (1867), etc., etc.

. . . The above letters give a good idea of the impression which was made upon the public by its appearance, which was heralded with delight by the world, and by the friends of the author, who were a host in themselves, for he was on terms of friendship with the whole of the Parisian aristocracy, whether of birth or of talent.

His profoundly witty and incisive pen took a pleasure in recalling the numberless recollections of his military life, and he has left his pages sparkling with strokes of wit and sarcasm. When I knew him—it was towards the end of his life,—he was a fine old man, full of energy and health, whose lips scarcely ever opened without letting fall flashes of exquisite satire. His whims often wounded, and his caustic humour often over-whelmed with sharply-pointed missiles, even his best friends, who were, however, perfectly willing to stand fire, reflecting that no one was exempt from the thick hail of his witticisms,—and besides they appreciated the wit which supplied the weapons.

He was an accomplished talker, and he con-versed with a gaiety which was absolutely juvenile. He had made a minute study of men, with the result that he came to the positive conclusion that women are infinitely superior to them ; it is not surprising, therefore, that he was extremely fond of women, and that they entirely reciprocated the sentiment. The noblest and most influential dames of Parisian society made it their study to surround the old age of this incorrigible cynic [who never abdicated his right to tyrannise over them] with delicate and touching attentions.

One evening he went peacefully. and quietly to

sleep, and in the morning they found him sunk in the perfect slumber from which there is no awakening on this side of the grave. And his lips had closed, wreathed as was their wont in a cynic smile.

[*From the French of*]

H. Gourdon de Genouillac.

Upon Cupid.

" *Love, like a gypsy, lately came,*
And did me much importune
To see my hand, that by the same
He might foretell my fortune.

He saw my palm ; and then, said he,
' I tell thee, by this score here,
That thou, within few months, shalt be
The youthful Prince d'Amour here.'

I smiled, and bade him once more prove,
And by some cross-line show it,
That I could ne'er be Prince of Love,
Though here the princely poet."

HERRICK, " *Hesperides.*'

TABLE OF CONTENTS.

LIST OF ILLUSTRATIONS.

HEAD-PIECES.—THE SEVEN AGES OF HANDS. THE CHEIRO-STEMON. THE LEGEND OF THE MONK RICARDUS. HAND-SHAPED ROOTS.

Rosamund Brunel Horsley, inbt. et delt.

Introductory Dissertation and Argument.

"Non ego ventosae venor suffragia plebis."—HORACE.

"Say that thou pour'st them wheat,
　　And they will acorns eat :
Twere simple fury still thyself to waste
　　On such as have no taste !
To offer them a surfeit of pure bread,
　　Whose appetites are dead !
　　No, give them grains their fill,
　　Husks, draff to drink and swill :
If they love lees, and leave the lusty wine,
Envy them not, their palate's with the swine !"

BEN JONSON.

AN INTRODUCTORY EXPOSITION OF THE BASES OF THE SCIENCE OF THE HAND — HAND CUSTOMS AND SUPERSTITIONS — AND A FURTHER ARGUMENT UPON THE CLAIMS OF CHEIROSOPHY TO RANK AS A PHYSICAL SCIENCE.

καὶ γὰρ πάντες οἱ ἄνθρωποι ἀνατείνομεν τὰς
χεῖρας εἰς τὸν οὐρανὸν εὐχὰς ποιούμενοι.
ARISTOTLE, ΠΕΡΙ ΚΟΣΜΟΥ.

THE philosopher to whom we are indebted for the above quotation spoke with truth when he described the hand as the member of the members—the ὄργανον πρὸ ὀργάνων.[7] From the earliest ages, and in all nations, homage has been paid to its importance by teachers in their writings, by priests in their ceremonies, and by the common people in their superstitions ; and this will not, I think, be wondered at when we reflect upon the part which is played by our hands upon the theatre of our existence, when we consider that there exists scarcely a single incident of our lives in which the hand is not the prime agent, the apparatus whereby we practically live, move, and have our being. It will not, I think, be taken amiss if as an introduction to this volume I devote a few moments to the consideration of, and the tabulation of a few notes upon, the superstitions

¶ 1.
University of hand superstitions and customs.

[7] ΠΕΡΙ ΖΩΩΝ ΜΟΡΙΩΝ Δ'., ί.

[¶ 1]

and customs which have grouped themselves round this all-important member.[8]

¶ 2.
Perfection of the hand.

When we consider the absolute perfection of the hand, whose entire structure, as Galen remarks,[9] is such that it could not be improved by any conceivable alteration, we cannot but realise the value of any study which draws our attention more closely to it, and we are impelled to ask with Sir Charles Bell,[10] "Is it nothing to have our minds awakened to the perception of the numerous proofs of design which present themselves in the study of the hand, to be brought to the conviction that everything in its structure is orderly and systematic, and that the most perfect mechanism, the most minute and curious apparatus, and sensibilities the most delicate and appropriate, are all combined in operation that we may move it?"

Sir C. Bell.

¶ 3.
Sudden sensations of the body.

It is not in any way strange that in days of old, as Steevens has observed, all sudden pains of the body which could not be naturally accounted for, were assumed to be presages of something that was shortly to happen,—an idea which we find expressed in the *Miles Gloriosus* of Plautus, in the line :—

Plautus.

"Timeo quod rerum gesserim hic ita dorsus totus prurit !"

How much more, therefore, should men have attached importance to sudden sensations of their all-important hands. Who has not heard of the "itching palm," which seems originally to have been regarded as a sign of coming fortune rather than, as

Itching palm.

[8] The following remarks upon "*Hand Superstitions and Customs*" form the substance of a lecture delivered on the 4th December, 1885, before the "SETTE OF ODD VOLUMES."
[9] "*De Usu Partium Corporis Humani,*" book i.
[10] "*The Hand, its Mechanism and Vital Endowments*" (London : 1832).

it is to-day, a sign of avarice?[11] and Shakespeare's lines, "By the pricking of my thumbs, something wicked this way comes" [*Macbeth*, iv., i.] have become a proverb.

This pricking of the thumbs as a warning of coming danger constantly recurs in old romances: perhaps one of the best known is the story of the Irish hero Fingal, whose Gargantuan master, having devoted many years to the attempt, at length caught a fish, the properties of which were, that whoso should first taste thereof should immediately be endowed with the gift of foresight. This finny phænomenon was handed to Fingal to be cooked; and during this operation, he having turned away himself, and in so doing having forgotten to turn the fish, a blister rose upon its side. Fingal, terrified at the prospective consequences of his inattention, pressed down the blister with his thumb: in so doing, the scorched fish adhered thereto and burnt it, whereupon our hero not unnaturally put his thumb into his mouth. The mischief [or rather the good—from Fingal's point of view] was instantaneous, *he* was the first to taste the fish, and consequently *he* was the depositary of the coveted power; he fled from the scene of his dereliction, and, of course, the giant followed vowing vengeance, but in vain, for whenever he approached his victim the pricking of the latter's thumb warned him of the coming danger, and

---

[1] JOHN MELTON, "*Astrologaster, or the Figure Caster*" (London: 1620):—"When the palme of the right hande itcheth it is a shrewde signe he shall receive money." This would seem to have originated in the East, judging by a custom quoted by G. Atkinson in his "*Customs and Manners of the Women of Persia*," translated from a Persian MS. (London: 1832), which reads: "If the palm of the hand itches, rub it on the head of a boy whose father and mother are still living, and a present of money will be the consequence."

[¶ 4]

Fingal pursued his way continually forewarned, and by consequence continually forearmed.

¶ 5.

Uses of the hand.

Successive generations of authors have called attention to the paramount importance of the hands in the human œconomy; with them man fashions all the implements and accessories which give him his vast superiority over every other created thing,

Galen.

"and lastly, by means of the hand man bequeaths to posterity the intellectual treasures of his own Divine imagination, and *hence* we who are living at this day are enabled to hold converse with Plato and Aristotle and all the venerable sages of antiquity;"[12] and indeed we need only reflect for a single moment to congratulate ourselves upon the *ipse scripsit* of the most famous men that the world has known, from the apostles and prophets even to the robber who said to

Don Quijote.

Don Quijote, "Know that I am Gines de Pasamonte, whose life is written with these pickers and stealers"[13] [Duffield's translation (London: 1881), vol. i., p. 292]. The writings of the wise, said

Chas. Lamb.

Charles Lamb, are the only riches our posterity cannot squander: and we might almost say in support of the theory of the descent of man from the ape, that man hangs monkey-like, by his hands, to the branches of the tree of knowledge.

¶ 6.

Loss of one or both hands.

Cervantes.

Cervantes, we are told, in the battle of Lepanto in 1571, had his left hand so much injured that he never recovered its use; I never remember this without wondering, almost with that sickening feeling that comes over one when a great danger has gone by, whether his immortal works would ever have been written with

[12] Galen, *op. cit.*, lib. i.

[13] "Sepa que yo soy Ginés de Pasamonte, cuya vida está escrita por estos *pulgares*."—"*Don Quijote*," part i., cap. xxii. As to the expression "*pickers and stealers*," compare *Hamlet*, Act III., sc. 2: "So do I still by these *pickers and stealers*."

[¶ 6]

Nelson.

his left hand ; whether, like Nelson, he would have overcome the loss of the right hand, or whether at this moment Don Quixote and Sancho Panza would have been sleeping in the Walhalla of unborn heroes of romaunt, waiting still to be galvanised into life by the touch of the magic quill of some latter-day genius. Whether Caius Mucius Scævola really went through *Mucius Scævola.* the traditionary performance of burning off his own right hand for having mistaken Lars Porsena's secretary for that "by-the-nine-gods-swearing" potentate himself, is, I think, highly problematical ; at all events, we know that the cutting off of hands has been a universal and dreaded punishment from the days of Horatius in Rome until within a couple of centuries in England.   I have recorded elsewhere the instance of the Roman poltroon of the time of Augustus Cæsar, *Poltroon.* who cut off the thumbs of his sons, lest they should be sent to fight [whence the word " poltroon "=*pollice truncatus*], and the gentle Norman barons would seem *Norman and* to have been impressed with the importance of this *Spanish tortures.* particular digit when they hung up their enemies thereby,—an operation which did not escape the observation of the chivalrous Spaniard who invented the instrument of persuasion known as the thumb screw.

Next after the thumb, the most important finger ¶ 7. has always been the third, which, as I have elsewhere *The third* pointed out,[14] was always alluded to as "*medicum*" *finger.* by the ancients   The *lustralis saliva* was always applied to the infant's forehead, as a preventive against the evil eye, with the third finger.

As late as the sixteenth century the thumbs of forgers ¶ 8. and seditious writers were cut off by the common *Cutting off* hangman, and much later than that the punishment *hands and* for drawing a sword in a court of justice or assaulting *thumbs in* *England.*

[14] *Vide* "*A Manual, etc.*," ¶ 37, and note 112, p. 189.

[¶ 8]

Ireby.

an officer of the Crown, or alderman of the City of
London, was amputation of the hand.   In Lord Chief
Justice Ireby's " *Notes to Dyer's Reports* " I recently
found the following delightfully quaint account of
the narrow escape of Sir Thomas Richardson, Chief
Justice of the Common Pleas, alluded to by Pepys
under date September 8th, 1667 :—" Richardson,
Chief Justice de C. Banc., al Assizes at Salisbury
in Summer 1631, fuit assault per prisoner la con-
demne pur felony ; que puis son condemnation ject
un brick-bat a le dit Justice, qui narrowly mist ; et
pur ces immediately fuit indictment drawn, per Noy
envers le prisoner, et son dexter manus ampute,
and fix at gibbet, sur que luy meme immediatement
hange in presence de Court."   Coustard de Massi,
in his " *History of Duelling*,"[15] tells us that among the
canons of duelling it was forbidden to the spectators
to sit, even on the ground, during a duel, on pain
of the loss of a hand.

¶ 8a.
The same in the
East.

I have elsewhere alluded to the Eastern punishment
of cutting off hands [*vide* ¶ 257 and note, and compare
" *A Manual of Cheirosophy* " ¶ 22] ; we find it continually alluded to in the Qur'án in passages like that
which occurs in the seventh chapter, and Mr. Sale [note
[214], p. 184] gives many very interesting notes on the
subject.

¶ 9.
The hands
in oratory.
Cicero.

The assassins of Cicero, when they killed him at
Caieta in 43 B.C., paid a rare tribute of homage to the
powers of the hand in oratory, when they cut off his
hands as well as his head and sent them to Rome to
be hung up in the Forum ; for what is more striking
than the action of the hands in speaking,—a function of

J. Bulwer.

the member which inspired Bulwer's curious opuscu-

[15] *Vide* "*History of Duelling in all Countries*,"
translated from the French of M. Coustard de Massi,
with introduction and concluding chapter by " Sir
Lucius O'Trigger " (London : *n.d.*), p. 22.

[¶ 9]

lum "*Chirologia*,"[16] and which called forth the glorious panegyric of Quintilian, "Nam ceteræ partes loquentem adjuvant, hæ [prope est ut dicam] ipse loquuntur. An non his poscimus? pollicemur? vocamus? dimittimus? minamur? supplicamus? κ.τ.λ.,"[17] which has been imitated by Montaigne in that celebrated passage of the "*Apologie de Raimond Sebond*," "Quoy des mains? nous requerons, nous promettons, appelons, congedions, menaceons, prions, etc., etc."[18]

Quintilian.

Montaigne.

The preference that is felt for a medium-sized or delicately-modelled hand over that hard, rough, and red paw which, as Sir Philip Sydney remarks, denotes "rude health, a warm heart, and distance from the metropolis," is, I think, universal in these latter days, when we prefer delicacy of mind shown by the former to the brute force indicated by the latter, though it was the latter that inspired Don Quijote with such respect in the Cave of Montesinos, when on the sepulchre of Durandarte he beheld that warrior extended "with his hand [which was somewhat hairy and of much muscle—a sign of great strength in its owner] laid across the side of his heart."[19] The rusé and elegant Villiers, Duke of Buckingham, and the Sultan Mahmoud II., renowned for his ghastly cruelties, had both of them small and delicate hands. We are told by Leigh Hunt in terms of sarcastic bad taste[20] that Lord Byron was similarly gifted, and successive historians have recorded that Queen Elizabeth prided herself to the like effect. Sir Henry

¶ 10.
Admiration
for small
hands.

Don Quijote.

Small-handed
tyrants.

Lord Byron.

Queen Elizabeth.

[16] J. BULWER, "*Chirologia; or the Naturall Language of the Hand*" (London: 1644).

[17] M. F. QUINCTILIANI "*De Institutione Oratoria Libri Duodecim*, lib. xi., c.iii.

[18] "*Essais de Montaigne*" (Paris: 1854), vol. ii., p. 282, book ii., ch. 12.

[19] "*Don Quijote*," parte ii., cap. 23

[20] *Vide* note [17], p. 127.

[¶ 10]

Ellis [21] quotes a letter from a Venetian minister, who, describing our virgin sovereign, says, " E sopra l' tutto bella mano de la quale fa professione " [*and above all the beautiful hand which she exposed to our view*]. The

Leo X.

celebrated Pope Leo X. had equally fine hands,[22] as may be seen by his picture in the Pitti Palace at Florence—hands which have been duly celebrated by Gradenigo in Cogliera's "*Nuova Raccolta degli Opuscoli*" (Venice: 1719).

¶ 11.
White hands:
Persia.

A certain mystical loveliness has always been attached to white hands : the Persians do not translate in the verse Exod. iv. 6 "*leprous*, as white as snow"' (R.V.), but take the " White hand of Moses" [a symbol of beauty and purity with them] to be a synonym for the white May-blossom which always blooms with them at their New Year (which begins with the vernal equinox). This, therefore, is the interpreta-

Omar-i-Khay-
yám.

tion of Omar-i-Khayyám's exquisite verse :—

"Now the New Year reviving old desires,
The thoughtful soul to solitude retires,
Where the WHITE HAND OF MOSES on the bough
Puts out, and Jesus from the ground suspires." [23]

The white hand has always been looked upon as

[21] "*Original Letters illustrative of English History*" (London: 1846).
[22] William Roscoe calls special attention to the whiteness and elegance of the hands of Leo X. in his "*Life and Pontificate of Leo X.*" (London: 1846, vol. ii., p. 377). *Vide* also note [29], p. 242.
[23] "*The Rubaiyát of Omar-i-Khayyám, etc.*" (London: 1879), quatrain iv. "*Les Quatrains de Khèyam,*" traduits du Persan par J. B. Nicolas (Paris: 1867), 186th quatrain.

وقتست كه از سبزه جهان آرایند

موسی صفتان زشاخ كف بنمایند

&c. &c.
Orientalists will note the signification of the phrase,
موسی صفتان &c.

the emblem of innocence, just as the red hand (red-handed) has always been a synonym for guilt; thus in Massinger's *Great Duke of Florence* (act ii., sc. 3) Sanazarro says: "Let this, the emblem of your innocence, give me assurance;" to which Lidia replies, "My hand, joined to yours without this super-stition, confirms it.[24] In the East the *white* hand is symbolical of open-handed and unasked generosity, the *black* hand being *per contra* the synonym for avarice and niggardliness. Thus we have it in the Arabian Nights, where Yahya says to the chamberlain in the story of the Caliph al Maamun and the Scholar:—"I owe thee a heavy debt of gratitude, and every gift the white hand can give," etc.[25] As an emblem of beauty we find it in such expressions as "rosy-fingered morn," of which no more beautiful use has been made than in the *Ecloghe de Mutio Justino Politiano*"[26] in the lines :—

Massinger.

Arabian Nights.

Justino Politiano.

> " Eran ne la stagion che l'aurea Aurora
> Con la rosata mano l'aurate porte
> Apre del cielo al rinascente giorno ! "

Among the ancient Egyptians the hand was the symbol of strength; among the Romans it was that of fidelity; and yet further back, how the mystic intensity of the warning to Belshazzar must have been tenfold increased by "the fingers of a man's hand," which the maddened monarch imagined that he saw writing its awful message in letters of living flame. Bacon has compared the will of the

¶ 12.
The Egyptians and Romans.

Bacon.

---

[24] On the subject of swearing by the hand *vide* note [202], p. 179, and "*A Manual, etc.*," ¶ 21. It is one of the commonest forms of oath in old English plays, as, for instance, in Ben Jonson's *Alchemist*, where Kastril says [act iv., sc. 2]: "By this hand you are not my sister, if you refuse!"

[25] Burton's translation [*vide* note [117], p. 112], vol. iv., p. 185.

[26] Venice 1555; Ecl. 2, Delle Marchesane.

¶ 12]

people to the hundred hands of Briareus,[27] but it is not necessary in this place to advert to the hand as a symbol of power ["*Manual*," ¶ 15, etc.]. We find the same *motif* in the expression "At a dear hand," signifying *expense*,[28] and "At an even hand," denoting *equality*.[29]

¶ 13.
Church ceremonies

Perhaps some of the most beautiful symbolisms connected with the hand are found in the use made of the member in the ceremonies of the Church, in the invocation of the blessing, in making the sign of the cross, and in the laying on of hands. We have the priestly blessing with the whole hand, and the episcopal blessing with the thumb and two fingers only extended, in the manner in which we see them arranged in the charms which are sold in Naples for the repulse of the evil eye;[30] and we find it laid down by the great authority Durand[31] that the sign of the cross is to be made with these three digits.

" The poets faigne, that the rest of the gods would have bound *Jupiter;* which he hearing of, by the Counsell of Pallas, sent for *Briareus* with his hundred hands, to come in to his Aid. An Embleme, no doubt, to shewe, how safe it is for Monarchs, to make sure of the good Will of Common People."—BACON, " *Of Seditions and Troubles.*"

[28] *Ibid.*, " *Of Despatch.*"
[29] *Ibid.*, " *Of Envy,*" and " *Of Expence.*"
[30] *Vide* " *A Manual of Cheirosophy* " (London: 1885), p. 32; and *vide* note [213], p. 183.
[31] GULIELMUS DURANDUS, " *Rationale Divinorum Officiorum* " (Venice: 1589), p. 140, verso, lib.. v. " Quid sit Officium," cap. i., sect. 12.*

* " Est autem signum crucis tribus digitis exprimendum, quia sub invocatione Trinitatis imprimitur. De qua propheta ait : Qui appendit tribus digitis molem terret [Es. 40]. Pollex tamen supereminet ; quoniam totam fidem nostram ad Deum unum et trinum referimus et mox post ipsam invocationem Trinitatis potest dici versus ille. . . . Secundo ad notandum, quod Christus de Judæis transivit ad Gentes. Tertio, quia Christus a dextra, id est, a patre veniens, diabolum, qui per sinistram significatur, in cruce peremit."

Eugene Schuyler, in his "*Peter the Great*,"[32] recounts that the orthodox make the sign of the cross with the thumb and two fingers, whereas the benediction is given with the first, second, and fourth fingers, the thumb meanwhile holding down the third. Havanski, in his interview [*vide loc. cit.*] with the Dissenter Sergius, calls attention to the fact that he makes the sign of the cross in the orthodox manner, as opposed to the Dissenters, who signed with the first and second fingers only.

¶ 14.
Knights of
Malta.

The most precious relic of the Knights of St. John of Malta was a mummied hand, said to have been that of St. John the Baptist, which had been given to the Grand Master d' Aubusson by the Sultan Bajazet. We are told that a dragon, who resided at Antioch, once displayed the want of foresight to eat a fragment of this relic, and was punished for his ill-advised temerity by the most unpleasant after-effects ; we are told that the said dragon swelled visibly to a preposterous size, and presently exploded with terrific violence. Whatever may be the credence rightfully attaching to this account, we know that this hand was carried through the Island in solemn procession once a year, and if the fingers opened, the harvests of the following season were plentiful and the year was prosperous; if, however, this miracle did not take place, the worst results were to be anticipated. At the dispersion of the order in 1798 by the French, it was taken away by the Grand Master, and subsequently restored to its original shrine.

¶ 15.
Touching for
the King's Evil.

The most interesting custom which has obtained in this country with regard to the laying on of hands has been, I think, that of touching for the King's Evil. Most interesting accounts of this ceremony may be

---

[32] *Scribner's Magazine*, vol. xix., 1880, chap. xi., p. 910.

found in various volumes of the *Gentleman's Magazine*, notably in that for 1747 (p. 13), that for 1751,
(p. 414), and that for 1829, pt. ii. (p. 499), the last containing a full and interesting account of the ceremony.
The custom is of the highest antiquity.   Suetonius
and Tacitus both record an instance of a blind man at
Vespasian.   Alexandria, who importuned the Emperor Vespasian
to anoint his cheeks and eyes with the royal saliva.
Vespasian having after some demur acceded to his
request, an instantaneous cure was effected, and
another supplicant, who had lost the use of his hands,
appearing at the same time, the Emperor touched him
also with the same beneficent result.[33]   I believe the
last instance of this having taken place was in 1712,
Queen Anne.   when Dr. Johnson was "touched" by Queen Anne.
Herrick.   Herrick, in his *Hesperides*, has a charming little poem
on the subject which closes with these lines :—

> " O lay that hand on me,
> Adored Cæsar ! and my faith is such
> I shall be healed, if that my King but touch
> The evil is not yours ; my sorrow sings,
> Mine is the evil, but the cure the King's."

Similar cures would seem to have been performed
in the seventeenth century, by one Valentine
Greatrakes.   Greatraks (or Greatrakes), who acquired an extremely
wide renown for the cures he performed by merely
"stroking" persons afflicted with disease,—a power
which, if the accounts of the thousands who flocked
to him to be cured be true, he must have found

[33] This account is given fully by Godwin in " *Lives of
the Necromancers* " (London : 1834, p. 155), and Hume
also has borne testimony to its veracity and probability
in section x. of part iii. of his " *Essays*." *

* " Ex plebe Alexandrina quidam, oculorum tabe notus, genua
ejus advolvitur, ' remedium cæcitatis' exposcens gemitu ; monitu
Serapidis dei, quem dedita superstitionibus gens ante alios colit :
precabaturque principem ' ut genas et oculorum orbes dignaretur

extremely irksome.[34]   I have before me a little
work, entitled "*A Brief Account of Mr. Valentine
Greatraks, etc., etc.*" (London : 1666), which gives the
history and progress of these phenomena, which would
appear to be merely an early instance of curative
mesmerism or of "faith-healing."

The custom of biting the thumb as a provocative of
strife is familiar to us all in the opening scene of
Romeo and Juliet; it seems to have originated in
France, where the practice was to bite a fragment
from the thumb nail, and draw it scornfully from between the teeth, this being the most deadly insult
that could be offered by one man to another, at least
so it is laid down at p. 44 of a most fascinatingly
quaint little work, entitled "*The Rules of Civility,*"
translated from the French (London: 1685). Connected with this custom it is interesting to note
the Arabic rite of biting the hands as an exhibition of
penitence, which we find cited in the Arabian Nights,
the Qur'án, and elsewhere. Thus, for instance, in
"The Story of Abu al Husn and his Slave-Girl"
[Burton's translation (*vide* note [117], p. 112), vol. v.,

¶ 16.
Biting the hand
or thumb.

Arabic customs

---

[34] I have given an account of this gentleman's cures
in my "*Discourse of Mesmerism, and of Thought-
Reading,*" in Pettitt's "*Early English Almanack*"
for 1886.

---

respergere oris excremento. Alius manuum æger, eodem deo
auctore, 'ut pede ac vestigio Cæsaris calcaretur,' orabat. Ves-
pasianus primo inridere, adspernari : atque illis instantibus modo
famam vanitatis metuere, modo obsecratione ipsorum et vocibus
adulantium in spem induci : postremo existimari a medicis jubet,
an talis cæcitas ac debilitas ope humana superabiles forent. Medici
varie disserrere : " Huic non exesam vim luminis, et redituram,
si pellerentur obstantia : illi elapsos in pravum artus, si salubris
vis adhibeatur, posse integrari." Id fortasse cordi deis, et divino
ministerio principem electum, etc., etc. Statim conversa ad
usum manus, ac cæco reluxit dies. Utrumque, qui interfuere,
nunc quoque memorant, postquam nullum mendacio pretium."
—TACITUS, "*Historiarum,*" lib. iv., c. 81.   *Vide* also the notes
to this passage in J. J. Oberlin's "*Tacitus*" (London: 1825),
vol. iii., p. 313, and compare SUETONIUS, lib. viii., cap. 7.

p. 191], in which we are told " all his goods went from him, and he bit his hands in bitter penitence ; " and again in the Qur'án, where we find the phrase, " The biting was fallen on their hands," which is generally translated, " they repented." [35]

¶ 17.
"Taking a sight."

The custom which obtains among vulgar little boys, known as "taking a sight," is, we are told, of incomparable antiquity. I have seen it stated (but I forget where) that the practice was known as a method of pantomimic derision among the ancient Assyrians. Now, if this is the case, I would submit that some of the curious and apparently strained

Assyria.

positions of the figures on the Assyrian bas-reliefs are fully explained ! Many years ago I used to think that the attitude of the exulting Assyrian warrior, with his hands extended one before the other on a level with his face, was strangely suggestive of the rude small boy, but if the explanation hinted at be the right one, why, history repeats itself—*et voilà tout !*

Rabelais.

We know, many of us, Rabelais' description of the meeting of Panurge and Thaumaste,[36] where we are told " Panurge suddenly raised in the air his right hand, and placed the thumb against the nostril on that side, holding his four fingers extended, and neatly

Bon Gaultier.

arranged in a line parallel to the end of his nose, entirely closing the left eye, whilst he made the other wink with a profound depression of his eyebrows and eyelids," and we have the other historic instance of the Lovelorn Youth erstwhile " coffee-milling care and sorrow, with a nose-adapted thumb." [57]

¶ 18.
Handshaking

After all, the hand-custom most familiar to all of us is that of hand-shaking, a custom to which I have

[35] سوقيتا فى آيديهم , chap. vii., 148.

[36] RABELAIS, "*Pantagruel*," chap. xix.

[57] SIR T. MARTIN and W. E. AYTOUN, "*Bon Gaultier Ballads*,"—"The Lay of the Lovelorn' (London : 1845, 1st edn.).

elsewhere alluded,[38] and which arose in "the good old days," when one could never be sure that one's dearest friend had not got a weapon concealed in his hand wherewith to take a mean advantage when one's back was turned.   It requires no more than a passing allusion here ; we have all experienced with regret the timid handshake of the sex which has for some unknown reason been termed "the weaker," and the eighty-one ton scrunch of the boisterous friend who cracks pebbles in his fists, and keeps in training for the performance by practising on the hands of his shrinking acquaintances (though in the two cases the causes of the regret are not identical).   Also, as W. S. Gilbert says, "the people who in shaking hands, shake hands with you like *that*," and lastly the haughty person of whom a poet, whose name I forget, said,—

> " With fingertips he condescends
>   To touch the fingers of his friends,
>   As if he feared their palms might brand
>   Some moral stigma on his hand,"

so that in time one actually begins to regret "the good old days," when it was not considered bad taste to slaughter people whose idiosyncrasies did not harmonise with one's own sense of the fitness of things.

A word, before I conclude these remarks upon Hand customs and superstitions, on the subject of finger nails, concerning which almost as many superstitions and customs exist, as there are concerning the whole hand.   Innumerable are the sayings and superstitions as to spots in the nails, "gifts" as our nurses were wont to call them, or the varied contingencies expressed by the couplet,—

¶ 19.
Finger nail superstitions and customs.

Spots in the nails.

> "A letter, a friend, a foe ;
>   A lover, a journey to go."

[38] "*A Manual of Cheirosophy*" (London : 1885), ¶ 4.

[¶ 19]

British Apollo.

A writer in that delightful repository of quaint information, " *The British Apollo* " (1708, vol. i., No. 17), says :—" Those little spots are from white glittering particles which are mixed with the red in the blood, and happen to remain there some time. The reason of their being called gifts is as wise an one as those of letters, winding sheets, etc., in a candle." Setting aside, however, as comparatively irrelevant, the ancient science of Onychomancy, or divination by the finger nails, we may note that the older cheiromants attached the greatest possible importance to these spots, concerning which they had a regular code of interpretations. I have gone into the question at length on p. 130 of " *A Manual of Cheirosophy*."

¶ 20.
Cutting the nails.

As to the cutting of nails the old wives are simply replete with dicta and dogmata, beginning with the old rhyme, " Cut them on Monday, cut them for health ; cut them on Tuesday, cut them for wealth, etc., etc. ;" but one and all are agreed on one vital point, and that is that " a man had better never been born, than have his nails on a Sunday shorn ;" and again in the same strain, " Better thou wert never

Holiday.

born, than on a Friday pare thy horn." Holiday, in his " *Marriage of the Arts* " (London : 1618), declaims

Lodge.

against such absurd superstitions ; and Lodge, in his " *Wit's Miserie* " (London : 1596), derides a young man for that " he will not paire his nails White-Monday to be fortunate in his love." One feels that

Tomkis.

Tomkis " hit out pretty freely all round him " when he says in " *Albumazar* " (London : 1615),

> " He puls you not a haire, nor paires a naile,
> Nor stirs a foot, without due figuring
> The Horoscope."

The Romans.

Nevertheless, superstitions concerning the paring of nails are of great antiquity. The Romans never performed this operation save upon the Feræ Nundinæ, which took place every ninth day, and the attention paid to such minor details as this is shown by the line of Ausonius,

" Ungues Mercurio, barbam Jove, Cypride crines." [39]

In Hazlitt's edition of Brand's " *Popular Antiquities of Great Britain* " (London: 1870) there is an account of an old woman in Dorsetshire, who always pared her children's nails over the leaves of the family Bible, in order that they might grow up honest ; and it is said to be with this object in view that for the first year of their lives children's nails are directed to be bitten off, and not on any account *cut*. And this notwithstanding the fact that Aristotle has clearly laid down, many centuries ago,[40] the obvious truth that " Nature has provided us with nails, not, as among the brutes, for purposes of offence and defence, but merely as a protection to the delicate tips of the fingers."

I might no doubt, had I time and space, say much upon Graphology or the science of detecting character from the handwriting. That the handwriting has certain marked characteristics in every individual, and that these characteristics, properly examined and interpreted according to given rules and method, will tell us much concerning the individual character of the writer, is an established fact, and the tabulation of these rules, and of this method, is now complete in the works of such acknowledged authorities as Rosa Baughan,[41] l'Abbé Flandrin, and Adolphe Henze.[42] Adrien Desbarrolles, in his major

¶ 22.
Graphology.

Rosa Baughan.
Flandrin.
Henze.
Des ba rrolles.

[39] " *Eclogarium*," 373. Hic versus sine auctore est.
[40] ΠΕΡΙ ΖΩΩΝ ΜΟΡΙΩΝ, Δ., ί.
[41] ROSA BAUGHAN, " *Character indicated by Handwriting*," 2nd edit. (London: 1886), revised and enlarged from a more elementary work on the subject.
[42] ADOLPH HENZE, " *Handbuch der Schriftgiesserei, etc.;*" in the "*Neuer Schauplatz der Künste*," Bd. 138, 1834 ; " *Die Chirogrammatomantie, oder*

[¶ 23]

work, " *Les Mystères de la Main: Révelations complètes, suite et fin* " (Paris : 1879), has devoted some 220 pages to this branch of the science of the hand, besides being the author of a standard work on the subject,[43] so that Graphology, or, as it is sometimes called, " Grammatomancy," boasts a literature of its own that it is not my intention to supplement in this place. " The more I compare different handwritings,"

Lavater. says Lavater in his " *Physiognomische Fragmente zur Beförderung der Menschenkenntniss,*" etc. (Leipzic : 1775-78), " the more am I convinced that handwriting is the expression of the character of him who writes. Each nation has its national character of writing, as the physiognomy of each people expresses the most salient points of character in the nation ;" and I may quote the remark of Rosa Baughan [*Op. cit.,* p. 2], " That the handwriting really reflects the personality of the writer is evident from the fact that it alters and develops with the intelligence, that it becomes firm when the character strengthens, weak and feeble when the person who writes is ill or agitated, and erratic when he is under the influence of great joy, grief, or any other passion."

¶ 23.
Origin of hand-
writing.

Handwriting is not, of course, in the present day —indeed, since the introduction of printing—of the vital importance that it was before the invention of the printing press, and the exquisitely-written manuscript is now, excepting on rare occasions, a thing of the past. Any remarks of mine on the early history of writing must necessarily be out of place here, but the curious in such matters should consult

Lehre den Charakter, etc., der Menschen aus der Handschrift zu erkennen, etc." (Leipzic : 1862) ; " *Das Handschriften Lesebuch* " (Leipzic : 1854), etc., etc.

[43] AD. DESBARROLLES and JEAN HIPPOLYTE, " *Les Mystères de l'Écriture ; Art de juger les Hommes sur leurs Autographes* " (Paris : *n.d.*).

the works of Astle and of Humphreys.[44]   Still, a
great interest must necessarily attach to fine pen
work ; to the elaborate script of Cocker, the originator   Cocker.
of the expression " according to Cocker " ;[45] whilst
we should many of us like to see a specimen of the
work of " Ricardus, Scriptor Anglicus," whose legend   Ricardus,
Scriptor Angli-
Miss Horsley has illustrated in the heading to this   cus.
introduction.[46]   Whilst, again, who is there of us
who has not profoundly objurgated the name of the   Hamlet.
friend who, like Hamlet, doth

> " hold it, as our statists do,
> A baseness to write fair, and laboured much
> How to forget [his] learning "—[Act v., sc. 2].

the incorrigible " kakographist," who is at once the
bane of his correspondents, and the chief thorn in the
uneasy chair of the hard-worked editor.   Perhaps
the most complete compendium of facts relating to
this subject is contained in a recent publication of the
" *Sette of Odd Volumes*" entitled, " *Pens, Ink, and
Paper, a discourse upon the Caligraphic Art with*

[44] ASTLE, " *The Origin and Progress of Writing*"
(London : 1802).   HUMPHREYS, "*Origin and Progress
of the Art of Writing*" (London : 1855).

[45] Edward Cocker [b. 1631, d. 1677], author of
" *Plumæ Triumphus, or the Pen's Triumph*" 1657 ;
" *Pen's Transcendencie*," 1657 ; " *The Artist's Glory,
or the Penman's Treasure*," 1659 ; "*Penna Volans*,"
1664 ; " *Vulgar Arithmetic: ACCORDING TO COCKER,*'
1677, and a quantity of other works on the same
subject.

[46] This English monk, by name " Richard," was a
writer of such perfection and elegance that, his tomb
having been opened twenty years after his death,
his right hand was found fresh and perfect as it had
been in life, though the rest of his body was reduced to
dust.   A full account occurs in " *Illustrium Mira-
culosum et Historiarum Memorabilium Lib. XII,
ante annos fere cccc à Cæsario Heisterbachiensi*"
(Cologne : 1591), lib. xii., cap. xlvii. ; and Miss
Horsley has illustrated the passage at the point above

D. W. Kettle.   *Curiosa,"* etc., by D. W. Kettle (London: *Oda Volume Opusculum*, No. X., 1885), which contains here-anent "marvellous riches in a little roome."

¶ 24.
Cheiromancy.

Nor is it necessary for me to enter here into the art of cheiromancy,—a subject which I have discussed at its fullest length in the volume to which this is supplementary, *"A Manual of Cheirosophy"* (1885). Of course, the tendency to seek for indications of the destiny upon different parts of the body is universal and of incomparable antiquity [*" Manual,"* ¶ 58], and though some nations have looked for interpretable signs in the sutures of the skull, and others [the

Authorities
pro and *con.*

Persians, for example],[47] have looked for them upon the forehead, like Subtle in Ben Jonson's *"Alchemist,"*[48] by far the most universal and antique form of divination has been by the hand : and notwithstanding the

---

named.* Another most interesting story of a hand which never putrefied is that one concerning the hand of King Oswald [A.D. 644], which, on account of his many good works, was blest by the Bishop Aidanus, and thereby rendered incorruptible. It may be found at folio 169, lib. i., of *" Flores Historiarum, Matthæus Westmonasteriensis monachus,"* 1567.

[47] *Vide*, for instance, the couplet in the 311th "Rubai" of Omar-i-Khayyám (Whinfield's translation, 1883),—

> "'Who wrote upon my forehead all my good
> And all my evil deeds ? In truth, not I."

[48] *Subtle.* " By a rule, captain,
> In metoposcopy, which I do work by ;
> A certain star in the forehead, which you see not.
> BEN JONSON, *The Alchemist*, act. i., sc. 1.

---

* " In Arnisberg monasterio ordinis Præmonstratensis, sicut audivi a quodam sacerdote ejusdem congregationis, scriptor quidam erat Ricardus nomine, Anglicus natione ; hic plurimos libros in eodem coenobio manu propria præstotansia coelis. Hic cum fuisset defunctus et in loco notabili sepultus, post viginti annos, tumba ejus aperta, manus ejus dextra tam integra et tam vivida est reperta, ac si recenter de corpore animato fuisset præcisa ; reliqua caro in pulverem redacta fuit. In testimonium tanti miraculi manus eadem usque hodie in monasterio reservatur," etc.

diatribes of such authors as Mason[49] and Gaule,[50] we can trace a strong inclination in favour of the science, not only among doctors, but among all classes of men who have made their fellow-creatures their study, from John of Gaddesden[51] and Ben Jonson[52] to Coleridge[53] and Ivan Tourgueneff.[54]   Among the Arabs, indeed, the physicians attach an immense importance to the condition of the hands; it is thus

Arabian physicians

[49] Mason drastically ridicules the science of palmistry, "where men's fortunes are told by looking on the palmes of the hande," in his "*Anatomie of Sorcerie*" (London : 1612), p. 90.

[50] JOHN GAULE, "Πυσμαντεία: *the Mag-astromancer or the Magicall-Astrologicall-Diviner posed and puzzled*" (London : 1652), p. 187.

[51] John of Gaddesden was, according to Freind, who gives a history of him in his "*History of Physick, from the Time of Galen to the Beginning of the Sixteenth Century*" (London : 1726, vol. ii., p. 277), a doctor of physick, the author of the famous "*Rosa Anglica*," who flourished at Merton College, Oxford, and subsequently at court, at the beginning of the fourteenth century.   He acquaints us [p. 284] with "his great skill in *Physiognomy*, and did design, if God would give him life and leisure, to write a treatise of *Cheiromancy;* but, to our unspeakable grief, this excellent comment upon *Fortune-telling* is lost."

[52] Ben Jonson seems [in *Volpone : or the Fox*, act i. sc. 1] to consider that *in articulo mortis* the hand is the last part in which any feeling remains; when Mosca says to Corvino :—

<blockquote>
" Best shew it, sir ;
Put it in his hand—'tis only there
He apprehends : he has his feeling yet !"
</blockquote>

[53] " A loose, slack, not well-dressed youth met Mr. ——— and myself in a lane near Highgate.   ——— knew him, and spoke.   It was Keats.   He was introduced to me, and stayed a minute or so.   After he had left us a little way, he came back, and said : ' Let me carry away the memory, Coleridge, of having pressed your hand !'   ' There is death in that hand,' I said to ———, when Keats was gone ; yet this was, I believe, before the consumption showed itself distinctly."—Coleridge's "*Table Talk*," August 14th, 1832.

[54] I refer to the passage in "*Monsieur François :*

that we have it in the "*Alf Laylah wa Laylah*" [Burton's trans., vol. v., p. 220], where the learned damsel says :—"A physician who is a man of understanding looketh into the state of the body, and is guided by the feel of the hands, according as they are firm or flabby, hot or cold, moist or dry."

¶ 25.
The Cheirostemon.

The reverence, therefore, which is paid to the hand being, as we have seen, practically universal, it is not surprising that the Cheirostemon [Χειρόστεμον] should have been worshipped as a sacred tree by the Indians. Of this extraordinary vegetable phenomenon only a single specimen existed, at Toluca, in Mexico, until 1801, because the Indians used to gather all the flowers as soon as they appeared, to prevent its propagation. In 1801, however, certain scientists obtained the fertilisation of a flower and propagated the tree at Toluca. It is described by early Spanish historians of the conquest of Mexico, by whom it was called *Arbol de Manitro*—the Indians called it the Hand-tree. Augustin de Vétancourt [55] describes it as "bearing in the months of September and October a red flower, having the appearance of a hand, formed with such perfection as to the palm, the joints, the phalanges, and the fingers, that the most expert sculptor could not reproduce it so exactly. When it is green it is closed like a fist, and as it becomes red it expands and remains half open." Miss Horsley has reproduced a drawing of it, given by Humboldt,[56] as a heading to Sub-section II., p. 101,[57] and I must con-

Vétancourt.

Souvenir de 1843," where M. François calls Tourgueneff's attention to the breaking of the Line of Life in his hand, and predicts his own violent death, which subsequently comes about.

[55] AUGUSTIN DE VETANCOURT, "*Teatro Mexicano*" (Mexico : 1698), fol.

[56] "*Voyage de Humboldt. Plantes Equinoxiales*" (Paris: 1808), vol. i., p. 85.

[57] Besides the cheirostemon Miss Horsley has given

[¶ 25]

fess that de Vétancourt seems to have been carried away by his artistic imagination in his description.

Now, with regard to the study of the hand, it is difficult, if not impossible, to say where it originated. M. Fétis — in tracing the origin of the violin,[58]—has, in more than one place, remarked that there is nothing in the West that has not come there from the East,— a remark in support of which Colonel Tcheng-ki-Tong has brought forward some very interesting evidence.[59] That this science was studied very many centuries ago in the East we have abundant proofs, from the writings of Philip Baldæus, who tells us of the daughter of a rajah who was told all her fortune by a Brahmin by an inspection of the lines of her palm [60]

¶ 26.
Oriental origin of cheirosophy.

Tcheng-ki-Tong.

Baldæus.

me as headings to Sections I. and VI. representations of two somewhat similar vegetable phænomena, taken from p. 25 of the first issue of Messrs. Cassell's publication " *The World of Wonders.*"   The radish [p. 169] grew in a sandy soil at Haarlem, and was painted from the life by Jacob Penoy, whose friend Zuckerbecker presented the picture to Glandorys in 1672.   From this picture an engraving was taken by Kirby, from which Messrs. Cassell's copy was taken.   The parsnep [p. 95] was bought of a market woman in the usual way, and is said to have represented the back of a hand so perfectly that it could not be surpassed by any .painter.   The article from which I quote gives many most interesting instances of a like nature.

[58] *Vide* ED. HERON-ALLEN, "*Violin Making; as it was and is*" (London : 2nd edition, 1885), p. 37.

[59] "Of such nature are the exact sciences which no Western nation can boast of having created; such are the alphabetic characters which have served to delineate sounds; the fine arts whose masterpieces date from remotest antiquity; modern languages themselves, whose roots are derived from a common origin, the Sanskrit; the properties of magnetism, imported from the East, and the foundation of the navigator's art; and such, lastly, the various descriptions of literary composition,—all of which, without a single exception, were created in the ancient world."—"*The Chinese painted by Themselves,*" p. 184.

[60] "Es begab sich dasz ermeldter Ragia sich einsmahls

[¶ 26]

Apollonius of
Tyana.

[notwithstanding the express prohibition contained in the laws of Manū!], down to Godwin, who tells us [61] that Apollonius of Tyana "travelled to Babylon and Susa in pursuit of knowledge, and even among the Brahmins of India, and appears particularly to have addicted himself to the study of magic."

¶ 27.
Universal iove
of divination.

Whatever reliance we can place upon either of these authorities, we have only to reflect for a moment upon the universal love of any knowledge which savours of divination, which appears to be deeply implanted in the human mind. The tendency of the human intellect is to be for ever progressing,—the fruit which is out of reach appears ever to us to be sweetest, and the first result of an intimate knowledge of what things are apparent and exoteric, is a burning

---

vor seinen Einwohnern sehen liesz u. nachdem er verstanden dasz unter andern ein erfahrner Braman angekommen, liesz er denselben für sich fordern u. sagte :— 'Narret' (denn also war sein Nahm) 'siehe doch meiner Tochter in die Hände u. verkündige mir ob sie glückselig oder unglückselig, arm oder reich sein, viel oder wenig Kinder bebähren werde, ob sie kurtz oder lange werde leben ; sag mir alles frey rund heraus u. nim kein Blat vors Maul.' Diese Manier in die Hände zu sehen is unter den Heyden sehr gebräulich, davon der hochgelehrte Vossius, l. 2., "Idol.," c. 47, 'Chiromantes etiam manus partes singulis subjecere planetis,' etc., etc. Der Braman wie er ihr in die Hand sahe hub an und sagte: 'Herr König, nach aller Anzeigung der Linien allein so stehets darauf dasz von ihr sieben Kinder sollen gebohren werden, nähmlich 6 Söhne u. eine Tochter, von welchen der letzte dich nicht allein deiner Krohn in Reichs, sondern auch des Haupts in Lebens berauben wird u. sich also dann auf deinen Stuhl sitzen.' " (And so it turned out, this being the eighth transformation of "Vistnum," beginning of the third period of time.) "Wahrhaftige Ausführliche Beschreibung der Ost-Indischen Küsten Malabar, etc." Philippus Baldæus (Amsterdam : 1672), cap. v., p. 513.

[61] WM. GODWIN, "Lives of the Necromancers" (London : 1834), p. 158.

desire to become acquainted, to extend our knowledge in the direction of things that are hidden and esoteric. How many are there of us who in all sincerity might say with Democritus, that "he had rather be the possessor of one of the cardinal secrets of nature, than of the diadem of Persia." Bearing, therefore, these things in mind, we need no longer express surprise at the necessity which appears to exist, that we should give credence to some cardinal error, abandoning very often the one, only, as Fontenelle said,[62] to fall into another.

Democritus.

The time is rapidly passing away when "wise men are ignorant of many things which in time to come every common student shall know,"[63] and every day sciences, such as the one at present under discussion, are establishing themselves more and more firmly, as physical and exact; and did Omar-i-Khayyám live in the present day, he would no longer have occasion to say, as he did in the eleventh century :—

¶ 28.
Ignorance of pretended wisdom.

> "These fools by dint of ignorance most crass,
> Think they in wisdom all mankind surpass;
> And glibly do they damn as infidel
> Whoever is not like themselves—an ass ! "[64]

Another great danger which at one time threatened the science is now passing away; I allude to the ill-directed enthusiasm of the ignorant, who, greedy of the marvellous, take up the science [which they are not qualified to understand], and *pretend* to believe in and to understand it, merely as an implement of

¶ 29.
Charlatans

---

[62] " Ils subiront la loy commune et s'ils sont exempts d'une erreur ils donneront dans quelque autre."—FON-TENELLE, " *Entretiens sur la Pluralité des Mondes* " (Amsterdam : 1701), Ent. iv.

[63] ROGER BACON, " *De Vigore Artis et Naturæ.*" ["*The Mirror of Alchemy. Also a most excellent and learned discourse of the admirable Force and Efficiencie of Art and Nature*" (London : 1597)].

[64] Whinfield's translation, p. 106, *vide* note [60], p. 70.

[¶ 29]

histrionic effect. I am reminded in such instances of a Chinese proverb which tells of a company of blind men who started to climb a mountain to admire the view, which they all described and extolled long before they reached the top.

¶ 30.
Value of such
studies.

Of the value of the study of such a science as this one, I hardly think there can be two opinions among people who have any right to an opinion at all. "By cultivation of the abstract-concrete sciences," says Herbert Spencer,[65] "there is produced a further habit of thought not otherwise produced, which is essential to right thinking in general. . . . Familiarity with the various orders of physical [and chemical] phenomena gives distinctness and strength to the doctrine of *cause and effect*."[66] And this remark applies not only to a science so fixed and physical as the one under discussion, but even, I contend, to its most legendary and traditional branches, even going so far as the gipsy cheiromancy, by which the coppers of the servant girl are diverted by the eloquence of the itinerant sorceress and peddler. "If we would know

Godwin.

man in all his subtleties," says Godwin in the preface to his "*Lives of the Necromancers*," "we must deviate into the world of miracles and sorcery. To know the things that are not, and cannot be, but have been imagined and believed, is the most curious chapter in the annals of man. To observe the actual results of these imaginary phænomena, and the crimes and cruelties they have caused us to commit, is one of the most instructive studies in which we can possibly be engaged."

¶ 31.
Necessity of
special training.

Of course, the training which is necessary as a precursor to the study of a science such as this, is somewhat special, must be in some way conducive to

[65] HERBERT SPENCER, "*The Study of Sociology*" (London : 11th edn., 1884), p. 318.
[66] Compare "*Manual*," ¶¶ 65, 66, 69, and 89.

a habit of analytical thought.   Different people will
study the question from different standpoints.   "There
are acquired mental aptitudes for seeing things under
particular aspects as there are acquired bodily apti-   Spencer.
tudes for going through evolutions after particular
ways.   And there are intellectual perversities produced by certain modes of treating the mind, as there
are incurable awkwardnesses due to certain physical
activities   daily   repeated"   [Herbert   Spencer, *op.
cit.*, p. 314].   A certain *sense of relation* is requisite
to the student who would study the science, a sense
which develops itself as the subject unfolds itself,—
" *viresque acquirit eundo.*"

It may perhaps be argued that were the science   ¶ 32.
really entitled to the encomia which I have bestowed   Value of Cheiro-
                                                         sophy.
upon it, it would long ago have been investigated and
brought to perfection ; but in answer to such an argument as this I may cite a hundred reasons for its non-
development, of which, however, a few will suffice.   As
long ago as 550 B.C., Confucius remarked that " the   Kung-fu-tzu.
study of the supernatural is injurious indeed ; " and
throughout the intervening centuries there has always
existed a strong prejudice against any study that can
give to the student such advantages as are to be
derived from this one.   Besides this it has been very per-
tinently remarked that a continued habit of self-analysis
has a strong tendency to lead one to self-deception :   Self-deception.
the arguments which arrive at persuading others
have passed that point, and have become exaggerated
and deceptive in our own cases, and from a knowledge of what we are, to a conviction that we are
what we are not, is with a feeble mind a very easy
and inappreciable transition.

Such causes as this, therefore, have warred against   ¶ 33.
the full development of our science, but now that   Establishment
                                                        of the Science.
mankind has reached a century in which a calm and
self-restrained habit of mind has become a leading

[¶ 33]

characteristic, I feel every confidence in submitting to the world the principia contained in the science, saying with the author of "*Hermippus Redivivus,*" [67] "These are my principles which I submit to the strictest examination ; if they can be demonstrated to be false or precarious, I shall be sorry for myself and for mankind, since undoubtedly they carry in them a strong appearance of truth, and of the most pleasing kind of truth—that which attributes glory to God by displaying His goodwill to man."

¶ 34.
Explanation

Longfellow.

I have endeavoured in the pages which follow to give a satisfactory explanation of the scientific bases of the science, [in defiance of Longfellow's opinion that explanations of a beautiful theory are supererogatory,[68]] an explanation more complete and minute, and in some measure supplemental to those which I have given in the Introductory Argument to "*A Manual of Cheirosophy.*"   I have thought it far better to adopt this course, relying on a certain amount of probability of success, than to shelter myself behind the arguments which I have advanced in ¶ 97 of that introductory argument, contenting myself in this

Drummond.

connection with merely quoting Henry Drummond's authority for saying that a science without mystery,

[67] [JOHN COHAUSEN], "*Hermippus Redivivus ; or, the Sage's Triumph over Old Age and the Grave*" (London : 1744), p. 14.   Recently reprinted by Mr. Edmund Goldsmid (Edinburgh : 1885).

[68] "And why should one always explain ?   Some feelings are quite untranslatable.   No language has yet been found for them.   They gleam upon us beautifully through the dim twilight of fancy, and yet, when we bring them close to us, and hold them up to the light of reason, lose their beauty all at once ; as glow-worms, which glimmer with such a spiritual light in the shadows of evening, when brought in where the candles are lighted, are found to be only worms like so many others."—H. W. LONGFELLOW, "*Hyperion,*" bk. iii., chap. 6 (Boston : 1881 edn., p. 253).

i.e., a fully-explained science, is not only unknown, but non-existent.[69] [*Vide* note [100], p. 91.]

¶ 35.
Mystery in science.
Bain.

"Mystery," says Bain [*vide* note [84], p. 66], "is correlated to explanation; it means something intelligible enough as a fact, but not accounted for, not reduced to any law, principle, or reason. The ebb and flow of the tides, the motions of the planets, satellites, and comets, were understood as facts at all times; but they were regarded as mysteries until Newton brought them under the laws of motion and gravity. Earthquakes and volcanoes are still mysterious; their explanation is not yet fully made out. The immediate derivation *of muscular power and of animal heat is unknown*, which renders these phænomena mysterious." And such is the case with the science which we have before us.

¶ 36.
Tabulation of facts and principles.

This being the case, though I have said in another place ["*Manual*," ¶ 93] that the tabulation and marshalling of facts alone will not by itself be sufficient for the establishment of this science upon a firm basis, still a proper observation of the *facts* of a case will very generally conduce to a great extent to an appreciation of its principles. "In the earlier centuries," says Drummond, in the opening paragraph of his "*Natural Law, etc.*," "before the birth of science, phenomena were studied alone. The world then was a chaos, a collection of single, isolated, and independent facts. Deeper thinkers saw, indeed, that relations must subsist between these facts, but the reign of law was never more to the ancients than a far-off vision." So it has been in the case of cheirosophy; the "palmistry" of the ancients was merely the interpretation of certain isolated facts arising from fortuitous concatenations of circumstances, but the

Drummond.

* HENRY DRUMMOND, "*Natural Law in the Spiritual World*" (London: 15th edn., 1885), pp. 28 and 88

[¶ 36]

cheirosophy of to-day is something more; it is the tabulation of certain received principles, upon certain rules approved by physical science, in obedience to a recognised system which has received the support of reason and of experience. Only, instead of treating our data as isolated truths, useful only by way of illustration and support, we are careful to verify our

Fontenelle.

facts as we go, like Fontenelle's "*true philosophers,* who are like *elephants,* who as they go never put their second foot to the ground until their first be well fixed." [70]

¶ 37.
Value of
investigation.

And after all, the more drastic the investigation to which such truths are submitted, the more clearly will their verity become apparent. " It is the great beauty of truth that the more we examine it, the more different lights in which we place it, the more pains we take in turning and twisting it, the more we perceive its excellence and the better the mind is satisfied about it " [" *Hermippus Redivivus,*" p. 45]; but the method of investigation pursued must be the right one, for, as Coleridge very justly remarked, to set up for a statesman upon historical knowledge only, is about as wise as to set up for a musician by the purchase of some score of flutes, fiddles, and horns. In order to make music you must know how to play; in order to make your facts speak truth you must know what the truth is which *ought* to be proved—the *ideal* truth,—the truth which was consciously or unconsciously, strongly or weakly, wisely or blindly, intended at all times. [71]

¶ 38.
Ridicule of the
science.

As for those persons who despise and ridicule the science, I have said enough concerning them else-

[70] " En fait de découvertes nouvelles, il ne faut pas trop se presser de raisonner, quoy qu'on en ait toujours assez d'envie, et les vrais philosophes sont comme les elephans, qui en marchant ne posent jamais le second pied à terre, que le premier n'y soit bien affermé."— FONTENELLE, *op. cit.,* vime soir, p. 139.

[71] "*Table Talk*" of S. T. Coleridge, April 14th, 1833.

where ["*Manual*," ¶¶ 71, 78] ; we must bear in mind Fontenelle's very just remark to the effect that "every species despises what it wants" [*Op. cit.*, vime soir], and it is of very little consequence that single sceptics make light of a science the importance of which is acknowledged by a vast body of their fellow-men,[72] for we can class them all with the Pococurante, of whom Candide said to himself, "What a surprising man, what a genius is this Pococurante! Nothing can please him!"[73]

Voltaire.

Under the same, or almost the same, category we may class those who declaim against the wickedness of the science ["*Manual*," ¶¶ 81—84], people who —to quote Candide once more—are like the father of Zenoida, who was wont to say, "Ill betide those wretched scribblers, who attempt to pry into the hidden ways of Providence" ["*Candide*," part II., ch. xiii.]. It is true that among the younger and abler minds of to-day there exists, as Drummond has said in the Preface to his "*Natural Law, etc.*," a " most serious difficulty in accepting or retaining the ordinary forms of belief. Especially is this true of those whose culture is scientific. And the reason is palpable. No man can study modern science without a change coming over his view of truth. What impresses him about nature is its solidity. He is there standing

¶ 39.
Wickedness of
the science.

Drummond.

---

[72] " There is no people, rude or learned, among whom" [such things as this] " are not related and believed. This opinion which prevails as far as human nature is diffused could become universal only by its truth ; those that have never heard of one another would not have agreed in tales which nothing but experience can make credible. That it is doubted by single cavillers, can very little weaken the general evidence, and some who deny it with their tongues confess it with their fears."— SAMUEL JOHNSON, " *Rasselas, Prince of Abyssinia*," chap. xxi.

[73] VOLTAIRE, "*Candide, ou l'Optimisme*," Ed. Originale: *reprint* (Paris: 1869), chap. xxv.

[¶ 39]

upon actual things, among fixed laws. And the integrity of the scientific method so seizes him that all other forms of truth begin to appear comparatively unstable." But this form of the free-thought of youth and ability is a very different thing to that against which Schlegel so drastically inveighs in the opening pages of his "*Philosophy of Life.*" A habit of thought engendered by a minute observation of the fixity of the laws of nature, so far from being pernicious or even undesirable, is calculated to advance us on rapid sails across what Professor Sir Richard Owen has called "*the boundless ocean of unknown truth.*" [74]

Schlegel.

¶ 40.
Fixed laws of nature.

Sir W. Scott.

Godwin.

Newman Smyth.

And of the actual fixity of these laws it is not necessary for me, in this place, to bring forward authorities. "The pursuers of exact science to its coy retreats," says Sir Walter Scott, "were sure to be the first to discover that the most remarkable phænomena of nature are regulated by certain fixed laws, and cannot rationally be referred to super natural agency, the suffering cause to which super-stition attributes all that is beyond her own narrow power of explanation;" [75] and Godwin has commenced his "*Lives of the Necromancers*" with the words, "The improvements that have been effected in. natural philosophy have by degrees convinced the enlightened part of mankind that the material universe is every-where subject to laws, fixed in their weight, measure, and duration, capable of the most exact calculation, and which *in no case admit of variation and exception.*" And it is upon this fixity of natural laws which Newman Smyth has described as the expression of

"SIR RICHARD OWEN, "*On the Nature of Limbs*" (London: 1849), p. 83. *Vide* "*Manual,*" ¶ 90.
[75] "*Letters on Demonology and Witchcraft,*" letter vi.

[¶ 40]

"the Divine veracity of nature"[76] that our science of cheirosophy has its foundation.

The books which contain the principia of this science are legion. The volumes catalogued in the " *Bibliotheca Cheiromantica,*" Appendix B, at the end of this book, dealing of course as they do principally with cheiromancy "*pur* [*mais pas*] *simple,*" contain a great mass of data which are of very little value to the cheirosophist.   It is not enough to buy a book upon the science, or half a dozen books, and devour them with unreflecting avidity,—a process which must inevitably result in intellectual indigestion.   We must bear in mind Bacon's excellent advice, in his essay " *Of Studies,*" to the effect that " some bookes are to be read onely in parts;   others to be read, but not curiously ; *and some few* to be read wholly and with diligence and attention," and so on.   Of course, the vast majority of books hereon are old and frequently obsolete, and great care is required in selecting and rejecting the data which they give.   Still, both Sampson and Don Juan pointed out to Don Quijote that there is no book so bad that it does not contain *some* good thing,[77] and a comparatively short study *and* practice of the science of cheirosophy will direct the student in his researches in this matter.

¶ 41.
Books on the science.

Bacon

Don Quijote

For it does not suffice merely to read books on the subject; it must be continually practised, that the principia laid down in the manual may be impressed upon the mind by their verification in actual experience.   " Personal experiment," said Coleridge [" *Table Talk,*" October 8th, 1830], "is necessary in order to

¶ 42.
Necessity of practising the science.

Coleridge.

---

[76] NEWMAN SMYTH, " *Old Faiths in New Light*" (New York : 1879) (London : 1882), p. 252.

[77] " No hay libro tan malo—dijo el bachiller—que no tenga algo bueno."—"*Don Quijote,*" parte ii., cap. 3. " Con todo eso—dijo el Don Juan—sera bien leerla, pues no hay libro tan malo, que no tenga alguna cosa buena."—*Ib.*, parte ii., cap. 59.

[¶ 42]

correct our own observation of the experiments which nature herself makes for us—I mean the phænomena of the universe. But then observation is, in turn, wanted to direct and substantiate the course of experiment. Experiments alone cannot advance knowledge without observation ; they amuse for a time, and then pass off the scene and leave no trace behind them." *Verb. sap.*

¶ 43.
Value of failures in the science.

The student must, of course, be prepared, in this as in every other science, to find occasional baffling inconsistencies.[78] I have laid down elsewhere the preparatory notes for the treatment of such contingencies ;—a young science possesses no more the perfection of the established science, than the human infant possesses that of the human adult, and the student will speedily learn to derive as much information from his failures as from his successes—just as in learning a language nothing impresses a phrase so firmly on the mind as to forget it suddenly, and to be obliged to dissect and re-acquire it.

¶ 44.
Necessity of learning good and evil in the science.

And finally you must be prepared to take, as it were, the rough with the smooth, to learn both the evil and the good which centres herein, for without the contrast of evil you cannot appreciate the good. It is continually argued to me that it is not good to know one's fellow-men as accurately and completely as one is enabled to do by means of this science ; that people are, as a rule, much more charming as they seem, than as they are. If such thoughts as this disturb you, lay down these volumes, O gentle-hearted reader, for the science of cheirosophy is not for you !

¶ 45.
Biassed arguments for one's own cause.

I may perhaps incur the charge of undue insistence, in advancing these arguments in favour of this science, as an introduction to this volume, but without wish-

[78] *Vide* ¶ 433 and note [337], p. 282.

ing it to be said in the words of Ben Jonson's Sir Epicure Mammon, that "if I take you in hand, sir, with an argument, I'll bray you in a mortar."[79] I am anxious to escape the censure bestowed upon Fontenelle by his marchioness, who complained of his convincing her only with his weaker arguments, and keeping his stronger ones in the background.[80]

I wish to avoid as much as possible the very easily fallen-into sin of advancing biassed or favourably exaggerated arguments in favour of my own case, and I wish to say everything which has to be said upon the subject, indifferently to its effect upon the reader's mind; not, like the Delphic pythoness,[81] to divide what it is judicious to say from what it is prudent to omit; to dwell upon one thing, and slur over another, as suits the purpose in hand; but that I may deal honestly with my subject and with the student, saying, as Drummond has said in the preface to his "*Natural Law*":—"And if with undue enthusiasm I seem to magnify the principle at stake, the exaggeration—like the extreme amplification of the moon's disc when near the horizon—must be charged to that almost

¶ 16.
The like.

Drummond

The Alchemist, act ii., sc. i.
⁸⁰ "Je vous ay pourtant pas dit la meilleure raison qui le prouve, repliquay-je. Ah ! s'écria-t-elle, c'est une trahison de m'avoir fait croire les choses sans m'en apporter que de foibles preuves. Vous ne me jugiez donc pas digne de croire sur de bonnes raisons ? Je ne vous prouvois les choses—répondis-je—qu'avec de petits raisonnemens doux, et accomodez a vostre usage ; en eussay-je employé d'aussi solides et d'aussi robustes que si j'avoîs eu à attaquer un Docteur ? Oui, dit-elle, prenez moy presentement pour un Docteur, et voyons cette nouvelle preuve du mouvement de la Terre." FONTENELLE, *op. cit.*, vi soir.—Placing, therefore, my reader in the place of the marquise, I propose to treat him as a doctor whom I labour to convince.
⁸¹ *Vide* HORACE, "*De Arte Poetica*," vers. 148 *et seq.*

[¶ 46]

necessary aberration of light which distorts every new idea, while it is yet slowly climbing to its zenith."

¶ 47.
Statement of facts.

Again, reflecting that arguments may often ruin the cause they desire to advance, by perplexing when they do not convince, I shall prefer to state the facts, the truth of the case, without comment, rather than to cover the light of the science with a bushel of argument.[82]

¶ 48.
Danger of recording only successes.

And lastly, I desire not to incur the charge of recording only successes and not failures of the system, so that the student may not paraphrase to me the words of the Roman to whom were pointed out the votive tablets of those who, "in consequence of their prayers to Neptune," had *not* been drowned in shipwreck, as a proof of the efficacy of such prayers, and who somewhat significantly replied, "Yes ; but where are the votive tablets of those who *have* been drowned?"

¶ 49.
Man's nature depends on his environment.

I would call particular attention to the now generally accepted fact that man is formed almost wholly by his environment,[83] that his organisation, like that of the animals, conforms to the necessities of his life. Perhaps the data upon which such a dictum as this may be amply supported, have never, in a condensed form, been put before the world more clearly than by the author of a now somewhat obsolete but most

---

[82] "Some in their discourse desire rather commendation of Wit, in being able to hold all arguments, than of Judgment in discerning what is true : As if it were a praise to know what might be said, and not what should be thought."—BACON, "*Of Discourse.*"

[83] "The Influence of Environment may be investigated in two main aspects. First, one might discuss the modern and very interesting question as to the power of Environment to induce what is known to recent science as variation. A change in the surroundings of any animal, it is now well known, can so react upon it as to cause it to change. By the attempt, conscious or unconscious, to adjust itself to the new conditions, a true physiological change is gradually wrought within the organism. Hunter, for example, in a classical

interesting opusculum, " *The Hand, Phrenologically Considered* " (London : 1848), chap. iii. of which is mainly devoted to a clear and progressive essay on the subject, entitled " Appendages to the Trunk, a Key to the entire Organisation and Habits of Animals" [pp. 22-43]. This essay terminates with the words:— " From this cursory examination of the animal world we may gather the important conclusion, that from the structure of an extremity we may obtain a complete insight into the entire organisation of an animal; and thus the paws furnished with sharp retractile claws of the lion, indicate at once to a naturalist its strong teeth, its powerful jaws, and its muscular strength of limb ; while from the cleft foot of the cow the complicated structure of its stomach, the definite peculiarities of its jaws, and its vegetable diet may with equal certainty be predicated " [p. 48].   A mass of interesting information, germane to this matter also, may be found in the chapter on " Relations," contained in Paley's " *Natural Theology.*"   At present   Paley. the point to be established is this, that man is formed by his environment, his actions, and his manner of life ; and that, therefore, by a discriminating examination of the man himself that environment, those

experiment so changed the Environment of a sea-gull by keeping it in captivity that it could only secure a grain diet. The effect was to modify the stomach of the bird, normally adapted to a fish diet, until in time it came to resemble in structure the gizzard of an ordinary grain-feeder, such as the pigeon."—DRUMMOND, " *Natural Law in the Spiritual World*," chapter on " Environment," p. 255.   Professor Drummond goes on to cite several other most interesting similar instances. *Vide* also KARL SEMPER [" *Die Natürlichen Existenz-Bedingungen der Thiere*, 1880] " *The Natural Conditions of Existence as they affect Animal Life* " (London : 1881), and C. DARWIN, " *The Variation of Animals and Plants under Domestication* " (London 1868, 2nd edn., 1875).

actions, and that manner of life may, with practical certainty, be inferred. "*I am a part of all that I have met*" ["*Odyssey*"].

¶ 50.
Body formed by the mind and vice versâ.

It may also be said contrariwise, that the mind is formed by the body as much as the body by the mind; *i.e.*, that a deformity of the body will very often result in a perversion of the mind, and that certain physical peculiarities will produce certain individualised habits of thought and action. The matter is entirely one of reciprocity,—a reciprocity which, so far from confusing us, renders our science more and more clear as we find more and more points of connection between mind and body.[84] "According to the definition of the physiologist Müller, the temperaments are peculiar permanent conditions, or modes of mutual reaction of the mind and organism, and they are chiefly dependent on the relation which subsists between the strivings or emotions of the mind and the excitable structure of the body. Even if we may be disposed to contend that they are not absolutely dependent on any particular constitution of the body, it must still be conceded that they are at least *associated* with certain peculiarities of outward organisation, by which they may be speedily recognised, so that the physical structure, the mental tendency, and the character of ideas are always intimately connected." [85]

¶ 51.
Bain on the relation of body and mind

"There is no example," says Bain [*vide* note [84]], "of two agents so closely united as mind and

[84] Dr. Alexander Bain's work, "*Mind and Body, the Theories of their Relation*" (London: 7th edit., 1883), is probably well known to many of my readers, as dealing exhaustively with the many theories which have been advanced in explanation with this relation.
[85] "*The Hand Phrenologically Considered*" (London: 1848), p. 6. *Vide* also Herbert Spencer's "*Study of Sociology*" (London: 1873), chap. xiii., "Discipline" [11th edit., 1884, p. 324].

body, without some mutual interference or adaptation. . . . On a theme so peculiar and so difficult the only surmise admissible beforehand would be that the two distinct natures could not subsist in their present intimate alliance and yet be wholly indifferent to one another ; that they would be found to have some kind of mutual co-operation ; that the on-goings of the one would be often a clue to the on-goings of the other. . . . We can begin at the outworks, *at the organs of sense and motion*, with which the nervous system communicates ; we can study their operations during life, as well as examine their intimate structure ; we can experimentally vary the circumstances of their operation ; we can find how they act upon the brain, and how the brain reacts upon them. Using all this knowledge as a key, we may possibly unlock the secrets of the anatomical structure ; we may compel the cells and fibres to disclose their meaning and purpose." [86]

I will close this section of my argument with this retrospective remark of Milne-Edwards,' "That the faculties of the *mammalia* are the more elevated in proportion as their members are the better constructed for prehension and for touch," [87]—an axiom the truth of which I think most of what has gone before has sufficiently established, an axiom which entirely clinches the data referred to in ¶ 49 and note [83], p. 64.

¶ 52.
Intelligence and Articulation of extremities.

It may perhaps be argued that these data may have been tabulated in error ; that the *real* nature of man

¶ 53.
Disclosure of man's real nature.

[86] " The form and posture of the human body, and its various organs of perception, have an obvious reference to man's rational nature; and are beautifully fitted to encourage and facilitate his intellectual improvement." —DUGALD STEWART, " *Outlines of Moral Philosophy*" (Edinburgh : 2nd edit., 1801), p. 68, sect. xi., ¶ 88.
[87] H. MILNE-EDWARDS, "*A Manual of Zoology*" (London : 2nd edit., 1863), §§ 306—343.

[¶ 53]

very seldom appears upon the surface; and that the deductions which are set down in the following pages may be founded upon misapprehension. I have only to say that, carefully watched, all men must at some time or another allow their true natures to appear,

Bacon.

"for Nature," as Bacon said ["*Of Nature in Men*"], "will lay buried a great time and yet revive upon occasion or temptation;" and again ["*Of Negociating*"], "Men discover themselves in trust, in passion, at unawares, and of necessities, when they would have somewhat done, and cannot finde an apt pretext;" and it is by watching for such occasions as these that we attain to the verification of our principia, my task being principally to state those principia clearly and correctly. By inapplicable phraseology many a question has been darkened and mystified to the point of despair. In the history of philosophy we find numerous instances of contradictions being brought about by inappropriate language, and in no case more fatally than in the tabulation of the data upon which to found such sciences as this one.

¶ 54.
Value of the
science in human
intercourse.

Of the value of this science it is not necessary for me in this place to say any more than I have already had occasion to say ["*Manual*," ¶¶ 85-6]. Alas! that, as Sir Walter Scott found it necessary to say,[88] truth should not be natural to man; were it so, one of the great arguments in our favour would be annulled; and again, one's first impressions being practically

---

[88] "The melancholy truth that 'the human heart is deceitful above all things, and desperately wicked,' is by nothing proved so strongly as by the imperfect sense displayed by children of the sanctity of truth. . . . The child has no natural love of truth, as is experienced by all who have the least acquaintance with early youth. If they are charged with a fault while they can hardly speak, the first words they stammer forth are a falsehood to excuse it."—"*Letters on Demonology and Witchcraft*," letter vii.

indelible,[89] is it not of vital importance that they should be correct, so that we may not start in our intercourse with our fellow-men at the prime disadvantage of having formed a mistaken idea of their characters and modes of thought? It is these two great advantages that I claim to confer upon you by the inculcation of the principles of this SCIENCE OF CHEIROSOPHY; it is this that I desire to impress upon thinking minds; it is this superiority that I claim to be existent in the HANDS as the outward indicators of the inner characteristics of man.

¶ 55.
Anatomy of the hand.

Before proceeding to consider the principles and practice of the science of cheirognomy, I think it is necessary that we should clearly understand what we are working upon, we should know thoroughly the anatomical construction of the hand, and that we should be in a position to say upon what development of what particular muscles and bones particular forms of hands depend.

¶ 56.
Man the most perfect created being.

I have already [¶ 52] quoted Milne-Edwards' remark upon the progression of intellect in a direct ratio to the articulation of the extremities of animals, a remark which finds a curious parallel in the second subsection of M. d'Arpentigny's work [¶ 99]. How obviously may we carry the observation further, and at length, contemplating the perfectly-articulated hands of MAN, say with the Persian tent-maker :—

" Man is the whole Creation's summary,
  The precious Apple of great Wisdom's Eye,
    The circle of Existence is a Ring
  Whereof the Signet is Humanity.

Ten Powers, and nine Spheres, eight Heavens made He,
And Planets seven, of six sides we see,
    Five Senses, and four Elements, three Souls,
  Two Worlds—but only ONE, Oh MAN ! like thee ! "[90]

[89] As Fontenelle remarks on the sixth evening of the " *Entretiens* " I have already quoted, *passim*.
[90] "*The Rubaiyát of Omar-i-Khayyám,*" Rubaiyát

¶ 57.
Observation of
environment.

But though a Power of which we can know *nothing* in the present finite condition of our intellects has brought us to this state of perfection, still we must not lose sight of the great principle continually laid down, to which I have before [¶ 49] referred, *i.e.*, that man is formed to a great extent by his environment; and the pre-natal conditions of his existence are only a further development of this same principle. The development of any organism in any direction is dependent upon its environment. A living cell cut off from all air will die. A seed-germ apart from moisture and an appropriate temperature will make the ground its grave for centuries. Human nature likewise is subject to similar conditions. It can only develop in presence of its environment.

¶ 58.
Early develop-
ment of the hand.

In the human embryo the hand is the first member which in any way suggests its final development. In the embryo of a month, though the hand resembles a fin more than anything else, it is still distinctly apparent, and the illustration given of a stage three weeks later on [p. 126 of vol. i of the eighth edition of Quain's "*Elements of Anatomy*" (London: 1876)] represents this elementary development very clearly and well. The hand is perfect at birth, save that the palm is always out of proportion with the fingers, itself an interesting factor in the establishment of the science, regard being had to the remarks of our author in ¶ 94. It is at about the age of fourteen, both in men and women, that the hand assumes the permanent form as the indicator of the characteristics which will follow a subject through life. [91] [*Vide* ¶¶ 65 to 67.]

340 and 120 of Whinfield's translation [note *"*, p. 48].
[91] We are not discussing the lines of the palm in this volume, but I beg to call attention to the following

It is now that the art of the cheiromant comes into requisition. "I can never see any difference in hands; I cannot tell whether my hands are big or small; I cannot tell whether my thumb is large or not; I do not know whether my fingers bear the right proportion to my palm." Such things as this are said to me every day, and it is of course only after a small amount of observation that the student is able to determine these points for himself, to distinguish differences which are imperceptible to the unpractised eye, just as the banker's clerk detects a forged banknote after it has deceived persons more clever intellectually, but less practised in this particular direction than he. The whole matter turns first upon the question of comparison or discrimination, then of agreement, and finally of memory. "When we have anything new to learn," says Bain [note [81], p. 66], "as a new piece of music or a new proposition in Euclid, we fall back on our previously combined combinations, musical or geometrical, so far as they apply, and merely tack certain of them together in correspondence with the new case. The method of acquiring by patchwork sets in early and predominates increasingly" [p. 87]. This then is the manner in which the science of cheirosophy is to be acquired.

¶ 59.
Differentiation of hands.

Stages of examination

paragraph which occurs on page 214 of vol. ii. of Quain's "*Anatomy*" :—" The free surface of the corium " [or horny layer of the skin] " is marked in various places with larger or smaller furrows, which also affect the superjacent cuticle. The larger of them are seen opposite the flexures of the joints, as those so well known in the palm of the hand, and at the jointure of the fingers. The finer furrows intersect each other at various angles, and may be seen almost all over the surface. *These furrows are not merely the consequence of the frequent folding of the skin by the action of muscles or the bending of joints, for they exist in the embryo.*"

¶ 60.
Composition of
the hand.

Of what elements, therefore, are the hands of man composed? Of bones, muscles, and nerves, the whole being vivified by a complete and complex arterial and venous system.

¶ 61.
Composition of
bone.

To begin with the basis of the fabric—the bone; let us consider what is its composition, and how does it play its part in the human œconomy. "Bone," says Holden,[92] "is composed of a basis of animal matter impregnated with 'bone earth' or phosphate of lime." The first ingredient makes it tenacious and

FIG. I.—TRANSVERSE SECTION OF BONE [*Radius*].

elastic; the second gives it the requisite hardness, the animal part forming about one-third, and the earthy two-thirds. The general construction of long and round, broad and flat, and short and irregular bones need not occupy us, but the microscopic structure is of interest to us in our present study. Fig. I represents a transverse section of a bone. [All bones present the same characteristics, and I have chosen a section from the radius, as its peculiarities are perfectly shown.] The big black spots are

[92] LUTHER HOLDEN, "*Human Osteology*" (London: 4th edit., 1879), pp. 2, 14, 15, 22.

sections of the canals [called Haversian canals], which transmit bloodvessels to the substance of the bone. The small dark spots are minute reservoirs, called "bone corpuscles," "bone cells," or "*lacunæ*." The Haversian canals vary considerably in size and shape : they are generally round or oval, and whilst those nearest to the circumference of the bone are small, towards the centre they grow gradually larger, at length opening into the hollow which occupies the centre of the bone, and is filled with the marrow. Round the Haversian canals are disposed the "*lacunæ*" in concentric rings, known as "*laminæ*," which are formed of concentric layers of bone, which have been developed within the Haversian canal. The *lacunæ* are microscopic tubes, connected spiderlike with one another by means of tinier tubes still, called "*canaliculi*." All these bunches, as it were, of canaliculi *anastomose, i.e.*, are connected with one another, so that by these means a constant connection is kept up between the Haversian canal and its concentric layers of bone, by means whereof the nutrient juices are continually distributed to every part of the bone. Each Haversian canal, with its layers of bone, *lacunæ*, and *canaliculi*, is called a " Haversian system," each system being to a great extent independent of the others. Besides these there are the triangular spaces between the systems [caused by their circularity], which, being filled up with bone layers and *lacunæ* similar to the Haversian systems, are known as Haversian interspaces. Fig. 2 represents a longitudinal section of a similar bone.

Between the articulations of the bone and the cartilage there exists another species of bone containing no Haversian canals ; the *lacunæ* are larger than in the subjacent bone, and have no *canaliculi ;* the result of this is, that, not being vascular, it is much less porous than common bone, and consequently forms a much

¶ 62.
Articular bone

[¶ 62]

stronger and unyielding surface for the support of the cartilage. With this bone, however, which is called *articular bone*, we need not deal in this place, as it does not concern the present discussion. It is not necessary either for us to enter into the structure and development of embryonic bone, as we are dealing with the fully-developed hand exclusively [*vide* thereon HOLDEN, *op. cil.*, p. 23].

¶ 63.
Skeleton of the hand.

The skeleton of the hand [Plate I., *Frontispiece*] consists of twenty-seven bones.[93] The first eight are

FIG. 2.—EARTHY MATTER OF BONE.
Longitudinal section showing the Haversian canals.

the little bones of the carpus; the five succeeding bones constitute the metacarpus; these support the bones of the fingers. Each finger has three bones, termed, in order from the wrist, the first, second, and third or ungual phalanx. The thumb has only two phalanges.

¶ 64.
The carpus.

It may be seen that the eight bones of the carpus are arranged in two rows of four each, so as to form a broad base for the support of the hand. The reason of this mass of small bones is dual: it is to

[93] Holden's " *Osteology*," pp. 148-157.

permit extended motion, and also to confer elasticity ; for each articular surface [*vide* ¶ 62] is covered with cartilage, which, by preventing jarring, reduces the risk of fracture or dislocation to a minimum. This beautifully-constructed and flexible " buffer " [as it were] is called the " carpal arch ; " all these bones articulate with one another by plane surfaces connected by strong ligaments, and the second row support, in like manner, the bones of the metacarpus, giving them by their varied shapes and positions different degrees of mobility, the thumb being the most, and the third finger the least, movable of the digits, on account of the arrangements of the carpus which give them motion. The carpus, which is entirely cartilaginous at birth, does not become fully ossified and perfect as we see it in Plate I., until the end of the twelfth year [*vide* ¶ 58]. No muscles are attached to the back of the wrist, but the palmar surface of the carpus gives rise to the main muscles of the fingers.

The metacarpus consists of the five bones that support the phalanges of the thumb and fingers. ¶ **65.** The metacarpus. The shafts of these bones are slightly hollow as seen from the palmar surface, the bases of the bones articulating, not only with the carpus, but laterally with one another. The metacarpal bone of the thumb is distinguished by a characteristic saddle-shaped surface at its base. There are no less than nine muscles to work the thumb, and its great mobility depends upon this saddle-shaped joint at its base. Each of the metacarpal bones have certain distinguishing characteristics by which they are known to the osteologist. It is interesting also for us to note that every metacarpal bone has an *epiphysis* [*i.e.*, a mass of bone ossified from a separate centre of ossification] at the lower end, and sometimes at the upper ; these appear at different ages up to the fourth year, but

[¶ 65]

none of them join on to the shafts of the bones until about the twentieth year, so that the skeleton of the hand cannot be said to be perfect and complete until that age.

¶ 66.
The fingers.

Each finger consists of three bones, termed the phalanges. The *two* of the thumb correspond to the *first* and *third* of the fingers. The shafts of the phalanges are concave on their palmar surface [like the metacarpal bones] for the convenient play of the flexor tendons, and on each side of the flat or concave surface there is a ridge for the attachment of the fibrous sheath which keeps the tendons in place. As in the metacarpus, the perfect ossification, and the attachment of the *epiphyses* of these bones, is not complete until about the twentieth year.

¶ 67.
Sesamoid bones.

Besides the above bones we often find small bones bedded in the tendons of the joints of the thumb, called *sesamoid* bones. These give greater leverage and strength to the joints, and where you can see or feel them in a hand, they are always a great sign of strength.

¶ 68.
Muscular system.

It would, of course, be beyond the limits of a work like the present to enter into anything like a complete survey of the muscles of the hand and the principal muscles of the arm which are connected with them, but, seeing that on the development of these muscles quite as much as upon the development of the bones the shapes of the hand depend, a few passing remarks will not, I think, be out of place. For any more complete survey the reader should refer to such works as Quain's "*Anatomy*," etc.[84]

¶ 69.
Composition of muscles.

Every muscle constitutes a separate organ, composed chiefly of a mass of contractile fibrous tissue, with other tissues and parts which may be called

[84] Quain's "*Elements of Anatomy*" (London: 8th edit., 1876), vol. i., pp. 183, 218-225.

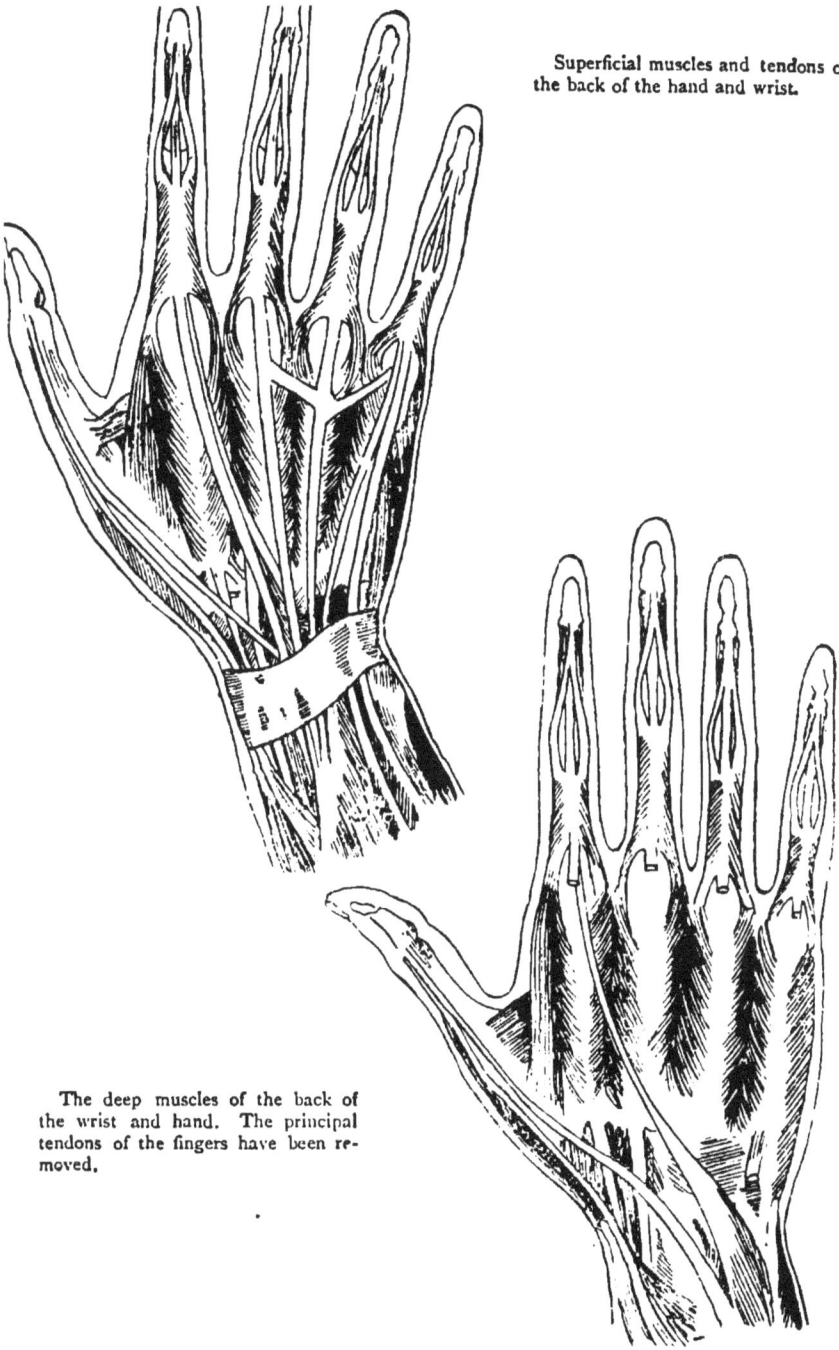

Superficial muscles and tendons on the back of the hand and wrist.

The deep muscles of the back of the wrist and hand. The principal tendons of the fingers have been removed.

PLATE II.—MUSCLES OF THE DORSAL ASPECT

accessory. Muscular fibres [*vide* Fig. 3] are connected together in bundles or *fasciculi*, and these
fasciculi are again embedded in, and united by, a
quantity of connective tissue, the whole "muscle"
being usually enclosed in a sheath of the same
material. Many of the muscles are connected at their
more or less tapering extremities with tendons by
which they are attached to the bones or hard parts,
and tendinous bands, as a rule, run into the substance,
or over the surface of a muscle. Bloodvessels are

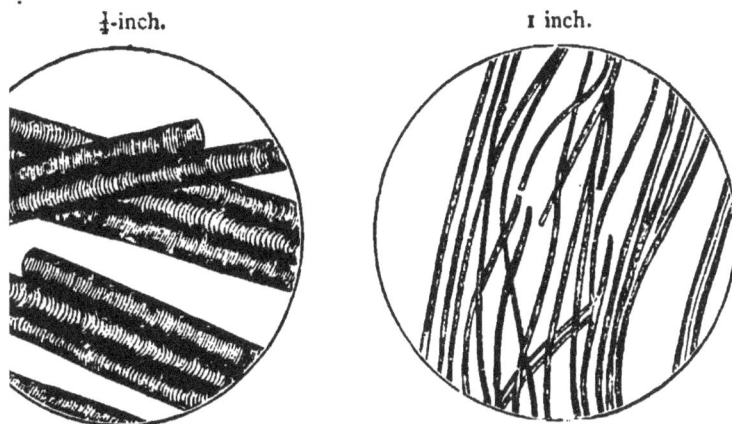

¼-inch. 1 inch.

A. Eyepieces.
FIG. 3.—VOLUNTARY MUSCULAR FIBRE.

largely distributed through the substance of a muscle,
carrying the materials necessary for its nourishment
and changes, and there are also lymphatic vessels.
Nerves are ramified through every muscle, by which
the muscular contractions are called forth, and a low
degree of sensibility is conferred upon the muscular
substance.

It is, of course, quite impossible, as it would be
inappropriate, to give in this place a full account of
the various muscles of the hand. They are as fully
represented as is necessary by Plates II. and III., and

¶ 70.
Muscles of the
hand.

they are there described. I wish merely to call particular attention in Plate II. to the great strength and variety of the tendons which support the muscles of the thumb and first finger, and in Plate III. to the enormous substance and importance of the muscles of the thumb and of the palm proper, which send off the most important digital muscles, an arrangement made especially clear to us by the differentiation in the two cuts of the superficial and of the deep-seated muscles. A glance at these two plates will be of more value to the student in cheirosophy than many pages of description.

¶ 71.
Arterial and venous system of the hand.

The arteries and venous system of the hand must necessarily interest us, as it varies continually in various individuals, and, as I have elsewhere pointed out [" *Manual*," ¶¶ 38 and 72], this very variation possesses of itself a very significant meaning for us. I have given a short account of the main arteries of the hand in " *A Manual of Cheirosophy;* " a fully detailed account may be found on pp. 410-419 of Quain's work already cited, containing practically all that is known concerning the various constituent parts of the superficial and deep palmar arches.

¶ 72.
Nervous system of the hands.

Nothing can be more interesting or significant to the student than the study of the nervous system of the hands,—a subject which I have discussed at length in my former volume, and to which it is not, therefore, necessary for me to do more than merely refer here. On the structure and arrangement of nerve fibres, cells, and centres, Bain [*vide* note ⁶⁴, p. 66] has given us some most interesting data on pp. 28-32 of his work already cited. "There are," says he, "some significant facts regarding the arrangement of the nerve elements. It is to be noted, first, that the nerve fibres proceed from the nerve centres to the extremities of the body without a break, and without uniting or fusing with one another ; so that each unfailingly

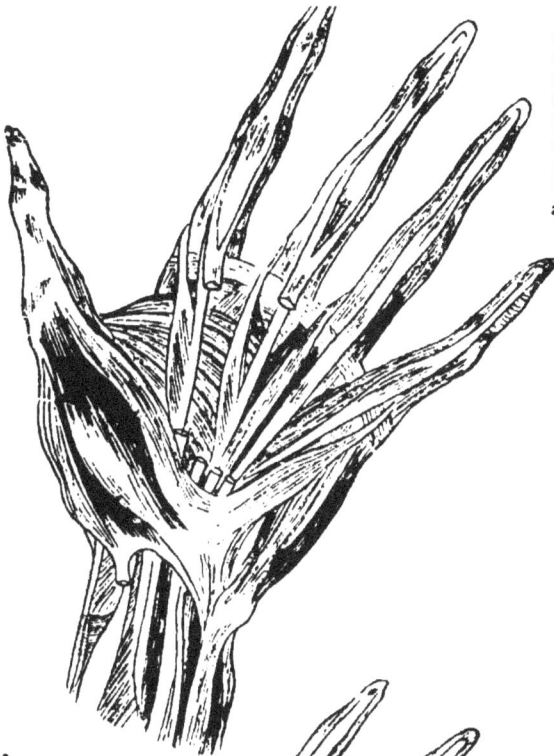

Muscular tendons of the palmar aspect. Parts of the superficial flexor muscles have been cut away to show the deep flexor and lumbrical muscles and their tendons.

Deep muscles of the palm; the abductor muscles of the thumb and little finger; the anterior annular ligament. The long flexor tendons in the palm have been removed. In the forefinger the tendons of both superficial and deep flexors remain; in the others only the deep flexors remain.

PLATE III.—MUSCLES OF THE PALMAR ASPECT.   6

delivers its separate message. Without this the greatness of their number would not give variety of communication. The chief use of the two coatings or envelopes appears to be to secure the isolation of the central axis.

"Remark, next, that the plan of communicating from one part of the body to another—as from the skin of the hand to the muscles of the arm—is not by a direct route from one spot to the other, but by a nervous centre. Every nerve fibre rising from the surface of the body, or from the eye or ear, goes first of all to the spinal cord or to some part of the brain, and any influence exerted on the movements by stimulating these fibres passes out from some nervous centre. As in the circulation of letters by post there is no direct communication between one street and another, but every letter passes first to the central office; so the transmission of influence from one member of the body to another is exclusively through a centre, or [with a few exceptions] through some part of the nervous substance contained in the head or backbone.

¶ 73.
Method of nerve communication.

"Every nerve ends in a corpuscle, and from the same corpuscle arises some other fibre or fibres, either proceeding back to the body direct, or proceeding to other corpuscles, whence new fibres arise with the same alternative. The corpuscles are thus the medium of connection of ingoing with outgoing nerves, and hence of communication between the outlying parts of the body. They are the crossings or grand junctions, where each part can multiply its connection with the remaining parts. *There is not a muscle of the body that could not be reached directly or indirectly by a pressure on the tip of the forefinger;* and this ramified connection is effected through the nerve cells or corpuscles." We are, therefore, in examining and dealing with the hands, concerned with the most

¶ 74.
Nerve terminations.

highly sensitive, the culminating point of this great
system by which practically our whole lives are
governed.

¶ 75.
**Principal nerve
trunks of the
hand.**
The hand is principally supplied by the two great
nerves of the arm, the *ulnar* and the *median*, the
latter doing most of the work. Branches of the
median nerve mainly supply the thumb, the first and

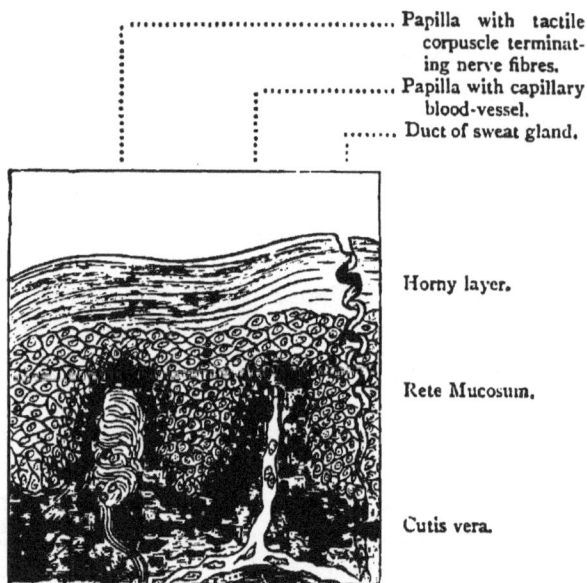

FIG. 4.—SECTION OF SKIN [*Diagrammatic*].
(*Highly magnified.*)

second, and one side of the third fingers; the other
side of the third and the little finger are supplied
mainly by branches of the ulnar. The dorsal surface
of the hand is mainly supplied by the *musculo-spiral*
and the *radial* nerves, principally by the latter, the
action of the former being confined principally to the
thumb and forefinger [*vide* further on this point
of neurology "*Manual*," ¶¶ 37, 39—41]. This
short sketch has, I think, been sufficient to give us

a general idea of the manner in which the nervous connection of the hand with the brain is carried out ; it follows, therefore, merely to consider the culminating point of this nervous system, the skin, which plays so important a part in the anatomy of the sense of touch [*vide* also " *Manual*," ¶¶ 40-51].

　　The skin is divided into the *cutis vera* [the true skin] and the *cuticle* [the epidermis or scarf skin]. These are shown diagrammatically by Figs. 4 and 5. The structure of the cuticle is best shown by Fig. 4,

¶ 76.
The skin
cuticle.

FIG. 5.—SECTION OF SKIN, SHOWING PAPILLARY SURFACE OF CORIUM, SWEAT GLANDS, AND PACINIAN CORPUSCLES.
[*1-inch objective A eyepieces.*]

and is as follows:—It is composed of cells agglutinated together in irregular layers, the lower ones of which, as may be seen in the figure, are arranged vertically, the cells, as they approach the surface, becoming more and more flattened till at last they lose their distinctive cellular formation, and become the flat horny scales of the scarf skin or cutis vera. This is clearly shown in Fig. 4 ; at the base we have the deep layer of vertical cells, above that the distinctive cells, known as the Malpighian layer or *rete mucosum*, and above this the *epidermis* or horny layer.

¶ 77.
Cutis vera and sensory apparatus.

The cutis vera or true skin is best seen in Fig. 5, and consists of a fibrous mass terminating in papillæ [of a pair of which Fig. 4 gives a magnified view], upon which the cuticle is moulded, and on the surface of which is spread an infinitesimally fine network of bloodvessels, generally running up into the papillæ, as in Fig. 4. The sensibility of the skin, its connection with the brain [*vide "A Manual"*] is due to the presence of nerve terminations, of which the largest are termed *pacinian corpuscles* [Fig. 5], and

FIG. 6.—HORIZONTAL SECTION OF SKIN, SHOWING PERSPIRATION PORES.

the smaller, which are found in the papillæ, *touch corpuscles* or *end bulbs* [Fig. 4].

¶ 78.
Sweat glands.

The sweat glands, of which 2,500 are to be found in every square inch of the skin of the palm, are shown in both figures, consisting of tubules coiled up in balls in the cutis vera, and proceeding by spiral ducts to the surface, where they can be seen with a strong magnifier, arranged in rows upon the curvilinear ridges [formed by the papillæ of the cutis], as shown in Fig. 6.

¶ 79.
The nails.

The nails are merely a thickening of the outer horny layer or cuticle, growing from matrices formed by folds

of that cuticle. They are formed from above as well
as from below at the root, so that the nail is continually pushed forward by the new growth at the
base. It is interesting to note that, if the nails are
neglected for years, they will grow as thick as they
are broad, and curl over towards the digit like a
claw.

Here, therefore, we have the complete physiology
of the hand so far as it is necessary for us to be con-
versant with it,—a physiology upon which we can base
that sense of touch which, as Sir Walter Scott has
said, it is impossible to deceive,[95] a quality which is
peculiar to it of all the senses of man, and a quality
which we find more highly developed in white-
skinned races than in any other;[96]—that sense of
touch which affords us perhaps the best evidence of

[95] " The sense of touch seems less liable to perversion than either that of sight or smell, nor are there
many cases in which it can become accessory to such
false intelligence as the eye and ear, collecting as they
do their objects from a greater distance, and by less
accurate inquiry, are but too ready to obey."—" *Letters
on Demonology and Witchcraft*," letter i.

[96] " As regards the darker-coloured races, we know
that they differ somewhat from the white in the texture
of their skin : it is coarse in its structure, provided with
a larger number of sebaceous glands, and covered by a
thicker layer of cuticle, so that the sentient terminations of the nerves being less exposed, its general sensibility must be considerably less than the skin of white
people."--[" *The Hand Phrenologically Considered*,"
p. 56.] Attention has been drawn by Dr. Wm. Ogle to
the fact that pigment occurs in the olfactory regions,
and he traces to this fact an increase in the acuteness
of smell. Dr. Ogle* attributes the acuteness of the
smell of the negroes to their greater abundance of pigment. Albinos and white animals neither see nor smell
so delicately as creatures that are dark-coloured [BAIN,
op. cit., p. 35].

* " *Anosmia*," by Dr. Wm. Ogle, in *Medico-Chirurgical
Transactions*, vol. liii.

[¶ 80.]

the connection of mind and body, and which is most highly developed in the human hand ["*Manual,*" ¶ 47].

¶ 81.
Conclusion.

The foregoing pages contain, I think, all that it is necessary for me to say by way of introduction to this volume, regard being had to the fact that I have already somewhat exhaustively argued the point [though along different lines] in the Introductory Argument to "*A Manual of Cheirosophy.*" If the above remarks are borne in mind during the perusal of the following pages, I think that it will be found that a new light is shed upon the *Science of Cheiro-sophy,* and that it has advanced yet another step towards a pre-eminence among the sciences which aim at a comprehension of man's inner nature by the observation of his physical peculiarities.

April 22nd, 1886.

Introductory Remarks

KNOW THYSELF. — Beautiful and wise maxim which the generality of men find it more easy to applaud and admire than to put into practice !

In Ionia, where the earth yields almost of her own accord all that is necessary to man, and where, by reason of the warm climate, a greater part of the needs which are indigenous to our latitudes are unknown, they were able to carve this maxim [97] upon the portico of their temple as a precept which every one in the

[97] " Know thyself." This maxim was engraved in letters of gold upon the front of the Temple at Delphi ; it was a dictum of the philosopher Socrates, according to the very interesting note at p. 1076 of Valpy's *Variorum Juvenal* (London: 1820) ; and, having been given as a maxim to Crœsus by the Delphic Pythoness, was inscribed, as above mentioned, on the front of the temple.* It is this circumstance which is referred to by Juvenal, when he says, " The maxim ' Know thyself ' descended from heaven ; "† and it is also recorded by Xenophon as a saying of Socrates, whom he causes to say (in the *Memorabilium*, bk. ii., cap. 2), " Tell me then, O Euthydemus, were you ever in Delphi ? " " Certainly," replies Euthydemus, " twice." " Do you remember seeing ' Know thyself ' inscribed upon the

* " Chilonis et Socratis præceptum tanquam oraculum e cœlo delapsum præ foribus templi Delphici aureis litteris scriptum fuit."—Valpy's " *Juvenal* " (London : 1820, p. 1076).

† " E cœlo descendit γνῶθι σεαυτὸν.—Juvenal, " *Satire* " xi., l. 27.

best interests of his or her own happiness is bound to put into practice : but in our own inclement lands, where we can get nothing from the soil excepting by the sweat of our brows, where the energies of our bodies and of our minds exhaust themselves in an endless struggle against the eternal aggressions of cold and damp, we have no time to devote ourselves to this beautiful esoteric study, recommended to us so long ago. At the same time it is in our foggy west, and it is since our population, augmenting itself in an ever-increasing ratio, has made manual labour more and more incumbent upon us, and more and more painfully exacting to us, that there have arisen amongst us the theories destined to reveal to us by the simple examination of a few physical signs the secrets of our inclinations and of our mental capacities.

¶ 84.
Phrenology and
physiognomy.
Gall.

Who has not read the works of Gall, the phrenologist, and of his enthusiastic disciples? [98] But their

altar?'' etc.*   Crœsus was a great benefactor of the Temple at Delphi; he recounts the reception of this oracle "Know thyself" to Cyrus, relating that the oracle said to him, "If thou knowest thyself, O Crœsus, thou shalt live happily."† Compare also Cicero, " *Tusculanum*," i., n. 52, and ii., n. 63 ; Pliny, vii., 32; and Persius, " *Satire* " iv., last line.

[98] Franz Joseph Gall was born in Tiefenbrunn, Baden, on the 9th March, 1758. He began lecturing on craniology in Vienna in 1796; but after six years of constant labour to inculcate his new theories was stopped by the Government in 1802. He lectured in various cities in company with Spurzheim, from 1807 till 1813, when he retired from public life.; he died in the year 1828.

---

* Καὶ ὁ Σωκράτης, Εἰπέ μοι, ἔφη, ὦ Εὐθύδεμε, εἰς Δελφοὺς ἤδη πώποτε ἀφίκου; καὶ δίς γε, νὴ Δί, ἔφη.   Καθέμαθες οὖν πρὸς τῷ ναῷ που γεγραμμένον τὸ ΓΝΩΘΙ ΣΕΑΤΤΟΝ ; Ἔγωγε. Πότερον οὖν οὐδὲν σοι τοῦ γράμματος ἐμέλησεν, ἢ προσέσχες τε καὶ ἐπεχείρησας σαυτὸν ἐπισκοπεῖν ὅστις εἴης; Μὰ Δί οὐ δῆτα ἔφη, καὶ γὰρ δὴ πανὺ τοῦ τό γε ᾦμην εἰδέναι· σχολῇ γὰρ ἂν ἄλλο τι ᾔδειν, εἴ γε μηδ' ἐμαυτὸν ἐγίγνωσκον.   Xenophon, ΑΠΟ-ΜΝΗΜΟΝΕΥΜΑΤΩΝ Βιβλ. Δ, 2.
† Σαυτὸν γιγνώσκων, εὐδαίμων, Κροῖσε, περάσεις.—Xenophon, ΚΤΡΟΤ ΠΑΙΔΕΙΑΣ Βιβλ. Σ'.

study is difficult, and their conclusions frequently con-
tradictory. Who has not read the works of Lavater

Johann Caspar Spurzheim, his most eminent pupil, was
born at Longrich, on the Moselle, on the 31st December,
1776. In the year 1804 he joined Gall as dissector and
demonstrator, and in 1808 (14th March) they presented
a Mémoire upon their science to the Institut de France,
which was published in 1809. In 1810 they published
the first edition of their joint work, "*Anatomie et Physio-
logie du Système Nerveux en général et du Cerveau en
particulier, avec des observations sur la possibilité de
reconnaître plusieurs dispositions intellectuelles et
morales de l'homme et des animaux par la configuration de leurs têtes*" (Paris : 1810-19). In 1813, on the
retirement of Dr. Gall, Spurzheim came to London, and
delivered a series of lectures on craniology, which attracted
much attention and criticism, both hostile and friendly.
Dr. Abernethy fully recognised the importance of the
new science, and delivered a most interesting address
thereupon before the Court of Assistants of the Royal
College of Surgeons in 1821, which was printed under
the title "*Reflections on Gall and Spurzheim's System
of Physiognomy and Phrenology*" (London : 1821). In
this opusculum he highly appreciates the theory, but con-
demns the arbitrary dogmata laid down by its exponents.
I have before me an amusing little satire, written by
Lord Jeffery and Lord Gordon, entitled "*The Craniad;
or Gall and Spurzheim illustrated*" (London : 1817),
which was one of the many publications which sought to
extinguish the new cultus. Whilst he was in London, J.
C. Spurzheim published "*The Physiognomical System of
Drs. Gall and Spurzheim, founded on an anatomical
and physiological examination of the Nervous System
in general and of the Brain in particular*" (London :
1814),—a work which was terrifically attacked by the press
Also "*A Sketch of the New Anatomy and Physiology
of the Brain and Nervous System of Drs. Gall and
Spurzheim*" (London : 1816). In the meanwhile he published in Paris his "*Encéphalotomie, ou du cerveau sous
ses rapports anatomiques*" (Paris : 1821). The great
work left behind us by Spurzheim's master, Gall, is entitled "*Sur les Fonctions du Cerveau et sur celles de
chacune de ses parties, avec des observations sur la
possibilité de reconnaître les instincts, les penchans,
les talens, ou les dispositions morales et intellectuelles
des hommes et des animaux par la configuration de
leur cerveau et de leur tête*" (Paris : 1825, 6 vols.). This

and of the other physiognomists?[99] But their indications are vague in their apparent precision, and their decisions are often entirely misleading.   Nevertheless, the one theory assisting the other, physiology has made a step in advance.   It is thus that the light in a crypt becomes more decided, and that the vault becomes more clearly illuminated, as, one by one, lamps are lit beneath its arching roof.   Yet another discovery,

work was translated into English by W. Lewis under the title " *On the Functions of the Brain and of Each of its Parts*" (London: 1835, 6 vols).   Since this time innumerable volumes have appeared in every language based upon the works above cited.*

[99] Johann Caspar Lavater was born at Zürich on the 15th November, 1741, and died on the 2nd January, 1800.   Educated originally for the Church, he went as a student to Berlin, where he made himself widely known by his theological and polemical works, many of which are recognised as standard authorities.   But he is, however, far better known as a physiognomist ; he took eagerly to this study at the age of twenty-five, and his *magnum opus*, " *Physiognomische Fragmente zur Beförderung der Menschenkenntniss und Menschenliebe* " (Leipzig : 1775-8, 4 vols.), is still the great leading work upon the science.   It was translated into French by Mme. de la Fite and MM. Caillard and Henri Renfner under the title, " *Essais sur la Physiognomie, destinés à faire connoître l'homme et à le faire aimer* " (La Haye : 1781-3, and 1803, 4 vols.), and an English translation was made by Dr. H. Hunter, entitled, " *Essays on Physiognomy destined to promote the Knowledge and the Love of Mankind*" (London : 1789-93, 3 vols.).   Lavater had preceded this work with a smaller one, and innumerable condensations of it have appeared, of which perhaps, in English, "*Physiognomy, or the Corresponding Analogy between the Formation of the Features and the Ruling Passions of the Mind*" (London : 1800 and 1827), and in French " *L'Art de connaître les Hommes par la Physionomie;* Nouvelle édition par M. Moreau," are the best and most handy.

* *Vide.*   A. CARMICHAEL : "*A Memoir of the Life and Philosophy of Spurzheim*" (Dublin : 1833), and N. CAPEN : "*Reminiscences of Dr. Spurzheim*" (New York : 1881).

and perhaps this science will reach a sufficient, if not a perfect degree of certainty.[100]   Very well then, the signs indicative of our inclinations and of our instincts, which Gall discovered in the bumps of the skull, and which Lavater discovered in the features of our faces, I claim to have found—not all of them, but those which concern *the intelligence*—in the formation of the hand.

And, indeed, after *speech* [which Charron calls *the hand of the spirit*],[101] is not the hand, particularly on the plane of materialism, the principal instrument of our intelligence ?   It can, therefore, reveal to us a great

¶ 85.
The hand the instrument of the mind.

[100] Herbert Spencer, in his *"Study of Sociology"* (London : 1873), has made some remarks which afford an interesting corollary to this passage. "Only a moiety of science," says he, "is exact science ; only phenomena of certain orders have had their relations expressed quantitatively as well as qualitatively. Of the remaining orders there are some produced by factors so numerous and so hard to measure that to develop our knowledge of their relations into the quantitative form will be extremely difficult, if not impossible. But these orders of phenomena are not therefore excluded from the conception of science. In geology, in biology, in psychology, most of the previsions are qualitative only ; and where they are quantitative, their quantitativeness, never quite definite, is mostly very indefinite. Nevertheless, we unhesitatingly class these previsions as scientific." Chap. vi., "Intellectual Subjective Difficulties." I have dealt with this passage at length in §§ 98-99 of *"A Manual of Cheirosophy"* (London : 1885). Henry Drummond also has said, "A science without mystery is unknown."*

[101] PIERRE CHARRON, *"De la Sagesse Livres trois."* (Bordeaux: 1601) p. 109. "Pour le regard de tous la parole est la main de l'esprit, par laquelle, comme le corps par la sienne, il prend et donne, il demande conseil et secours, et le donne." Chap. xiii. " Du voyr, ouir, parler."

* HENRY DRUMMOND, *"Natural Law in the Spiritual World"* (London : 1884), at p. 28 of his " *Natural Law*," and again at p. 88, "Even science has its mysteries, none more inscrutable than around this Science of Life."

deal concerning the attractions and repulsions, and the intellectual aptitudes of each individual.

¶ 86.
The hand adjusted to the uses man makes of it.

For, just as animals have organisations conformable to their several instincts ; just as the beaver and the ant, who are endowed with the instincts, in the one case of building, and in the other of burrowing in wood, have at the same time been furnished with the instruments, in the one case for building, and in the other for burrowing ; just as we find that among the animals of a particular family, whose instincts are identical, the organisation is also identical, whilst among animals of the same *species* [such as, for instance, dogs or spiders], whose instincts are partially different, the organisation is also commensurately different,—so, I opine, God in giving us men different instincts, has logically given us *hands* of diversified formations.[102] The hand of a poet cannot resemble that

---

[102] This is the tenour of the long preamble which introduces Galen's great work, " *Claudii Galeni Pergameni, secundum Hippocratem Medicorum Principia, Opus de Usu Partium Corporis Humani, magna cura ad exemplaris Græci veritatem castigatum universo hominum generi apprime necessarium Nicalao Regio Calabro interprete*" (Paris : 1528), in which he points out that : " The hands are themselves the implements of the arts, as are the lyre to music, and the tongs to the smith. And just as the musician was not instructed by the lyre, nor the smith by his forceps, but each of them is skilled in producing works which he could not produce without these accessories, no one can do the work he is born to do without the necessary tools for doing it, . . . but every animal performs the functions of his own peculiar instinct without requiring any instruction therein. . . . On which account it seems to me that animals perform their various functions by nature rather than by reasoning (as for instance the complicated habitations and occupations of bees, of ants, and of spiders), and to my mind without any instruction."* *Vide* also " *A Manual of Cheirosophy*," §§ 6-7.

---

* " Manus autem ipsæ sunt artium organa sicut lyra musicis, et forceps fabri. Sicut igitur lyra musicum non docuit nec

[¶ 86]

of a mathematician, nor can the hand of a man of
action resemble that of a man of contemplation.
When I lay this down as an axiom I must not be
misunderstood ; I do not mean the hand of a poet or
mathematician by cultivation, but of a poet whom
nature has made poetic, and of a mathematician to
whom nature has given a mathematical mind. And
again, it would argue a very weak idea of the prevision
of the Omnipotent Creator, of His justice and of His
power, to believe that the instruments with which He
has furnished us are not appropriated by the variety
of their forms to the variety of our intelligences.[103]

It is upon this great truth, and starting from this
point, that I have based my system. Cleverer men
than I, if they consider it worthy of their attention,
starting again *ab initio*, will enunciate and develop it
better than I have been able to do. For myself I
claim only the honour of having been the first to
catch a glimpse of the fertile regions of this new
science.

¶ 87.
Bases of
cheirognomy ·

[103] " There is as much diversity and want of resemblance between the forms of the hands as there is
between varied physiognomies. This truism is founded
on experience, and requires no proof ; . . . the form of
the hand varies infinitely, according to the relations,
the analogies, and the changes to which it is amenable.
Its volume, bones, nerves, muscles, flesh, colour, outlines, position, movement, tension, repose, proportion,
length, curvature,—all of them offer you distinctions
which are apparent and easy to recognise." " *L'Art
de connaître les Hommes par la Physionomie*," par
Gaspard Lavater (Paris : 1806), vol. iii., p. 1.

forceps fabrum, sed est uterque ipsorum artifex per eam qua pro-
ditus est rationem agere autem non potest absque organis, ita et
una qualibet anima facultates quasdam a sua ipsius animæ facultates, ac in quos usus partes suæ polleant maxime nullo doctore
præsentit . . . Qua propter cætera quadam animalia mihi naturâ
magis quam ratione artem aliquaquam exercere videntur (apes,
videlicet, plasmare fingereque alveolos, thesaurus vero quosdam
et labyrinthos formicæ fabricare, nere autem et texere araneæ)
ut autem conjecto sine doctore," lib. i.

AUTHOR'S NOTE.—Perhaps I should say "*rediscovered*," for Anaxagoras is said to have seen significant signs of the tendencies of the mind in the formations of the hand.[104] The Greeks have, almost without exception, been gifted with the faculty of prevision.

[101] Anaxagoras was the first, unless I am mistaken, who is said to have pointed out the fact that man was the wisest of all animals, *because* he had hands. I do not know where he made the remark, nor do I *think* it occurs in any of the works which remain to us of him; but Aristotle alludes to the statement, correcting him by saying that it was *because he was the wisest* of all animals that he was given hands.* Galen in the work cited above, agrees with Aristotle.†

---

* " 'Αναξαγόρας μὲν οὖν φησὶ διὰ τὸ χεῖρας ἔχειν φρονιμώτατον εἶναι τῶν ζῴων ἄνθρωπον· εὔλογον δὲ διὰ τὸ φρονιμώτατον εἶναι χεῖρας λαμβάνειν."—ARISTOTLE, ΠΕΡΙ ΖΩΩΝ ΜΟΡΙΩΝ, Βιβλ. Δ'., κεφ. ί.

† " Ita quidem sapientissimum animalium est homo. Ita autem et manus sunt organa sapienti animalia convenientia. Non enim quia manus habuit, propterea est sapientissimum, ut Anaxagoras dicebat ; sed quia sapientissimus erat, propter hoc manus habuit, ut rectissime censuit Aristoteles."—"*De Usu Partium Corporis Humani*," lib. i.

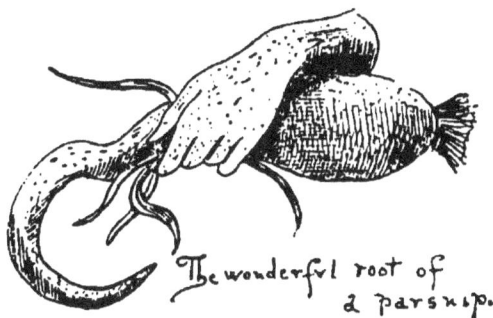

The wonderful root of
a parsnip.

SUB-SECTION I.

THE CLASSIFICATION OF HANDS.

HANDS may be divided into seven classes or types, which are sufficiently distinct from one another in their peculiar formations to be clearly and distinctly described. I have classified them as follows :--

¶ 83.
The seven types of hands.

The Elementary, or Large-palmed hand.
The Necessary, or Spatulated hand.
The Artistic, or Conical hand.
The Useful, or Squared hand.
The Philosophic, or Knotted hand.
The Psychic, or Pointed hand.
The Mixed hand.

These types, like the separate breeds of the canine race, cannot alter or modify themselves beyond a certain point, in obedience to an occult force similar to that which brings about the fact that the man of to-day is the prototype of the man of the patriarchal times, and which continually brings them back to their original purity and distinctness of characteristic.[105]

¶ 89
Recurrence of the types.

From the various ways in which these types strike

¶ 90.
Their influence

[105] Desbarrolles has said upon this paragraph :—"We will not follow M. d'Arpentigny in his classification, because we consider it to be useless. Hands may resemble one another, but nature never repeats herself,

out in new lines and mix themselves together, result the various civilisations which succeed one another upon the surface of the globe.

¶ 91.
Illustrations.

In his primeval forests, with their alternate lights and shadows, Pan on his pipes, which never alter their primitive form, is constantly ringing the changes upon new tunes.

A nation consisting exclusively of two, or three, types, would resemble a lyre with only two or three chords.

Humanity is an argosy of which God is the pilot ; and man is a passenger on board this argosy, a passenger governed by his own instincts and inclinations.

He obeys, like the little planet which he inhabits, two great forces ; the one general and exoteric, and the other particular and esoteric.

¶ 92.
Origin of laws.

Laws are evolved by the knowledge that we have of the *abusive* powers of our instincts ; but they demonstrate our freedom in this sense, that they sum up and codify the *reflective* forces which reason opposes to the *spontaneous* forces of our instincts.

¶ 93.
Recognition of the types.

Each type asserts itself by the invincible persistence of the tendencies which it exhibits.

From the day that he ceased to give utterance to the sighing harmony which testified to his divinity, Memnon ceased to be looked upon as a god[106].

and in objects apparently the most similar she places, sometimes by an imperceptible touch, a complete diversity of instincts."—"*Les Mystères de la Main*" (Paris : 15th edn., *n.d.*, p. 176). But I do not think that the remark is called for, because it is sufficiently guarded against by the minuteness of M. d'Arpentigny's analysis and differentiations.

[106] This is the statue referred to by Juvenal ("*Sat.*," xv., l. 4), when he says :—

"Effigies sacri nitet aurea cercopitheci,
Dimidio magicæ resonant ubi Memnone chordæ,
. Atque vetus Thebe centum jacet obruta portis."

In John Pinkerton s "*Voyages and Travels in All*

Parts of the World" (London : 1814, 17 vols.), in the fifteenth volume, containing "Pocock's Travels in Egypt," he says :—"Strabo, speaking of Thebes, says that there were in his time several villages on the site of it, part of them on that side which was in Arabia, where the city then was, part on the other side, where the Memnonium was. Here were two colossal statues of one stone near one another, one being entire ; the upper part of the other was fallen down from the seat, as it was said occasioned by an earthquake. It was thought that once a day a sound was heard as of a great blow from that part which remained on the seat and base. When he was there with Ælius Gallus and others he heard the sound, and whether it came from the base or the statue or the people around it he could not tell, the cause not appearing ; he would rather believe anything than that a sound could be occasioned by any particular manner in which the stone is composed."* Pausanias says that Cambyses broke it, and hat when the upper part from the middle was seen lying on the ground, the other part every day at sun-rising uttered a sound like the breaking of a string of a harp when it was wound up.† Philostratus, in his "*De Vita Apollonii Tyanei*," lib. vi., c. 3, describes the statue as being of black stone, with its feet set together, in the manner, I presume, usual to sitting colossi ; and Pliny, in speaking of basalts, reckons among celebrated statues of this stone that of Memnon now under consideration.‡ Sir J. Gardner Wilkinson, in his "*Topography of Thebes*" (London : 1835, p. 36), gives some very interesting notes on the statue, and makes some very

---

* STRABO, ΓΕΩΓΡΑΦΙΚΩΝ, Βιβλ. 17.—"Μέρος δ'ἐστὶν ἐν τῇ περαίᾳ, ὅπου τὸ Μεμνόνιον· Ἐνταῦθα δὲ δυοῖν κολοσσῶν οὐ τῶν μονολίθων ἀλλήλων πλησίον, ὁ μὲν σώζεται, τοῦ δ'ἑτέρου τὰ ἀνώμερη τὰ ἀπὸ τῆς καθέδρας πέπλωκε σεισμοῦ γενηθέντος ὥς φασι. Πεπίστευλαι δ' ὅτι ἅπαξ καθ' ἡμέραν ἑκάστην ψόφος, ὡς ἂν πληγῆς οὐ μεγάλης ἀποτελεῖται ἀπὸ τοῦ μένοντος ἐν τῷ θρόνῳ καὶ τῇ βάσει μέρους ... Διὰ τὸ ἄδηλον τῆς αἰτίας πᾶν μᾶλλον ἐπέρχεται πιστεύειν ἢ τὸ ἐκ τῶν λίθων οὕτω τεταγμένων ἐκπέμπεσθαι τον ἦχον."

† PAUSANIAS, book i., c. 42.—"Ἔστι γὰρ ἔτι καθήμενον ἄγαλμα Ἡλεῖον, Μέμνονα ὀνομάζουσιν οἱ πολλοί ... ὁ Καμβύσης διέκοψε καὶ νῦν, ὁπόσον ἐκ κεφαλῆς ἐς μέσον σῶμα ἦν ἀπερρίμένον, τὸ δε λειπὸν κάθηται τε καὶ ἀνὰ πᾶσαν ἡμέραν ἀνίσχοντος Ἡλίου βοᾷ, καὶ τὸν ἦχον μάλιστα εἰ κάσει τὶς κιθάρα ἢ λύρας ῥαγείσης χορδῆς."

‡ PLINY, "*Naturalis Historia*," lib. xxxiv., c. 7, "Non absimilis illi narratur in Thebis delubrio, ut putant, Memnonii statua dicatur, quem quotidiano solis ortu contactum radiis crepare dicunt."

[¶ 93]

pertinent observations on the probable solution of the mystery, which is probably explained by the heat of the sun acting upon the cracked and porous stone. Memnon himself was the son of Tithonus and Aurora, or, as Diodorus says, of some Eastern princess. He was evidently one of the most celebrated rulers of Egypt, and one of the most respected ; for we find long accounts of him in Suidas, Diogenes Laertius, and Virgil. We also find mention of him in the works of Dictys Cretensis, Simonides, and Josephus. Philostratus tells us that he assisted Priam at the siege of Troy, and was killed by Achilles.

SECTION II.

The Hand in General.

The
Cheiros:
tem·
on
or
hand:
flower.

SUB-SECTION II.

THE SIGNIFICATIONS FOUND IN THE PALM OF THE HAND.

BEFORE tabulating the deductions I have drawn from the observation of the various types, I propose to say a few words upon the significations to be traced in the various parts of the hand.

On the palm of the hand are found the indications of the physical appetites of men, and, up to a certain point, those of the intensity of the intellectual aptitudes which these appetites determine.

¶ 94.
Indications of the palm.

Too slim, too narrow, too meagre, it indicates a feeble and unfruitful temperament, an imagination lacking warmth and force, and instincts without any settled object.[107]

¶ 95.
Narrow and small.

[107] Aristotle, in his treatise upon physiognomy, points out this indication of a long and graceful hand, in a passage commented upon at much length by Camillo Baldi in his "*In Physiognomica Aristotelis Commentarii,*" a Camillo Baldo . . . . lucubrati (Bononiæ: 1621, fol.), p. 69 (33). *Vide* also "*A Manual, etc.,*" ¶ 112.

¶ 96.
Medium and normal.

If your palm is supple, of a medium thickness and consistency of surface,—that is to say, if it is in proper proportion with the size of the fingers and thumb,—you will be capable of enjoying all pleasures [incalculable privilege!], and your senses, easily excited, will keep pace with the faculties of your imagination.

¶ 97.
Large.

If, whilst still supple, its developments are too pronounced, egoism and sensuality will be your dominating instincts.

¶ 98.
Excessive development

Finally, if its amplitude is utterly out of proportion with the rest of the hand, and if it is at the same time excessively hard and excessively thick, it will indicate instincts and individualities verging upon an animality which is destitute of ideas.

¶ 99.
Articulation and intelligence in proportion to one another

Look, for example, at the animals whose solid and rounded feet are formed of a single nail, cloven or solid-ungulous, as, for instance, the ox, the horse, the ass, the camel; does not the fact that we men make use of their powerful strength, of which God has withheld from them the knowledge and the power to use it for their own advantage, afford ample evidence of their want of intellect? It is not the same thing with regard to the animals whose feet are articulated, like those of lions, tigers, and so on; the superiority of their organisation is proved and verified by the superiority of their intelligence, which is demonstrated by the state of liberty in which they live.[108]

[108] Helvetius has made some very interesting and analogous observations upon this point in his treatise "*De l'Esprit*" (Londres: 1776), ch. 1, which are as follows:—"The human faculties which I regard as the productive causes of our thoughts, and which we have in common with the animals, would supply us with but a very small share of ideas, if they were not, with us, combined with a certain external organisation. If nature, instead of with flexible hands and fingers, had terminated our anterior extremities with a horse-like hoof, what doubt is there that men, without art, without

By the amount of liberty which they enjoy one can also form an estimate of the moral forces of a nation ; for liberty pre-supposes morality.

¶ 100.
Liberty indicating morals.

You will weigh, if you please, the importance of these comparisons and similes, notwithstanding the succinctness with which they are placed before you, and then we will go a step further. You will close this volume at once if your mind cannot grasp anything which is not amplified and minutely developed in the exposition.

¶ 101.
Importance of these preliminaries.

The indications furnished by the palm are, of course, modified or confirmed by the indications furnished by the other parts of the hand.

¶ 102.
Modification of indications.

habitations, without defence against animals, entirely occupied by the cares of providing their nourishment and of avoiding wild beasts, would be still wandering in the forests like wild herds ?" And to this he appends the following note :—"All the feet of animals terminate either with a horn, like those of the ox or stag, or in nails, as in the dog and wolf ; or in claws, like the lion and cat. Well, this difference of organisation between our hands and the paws of animals deprives *them*, not only, as M. de Buffon says, almost entirely of sense or tact, but also of the skill necessary to handle any tool, or to make any discovery which requires hands for its development." *Vide* ¶ 52 and note ⁴⁶, p. 67.

THE SIGNIFICATIONS FOUND IN THE FINGERS OF THE
HAND.

¶ 103.
Principal indica-
tions.

PUTTING aside the doubtful or unimportant signs, I
will concern myself only with the leading significations, the practically infallible indicators of the leading
instincts.

¶ 104.
Fingers
generally.

Fingers are either smooth, or knotty [*i.e.*, with developed joints].

¶ 105.
Joints.

Of the latter class some fingers have only one
joint developed, whilst others have two. Joints which
carry with them meanings for us, are not those which
are only apparent to the sense of touch, but those
which the eye easily perceives at the first glance.

¶ 106.
The finger-tips.

Our fingers terminate either (*a*) in a spatule, *i.e.*,
are slightly enlarged at the tips ; (*b*) squarely, *i.e.*, by
a phalanx whose lines extend parallel to a more or
less square tip ; and (*c*) in a cone, whose rounded
tip is more or less accentuated.

¶ 107.
The tips and
joints.

To these different formations belong as many
different interpretations ; but before interpreting them
let us say a few words concerning the joints.

¶ 108.
Significations of
the joints.

If the joint which connects the third phalanx [the
outer or nailed phalanx] with the second is prominent,
you have order in your ideas, *i.e.*, a well-regulated
mind ; if the joint which connects the second [or
middle] phalanx with the first [or lower] one is

prominent, you have a remarkable gift of material order and of method in worldly affairs.

With both joints developed, at the same time that you have the instincts of arrangement, symmetry, and punctuality, you will proceed by reflection; there will occur in your mind an appreciable interval between thought and action. You will have an innate faculty for the pursuit of science.

¶ 109.
Both joints developed.

Smooth fingers, on the contrary, are endowed with the faculty of art. However practical and positive the end towards which they are goaded by material interests, they will always proceed by inspiration rather than by reason, by fantasy and sentiment rather than by knowledge, by synthesis rather than by analysis.

¶ 110.
Smooth fingers.

Taste [from the intellectual point of view] resulting as it does from consideration, belongs essentially to knotty fingers; and *grace*, unreasoning and instinctive as it surely is, belongs essentially to smooth fingers.

¶ 111.
Taste and grace.

There are people who sacrifice superior to inferior orderliness; they ruin themselves so as to have a well-ordered household. Louis XIV. sacrificed well-being to symmetry, merit to rank, the State to the Church.[109] He probably lacked the upper joint [that of philosophy].

¶ 112.
Excessive order.

Louis XIV.

[109] A better illustration of a mind absorbed by an attention to trifles and punctilios, which blinded it to all the great considerations which should have occupied it, than the Grand Monarque who immortalised himself by the sentence "L'Etat, c'est moi!" could not have been found. This pettiness of spirit in the midst of his grandeur is especially noted by MM. A. Roche and P. Chasles in their "*Histoire de France*" (Paris: 1847, vol. ii., p. 274); and Voltaire's "*Histoire*," tome iv., "Siècle de Louis XIV." (Paris: 1856, chap. xxv., p. 176), abounds in illustrations and instances of the trait we have under consideration. "He had a manner of bearing himself," says Voltaire, "which would have sat ill upon any one else; the embarrassment which he caused to those with whom he conversed, secretly

¶ 113.
Impulse.
Rabelais.

After the capture by Picrochole from Grandgouzier of some paltry little town which he had conquered, Picrochole, comparing himself to Alexander of Macedon, proposed to march against Angoumois, Gascony, Galicia, Spain, Tunis, Hippes, Corsica, Rome, Italy, and Jerusalem. His counsellor, Captain Merdaille, baring his head to speak to him, "Be covered," said

flattered the complaisance with which he felt his superiority.* The magnificence of his banquets, and his lavish generosity with the money of the nation, have become historic, as also have become his rules of etiquette, etc. To distinguish his principal courtiers he had invented a blue tunic, embroidered in gold and silver. They were as much sought after as the collar of the order of St. Louis. He established in his household a *régime* which exists to this day ; regulated the orders of men and the functions of the court ; and created new posts of honour among his personal attendants. He re-established the tables instituted by François I., and increased their number, . . . and all were served with the profusion and ceremony of a royal board." Voltaire adds the following very pertinent note :—" All this profusion, made with the money of the nation, all these posts and sinecures, were absolute injustice, and certainly were a far greater crime (save in the eyes of the priests) than any he could commit with regard to his mistresses."† A very interesting collection of notes upon this reign, fully illustrating the above, will be found in M. l'Abbé de Choisy's eccentric and egotistical little work " *Mémoires pour servir à l'Histoire de Louis XIV.* " (Utrecht : 5th edn. : 1727, 2 vols.), especially in the second volume. The minutiæ of his household is demonstrated by passages like the following :—" Le Roi fit un grand plaisir à M. le Duc en lui accordant les grandes entrées, c'est a dire le droit d'entrer le matin

---

* " Il avait une démarche qui ne pouvait convenir qu'à lui et à son- rang, et qui eut été ridicule en tout autre.   L'embarras qu'il inspirait à ceux qui lui parlaient flattait en secret la complaisance avec laquelle il sentait sa supériorité."

† " Toutes ces profusions faites avec l'argent du peuple, toutes ces créations de charges inutiles, constituaient des véritables injustices, et certes un beaucoup plus grand péché (sauf aux yeux des Jésuites) que ceux que le roi pouvait commettre avec ses maîtresses."

Picrochole. "Sire," replied Merdaille, "I attend to my duty, but do not be so sudden in your enterprises. [110]

This advice of the gallant captain I warmly recommend to the consideration of smooth-fingered subjects. They are too passionate, too hasty. In study, in love,

¶ 114.
Evils of overhaste.

dans sa chambre en même tems que les premiers Gentils-hommes de la Chambre, dès qu'il est éveillé, avant qu'il sort du lit. Car quand il se lève, et qu'il prend sa robe de chambre et ses pantoufles, les Brevets entrent, et ensuite les Officiers de la Chambre et les Courtisans pour qui les Huissiers demandent d'abord, etc. etc." (p. 55, vol. ii.). On p. 138 a minute description is given of the *nuances de cérémonie*, exhibited to the Papal Nuncio Ranuzzi. I have elaborated this note, as it will be found during the perusal of the following pages that M. d'Arpentigny draws largely on this exaggerated methodism of Louis XIV. in illustrating his pages.

[110] Our author has dealt somewhat freely with his original in this paragraph. The conversation takes place when at the close of the day le Duc de Menvail, Comte Spadassin, and Capitaine Merdaille greet him with the words:—"Sire, to-day we make you the happiest, the most chivalrous prince that has lived since Alexander of Macedon." "Be covered," replied Picrochole. "Many thanks," said they; "sire, we attend to our duty;" and after some conversation, Picrochole having unfolded his great plans, "No," reply they, "wait a little ; never be so sudden in your enterprises."*

---

* RABELAIS, "*Gargantua*," liv. i., chap. 33. "Cyre, aujourd-hui nous vous rendons le plus heureux, plus chevaleureux prince qui encques feut depuys la morte de Alexander Macedo." "Couurez, couurez vous," cist Picrochole. "Grand mercy," dirent-ils; "Cyre, nous sommes a nostre debuoir" .... "Prinse Italic, voyla Naples, Calabre, Apoulle, et Sicilie toutes en sac, et Malthe avec. Je vouloys bienque les plaisans cheualiers iadiz Rhodiens vous resistassent. . . . Je iroys" (dist Picrochole) "voulentiers à Lorette." "Rien, rien," dirent ilz," "ce sera on retour." "De la prendons Candye, Cypre, Rhodes, et les Isles Cyclades, et donnerons sur la Morée. Nous la tenons Sainct Treignan, Dieu guard Hierusalem, car le Soudan n'est pas comparable à vostre puissance. Je," dist il, "feray donc bastir le temple de Salomon." "Non," dirent ilz encores "attendez ung peu ; ne soyez jamais tant soubdain dans vos entreprinses."

[¶ 115]

in business, they often fail to attain their ends by aiming at them with too much vigour.[111]

¶ 115.
The finger tips.

I shall, however, return to this subject. Let us proceed to consider the interpretation of the external phalanges.

¶ 116.
Seven hands.

We have before us [let us suppose] seven hands belonging to as many different individuals. They are held towards us, resting upon nothing, the fingers slightly opened.

¶ 117.
Spatulate.

The first hand has smooth fingers, terminating in a spatule ; the second has knotty or jointed fingers, also terminating in a spatule.

¶ 118.
Indications of
the spatulate.

Both, by reason of the spatulated tips of the fingers, are characterised by an imperious desire for corporeal exertion, for locomotion as a rule, and for manual labour ; they act more upon the promptings of their organisations than of their heads; they cultivate the science of things according to their useful and physical aspects, and are inspired with a love of horses, dogs, and hunting, navigation, war, agriculture, and commerce.

¶ 119.
Their practical
instincts.

Both are gifted with an innate sense of tangible realities, with an instinctive appreciation of real life, a tendency to cultivate physical power, the talent of calculation, of industrial and mechanical arts, and

//

---

[111] The moral pointed in these two latter paragraphs receives a striking confirmation in the case of Gustave Doré, the French artist. As a painter he was comparatively a failure, probably by reason of the fact that he always insisted on "running before he could walk," so to speak, with the result that his paintings always remained crude and unfinished in appearance. This characteristic is vividly pourtrayed throughout Blanche Roosevelt's "*Life and Reminiscences of Gustave Doré*" (London: 1885), and shows itself especially in his favourite aphorism, "Never be modest in your undertakings, but always be modest in the day of success." *

---

* "Ne soyez pas modeste dans vos entreprises, mais soyez-le toujours dans le succès."—*Op. cit.*, p. 352.

of exact, applied, natural, and experimental sciences, graphic arts, administration, jurisprudence, and so on. Jacquard,[112] Vaucanson,[11] and Constantin Périer had extremely spatulate fingers.

BUT, as fingers which are smooth proceed, as I have remarked before, by inspiration, passion, instinct, and intuition, and as knotty fingers [*i.e.*, those with both joints developed] proceed by calculation, reason, deduction, and by a balancing of probabilities; the

¶ 120.
Variations of their procedure.

[112] Joseph Jacquard was the George Stevenson of the Lyons silk trade; he immortalised himself by the invention of many labour-saving appliances in connection with the silk-weaving industry. Born, about the year 1750 at Lyons, he was first of all employed under his father in a silk factory, but the work proving too arduous for his delicate health, he was apprenticed first to a binder, and then to a hatter, which latter trade he subsequently took up. It was whilst he was thus employed that he constructed, from a few sticks of firewood, one evening the rough model of the Jacquard loom, the which, having attracted the attention of the authorities, he was taken to Paris to complete and expound it in the Ecole Polytechnique. He entirely revolutionised the silk trade by his invention, which caused him to be abhorred by the workmen of Lyons; his machine only became universally used in 1809. Whilst in Paris he completed and perfected the half-finished spinning machine of Vaucanson, of which he found a model in the school. He died in August 1834. For further particulars of Jacquard and of his work see Madame Grandsard's work " *Jacquard, sa Vie et son Œuvre* " (Lille : 1869), or the short biography contained in A. du Saussois' " *Galérie des Hommes Utiles* " (Paris : 1875, etc.).

[113] Jacques de Vaucanson was a man of character similar to that of Jacquard. Born at Grénoble in 1709, he devoted his energies to the improvement of the silk-spinning machinery used in Languedoc; but he is better known as the constructor of several marvellous automata, of which the most celebrated were a duck that swam about upon a pond, and ate grain which was thrown to it, and the world-famous automaton flute-player, which created such an excitement in 1738. All the journals of the time are full of it, but the best account

8

hand with smooth fingers will excel particularly in
art by locomotion, by activity, and in applied sciences,
where spontaneity and address, and the talent of
grasping a subject with promptitude, excel over the
mere capacity for combination and deduction.[11]

¶ 121.
The spatulate
type.
Prince Jules de
Polignac.

Prince Jules de Polignac was devoted to the chase,
to travelling, to horses, and to all forms of bodily
exercise. His was a *sanguine* temperament, and it
was shown by his aquiline nose and high colour.
His shoulders were broad, his chest deep, and his
figure was well proportioned, but his lower limbs did
not come up to the standard of the rest of his body ;
with large feet and bowed legs he had a general
appearance of boorishness, which gave one an idea
of a swan out of the water. In his youth so great
was his muscular strength and his agility that on one
occasion, having been attacked by a bear, he succeeded
in throwing it on the ground and killing it. On
another occasion he held his own against a pack of
the huge mountain dogs which the cowherds on the
Ural slopes had set at him. As a final instance we
may cite that one day, before he knew how to swim,
he laid a wager that he would cross the Volga at a

is to be found in a little work entitled, "*Le Méchanisme
du Fluteur Automate, avec la Déscription d'un Canard
Artificiel et celle d'une autre Figure jouant du Tam-
bourin et de la Flute*" (Paris : 1738). A translation of
this work by J. T. Desaguliers appeared in 1742,
entitled, "*An Account of the Mechanism of an
Automaton playing on the German Flute, etc.*"
(London : 1742). Vaucanson died in the year 1782
(November).

[11] I may add here a comment of Adrien Desbarrolles on
the above paragraphs. "Let us add a most important
remark, which M. d'Arpentigny has not made, *viz.*, that
exaggerations in the forms of the tips of the fingers, or in
the development of the joints, announce always an *excess*,
and, therefore, *a disorder* of the qualities or instincts
represented by those developments." "*Les Mystères
de la Main*" (Paris : 15th edn., *n.d.*), p. 160.

point of its greatest width, sustained only by a hand placed beneath his chin, and this he succeeded in doing.[115] Such athletes as this are but little fitted to govern nations of varied individualities; equally ready to devise sudden expedients, to make unpremeditated plans and to put them into execution, they are very prone to lose themselves in the unlimited wastes of abstract *ideas*.[116] Their fingers are spatulated and smooth. If their empire requires a prime minister they burden it with a Wazir;[117] if their sails require a gentle breeze, they let loose a hurricane.

Spatulated hands with the joints developed have the talents of practical and mechanical sciences brought to perfection, such as statics, dynamics,

¶ 122.
Spatulate hands with joints developed

[115] The eminence of Prince Jules de Polignac in all feats of athletic skill was a fruitful theme for the satirists of the empire. The following is from a political squib, entitled, " *Feu partout, voilà le Ministère Polignac* " (Paris: 1829, p. 10) :—

"Qu'il était fort—surtout au jeu de paume,
Nul n'eut osé lui disputer le prix ;
Depuis Nemrod quel chasseur du royaume
Abattait mieux au vol une perdrix ?
Voilà comment sa grandeur se fit homme !"

[116] " Le prince était un de ces hommes comme les gouvernements savent en choisir aux jours de leur décadences et qui ne font que hater leur chute et précipiter les révolutions. Loyal et conscientieux, mais d'une profonde incapacité, aveuglé par ses préjugés de caste et ses opinions retrogrades, il ignorait absolument l'esprit, les tendances, et les besoins de la France nouvelle, et marcha en sens contraire de l'opinion publique."— LAROUSSE, "*Dictionnaire Universel du XIXe Siècle*" (Paris: 1866-77), Art. " POLIGNAC."

[117] The author has, I think, hardly appreciated the true signification of the word وزير [pronounced by various authors vizier, wuzeer, wezeer, vizír, vizir], being derived from a root وزر (*wizr*), meaning "*a burden*,' or "*load*" and the وزارة [*grand-wiziriate*], or premiership, is in no sense an autocratic office. Sale, in his translation of the Qu'rán,* translates the word "Coun-

---

* " *The Koran, commonly called the Alcoran of Mohammed*, translated by G. Sale (London : 1865 .

[¶ 122]

Illustrations.

navigation; military, naval, and utilitarian architecture, as for instance bridges, streets, and so on,
great industries and combined strategies, κ.τ λ.   I
might quote, as examples of the type, Vauban, Monge,
Carnot, Cohorn, and Arago.

¶ 123.
Square hands.

Here we have a hand whose smooth fingers terminate squarely, *i.e.*, by a nailed phalanx whose lateral
sides prolong themselves parallel to one another; and
again, this other hand which we have here [which is
also square as to its finger-tips] has its joints apparently developed.

¶ 124.
Indications of
the square tip.

Both of them, by reason of the square tips, are
endowed with the tastes for moral, political, social,
and philosophic sciences; didactic, analytic, and dramatic poetry; grammar, form, languages, logic,
geometry; love of literary exactitude, metre, rhythm,
symmetry, and arrangement, strictly-defined and conventional art.   Their views of things are just rather
than wide; they have great commercial talent; they
are great respecters of persons, and have positive but
moderate ideas; they have the instincts of duty and
of the respect due to authority, of the cultivation of
practical truths and of good behaviour, with a strong
paternal instinct; in fact, generally speaking, having
more brains than heart, they prefer what they discover to what they imagine.

sellor," and appends a note:—"Wazîr: one who has the
chief administration under a prince," a rendering in
which he follows J. M. Rodwell ["*Koran,*" chap. xx.,
30, p. 256, ed. 1865].   Captain Sir R. F. Burton, in his
recent translation of the Arabian Nights,* gives an
extremely interesting note upon the word.   It must not
be understood in the sense implied by the above
passage.

---

* "*The Book of the Thousand Nights and a Night: A plain
and literal translation of the Arabian Nights' Entertainments,
with introduction, explanatory notes . . . and a terminal essay,*"
by Richard F. Burton (Benares: 1885, FOR PRIVATE CIRCU-
LATION), vol. i., p. 2.

Square fingers are responsible for theories, for methodical registration of facts : and not for the higher flights of poetry, to which they never attain, but for literature, sciences, and some of the arts. The name of Aristotle is embroidered upon their banner,[118] and they march in the van of the four faculties.

¶ 125.
Their tastes.
Aristotle.

This type does not shine by the effulgence of its imagination as the term is understood by poets. That which results from this faculty belongs essentially to smooth fingers, as for instance literature properly so called, literature whose sole aim is its own perfection; whilst that which results from reasoning and from combination [as for instance social science and history] belongs essentially to knotted fingers. The fingers of Descartes and Pascal were jointed, whilst those of Chapelle and Chaulieu were smooth.

¶ 126.
Their want of
imagination.

Illustrations.

To fingers which terminate in a spatule belongs action, instinctive tact, and knowledge. There are in France more square than spatulate hands, i.e., more talkers than workers, more brains particularly adapted for the evolution of theories than men fit to put those theories into practice.

¶ 127.
Spatulate
fingers.

The hand of the ex-minister M. Guizot is large,

¶ 128.
The square type.
M. Guizot.

[118] Of all the classic authors none could have afforded a more perfect illustration of the habits and instincts of the square-handed subject than the Stagyrite philosopher. Those of my readers who are familiar with his works cannot fail to have been struck, not only by the astounding extent of his knowledge, but also by the marvellous symmetry and order exhibited in the way in which he marshals his facts and unrolls his theories with all the exactitude and terseness of a proposition of Euclid ; never repeating himself, excepting, as Bacon says, "to gain time."*

* "*Iterations* are commonly losse of time ; but there is no such gaine of time, as to *iterate* often the *state* of the *question*. For it chaseth away many a frivolous speech, as it is coming forth."— FRANCIS BACON. Essay on "*Despatch*." 1625.

with highly developed joints, and large square terminal phalanges to the fingers. His is one of those
retrospective minds whose light casts its rays in a
backward direction only, which seeks to obtain from
the dead the secrets of the living, and for whom the
present is obliterated by the past. Bred for the pro-
fessorate, he has acquired the disdainful bearing and
pedantic manner of the professor ; two things have
always been particularly objectionable to him,—war,
because it throws into shadow talkers who do not act ;
and the people, because it is not enough in his eyes
that a man should be endowed with a high spirit for
him to be great. Then, biliously complexioned, his
head large, and well-filled rather than well-made, as
Montaigne has expressed it,[119] with large features, and
clever at excusing his defalcations by specious maxims,
he has made himself by words, and has sustained himself by corruption. Seeing that one only knows what
one really *loves*, he knows by heart his legal, mechanical England ; but our France, which is as variable as
its own climate, diversified like its various Departments, eager for lofty emotions, fatigued by uniformity,
impregnated by storms, which to sophists without
patriotic or national emotions are repugnant,—he has
never understood her, and he never will.[120]

[119] This is the text upon which Montaigne bases two of
his most celebrated essays: *" Du Pédantisme "* and
" De l'Institution des Enfants."—ESSAIS.*
[120] M. le Capitaine d'Arpentigny in this paragraph
reflects the opinions of many contemporary and recent
writers upon Guizot ; the following passages may serve
as examples : " C'est à Genève qu'il a pris ces manières
gourmées, ce ton pédant, ces mœurs roides et cas-
santes."—*Eugène de Mirecourt*.† " La figure toujours
grave jusque dans son sourire . . . tel il apparaissait

* *" Essais de Montaigne, suivis de sa Correspondance, etc."*
(Paris : 1854. 2 vols.), livre i., chap. 23-4.
† *" Les Contemporains "* (Paris : 1857), Art. "GUIZOT,"
p. 12.

With more talents and less chivalry than the prince of papal creation, whose influence was so fatal to the elder branch of the Bourbons,[121] Guizot has become the Polignac of the younger branch, with this difference, however, between them,—M. le Prince de Polignac, a man of *action*, fell with his sword in his hand, whilst M. Guizot, a man of *words*, fell with an oration on his lips.   On the one hand fingers smooth and spatulate, on the other fingers knotty and square.

There is more simplicity but less politeness, more

¶ 129.
Guizot and
Polignac
contrasted.

¶ 130.
Manners of
square and
spatulate
fingers.

dans toute la raideur de son dogmatisme austère, laissant tomber de sa lèvre dédaigneuse des paroles, tour à tour mordantes et glacées."—*E. Langeron.* *   "Le lourd pédantisme de son procédé . . . j'aime à trouver dans un critique un homme qui me fait part de ses impressions, et non pas un pédagogue."—*Hippolyte Castile.*†

[121] It was by the counsels and frightful extravagances of the father of Prince Jules de Polignac that the revolution of 1789 was hastened on.   Mirabeau is reported to have said of him :—" Mille écus à la Famille d'Assis pour avoir sauvé l'Etat ; un million à la Famille Polignac pour l'avoir perdu ! "   The article in the " *Dictionnaire Universel du XIXe Siècle*, quoted in note [116] , continues, "On sait le résultat ; les ordonnances de Juillet 1830, contre-signées par lui, firent éclater une révolution qui consomma la ruine de la branche ainée ! "—quite a parallel passage to the above.   Eyre Evans Crowe, in his " *History of France* " (London : 1868, vol. v., p. 381), says : " The very name of Polignac as minister was a declaration of war against the nation," and Bertin, in an article in the *Journal des Debats* (1st August, 1829),—a paper described by Martin ‡ as " un journal attaché aux Bourbons par des liens que son ardente opposition n'avait point brisés jusque là," — says gloomily, " The glory of the dynasty was its moderation in the exercise of authority.   But moderation is henceforth impossible ; the present ministry could not observe

* " *Portraits Contemporains* " (La Rochelle : 1875), p. 15.
† " *Les Hommes et les Mœurs en France sous le Règne de Louis Philippe* " (Paris : 1853), pp. 47-53.
‡ HENRI MARTIN, " *Histoire de France* " (Paris : 1879), vol. iii., p. 408.

[¶ 130]

freedom but less elegance, among people whose hands are principally spatulate than among those in which the square type predominates.

¶ 131.
Conic fingers.

This fifth hand has smooth fingers whose terminal phalanges present the form of a cone, or of a thimble. Plastic arts, painting, sculpture, monumental architecture, poetry of the imagination and of the senses [Ariosto],[122] cultivation of the beautiful in the solid and visible form, romantic charms, antipathy to rigorous deduction, desire of social independence, propensity to enthusiasm, subjection to phantasy.

¶ 132.
The same with joints.

This same hand, but with jointed fingers, betrays the same instincts, but with more combination and moral force.

¶ 133.
Philosophic fingers.

This other hand has knotty fingers with the external phalanges partaking of the natures both of the

it, however much they might desire it." The author of "*Les Omnibus du Nouveau Ministère*" (Paris: 1829, p. 98), speaking of Prince Jules, makes the portentous remark: "Il y a des noms qui sont fatals à des états!" These notes will show that "the younger branch" had already a Polignac, and, therefore, Guizot was simply an aggravation.

[122] Ludovico Ariosto [born at Reggio, in 1474, died at Ferrara in 1533,] was perhaps the most romantic, enthusiastic, and phantastic poet that Italy has ever seen, and is, therefore, very happily introduced here as an illustration. He was bred for the law, but abandoned it to become a poet. In 1503 he became attached to the court of Cardinal Hyppolytus d'Este, at Ferrara, and, after a labour of about ten years, produced his "*Orlando Furioso*," of which the first edition was printed at Ferrara in 1516, 4to, and which appeared in its present completed form in 1532 (46 cantos). The poem is described by a writer in "'*Chambers' Encyclopædia*" as "a romantic imaginative epic, marked by great vivacity, playfulness of fancy, and ingenuity in the linking together of the various episodes." The best English rendering of the "*Orlando Furioso*" was made by W. S. Rose, in 1823; the translations of Sir John Harrington [1634], and John Hoole [1783], being of doubtful merit.

square and of the conic, the upper joint giving the terminal phalanx almost an oval formation.

It indicates a genius inclining towards speculative ideas, meditation, the higher philosophical sciences, and the rigorous deductions of verbal argument, love of the absolutely *true*, poetry of reason and of thought, advanced logic, desire of political, religious, and social independence ; deism,[123] democracy, and liberty.

¶ 131. *Their instincts.*

This is the Philosophic hand ; it examines *itself* rather than its surroundings, and is more taken up with ideas than with things. It hates the soldier and the priest, the former because his existence is antipathetic to liberty, and the latter because he is a stumbling-block athwart the path of progress.

¶ 135. *Their tastes.*

Finally, this last hand has smooth fingers terminating in a long-pointed cone. Contemplation, religious feeling, and idealism ; indifference to material interests, poetry of soul and of heart, lyric inspirations, desire of love and liberty, cultivation of all things beautiful, by their form *and* by their essence, but particularly by the latter. I have given to this hand, by reason of its attributes, the appellation " Psychic."

¶ 136. *Pointed hands.*

[123] I wish this *word* to be noted as a characteristic of the philosophic type. There is a strong tendency in the present day to regard *everything* un-absolutely-ortho-dox as *atheistical*, and atheistical has become a synonym for *unfamiliar*,*—a state of things recognised by Bacon, and concerning which he says, " For all that impugne a received *religion* or *superstition*, are by the adverse part branded with the name of *Atheists*" (Essay on " *Atheisme*," 1625) ; and I have emphasised the word " deism " in the above paragraph, for it will be immediately apparent to the reader how a philosophical mind may come to true reverent deism, when it cannot accept any recognised dogma.

* A friend of mine once tried the experiment of wearing a green hat and announcing that it was the symbol of his religious opinions ; in four days from its first appearance a well-meaning acquaintance deplored to me his atheistical views !

¶ 137.
Spatulate and square, or conic and pointed fingers.

Thus God has given to fingers which are spatulate or square, *matter*, *i.e.*, materialism and the appreciation of things real, as exemplified by the useful and the necessary arts, action, theory in undertakings, the comprehension of actual facts, and the pure sciences ; and thus to fingers which are conic and pointed He has opened the gates of the illimitable ideal : to conic fingers, by giving them the intuition of the beautiful according to its outward aspect, *i.e.*, *Art;* and to pointed fingers by endowing them with the intuition of the true and of the beautiful, according to their inner meanings, as exemplified by the higher forms of poetry, by ideal philosophy, and by lyric abstraction.

¶ 138.
Hard, stiff hands.

The hand which is hard and stiff, and which finds a difficulty in extending itself to its utmost limit of extension, indicates a stubborn character and a mind without versatility or elasticity.

¶ 139.
Importance of *synthesis* in the study.

I shall state further on, what must be, in each type, the proportions of the various parts of the hand to one another, but meanwhile you must not forget what I have said concerning the palm and the joints ; concerning the palm, which tells us all we want to know concerning the temperament and the intensity of the developed instincts; and concerning the joints, whose influence is always in harmony with the genius indicated by the outer phalanx, and which announce at once to the Cheirosophist the existence in the subject of a spirit of calculation and combination.

¶ 140.
Large hands and minutiæ.
Frederick William.

Large hands, therefore, are endowed with a spirit of minutiæ and detail. From the love of trifles which he displayed to his dying day, we know that Frederick the First of Prussia, known as the Sergeant-King, who reigned with a scourge in his hand, who used to cudgel his son when he was displeased, and into whose graces a pair of well-polished boots would carry

a man a long way, had very large hands.[124]   In the
same way, from the surname of "Long-handed," which
was applied to one of the kings of Persia [whose
name I do not know], one may infer that this sovereign,
whose politics were shuffling and petty rather than
grandly arranged, had essentially the spirit of detail.[125]

Artaxerxes
Μακρόχειρ.

[124] Carlyle, in his " *History of Frederick II. of
Prussia* " (London : 1858, vol. i., p. 579), speaks of
Frederick William I. as the " great drill sergeant," and
gives many instances of the manner in which he ill-
treated his son and the rest of the family.   In vol ii.
[chap. 8, p. 113,] we find an account by Dubourgay—
under date November 28th, 1729—of his " raining
showers of blows upon his son." (*Vide* also vol. ii.,
pp. 61, 71, 87, 253.)   Compare L. P. de Ségur's " *Geheime Nachrichten über Russland*" (Paris : 1800),
Noten zum Fünften Heft, vol. i., p. 430.

[125] Artaxerxes I. [who succeeded his father Xerxes,
after having slain Artabanus, his father's murderer]
was surnamed " Longimanus," or " μακρόχειρ," from
the fact that one of his hands—the right—was longer
than the other.*   Strabo tells us that when standing
upright, he could touch his knees without bending his
body [like Rob Roy].   A writer in the " *Encyclopædia
Britannica*," says, " His surname μακρόχειρ, first mentioned by Dinon, has no doubt a symbolical meaning of
'far-reaching power,' but later Greek writers took it
literally."   I see no grounds for this assumption.
Neither Thucydides, Diodorus, nor Herodotus pays much
attention to him, save as the father of the Great Artaxerxes (Μεμνὸν), and it is only in this capacity that he is
noticed by Plutarch.   Cornelius Nepos tells us that he
was famed for his beauty of person and of character ;†
and Plutarch describes him as " the first Artaxerxes,
who, of all the Persian kings, was most distinguished
for his moderation and greatness of mind " (*Langhorne*).   These accounts, it will be observed, hardly
tally with M. d'Arpentigny's treatment of the name.

---

* PLUTARCH, BIOI:—ΑΡΤΑΞΕΡΞΟΤ. "'Αρταξέρξης, ὁ Ξέρξου,
ὁ μακρόχειρ προσαγορευθεὶς διὰ τὸ τὴν ἑτέραν χεῖρα μακροτέραν
ἔχειν," etc.
   † C. NEPOS, " *Reges*."—"At Macrochir præcipuam habet
laudem amplissimæ pulcherrimæque corporis formæ, quam in-
credibili ornavit virtute belli. Namque illo Persarum nemo fuit
manu fortior."

Louis XVI., born a locksmith ;[126] Paul of Russia, born a corporal ;[127] and the plausible Francis II. of Austria, born a sealing - wax manufacturer,[128] — sovereigns whose innate tendencies led them to adopt these mean characteristics and pursuits, — had in like manner very large hands. They were endowed with

---

Artaxerxes Macrochir reigned for thirty-five years, and died B.C. 425.

[126] Madame Campan, in her charming book *"Mémoires sur la Vie Privée de Marie Antoinette"* (London : 1823), says upon this point :—" Unfortunately the king displayed too pronounced a taste for mechanical arts. Masonry and lock-making pleased him to such an extent that he admitted to his house a locksmith's apprentice, with whom he forged keys and locks, and his hands, blackened by this work, were often, in my presence, a matter for expostulation, and even for re-proaches, from the queen, who would have preferred other amusements for the king." — Vol. i., chap. v., p. 112. And Martin, in his *"History of France"* (Paris : 1878, vol. xvi., bk. 103), says of this monarch, " Il n'était à son aise qu'au milieu de ses livres . . . ou mieux encore dans son atelier de serrurerie," etc. Compare also the passages to be found in J. Michelet's *"Histoire de France,"* vol. xiv., chap. 12, and the *"Histoire Parlementaire de la Révolution Française,"* by B. Buchez and P. Roux (Paris : 1834, vol. iv., p. 198),—a passage quoted by Carlyle on p. 3 of the second volume of his *"History of the French Revolution."*

[127] Of the taste displayed by Paul I. of Russia for the minutiæ of military life, we are told by Alfred Rambaud [*"History of Russia,"* translated by L. Lang (London : 1879), vol. ii., ch. xi., p. 183]. · *Vide* also L. P. de Segur's *" Geheime Nachrichten über Russland"* (Paris : 1800), p. 341, 342.*

[128] I cannot find any authority for this statement, that Francis I. of Germany [and II. of Austria] was in any way interested in the sealing-wax industry. Dr. Hermann Meynert, in his *" Kaiser Franz I."* (Vienna : 1872, p. 6), tells us that Francis II. of Austria took a

---

* " Ihre entschiedene Abneigung für alles was Studium und Nachdenken erfordert, flösste ihnen beiden die sonderbare Leidenschaft für militärische Kindereien ein. . . . [Er war] ein Mann der nichts liebte als Soldaten, Wein und Taback " (!!)

the genius of their aptitudes, that is to say, of their respective *natures*, but they were devoid of that of their kingly stations. They *reigned* because they were the scions of royal families : and they would have reigned *well* if they had been gifted with royal natures. These instances afford us good examples of the theory of Joseph le Maistre.[129]

The theory of J. de Maistre.

With medium-sized hands we find the spirit of synopsis, *i.e.*, a capacity for comprehending at one and the same time the details and the mass of a subject.[130]

¶ 141.
Medium-sized hands.

great interest in the industries of the empire, and founded many manufactories ; but they were of the soap boiling and glass blowing order, and I cannot find any mention of the " cire à cacheter " of which d'Arpentigny speaks.*

[129] The theory of Joseph de Maistre, thus vaguely alluded to, refers to the office of kings and of kingly power, and is to be found enunciated in the second chapter of the fourth book of his work, " *Du Pape* " (Lyon : 1830), entitled " Institution de la Monarchie Européenne." His statement of the theory is as follows :— " Kings abdicate the power of judging for themselves, and, in return, nations declare their kings to be infallible and inviolable. Nothing can happen, nothing exists, without a sufficient reason : a family cannot reign excepting for the reason that it has more vitality, more royal spirit,—in a word, more of that quality which makes a family more fitted to reign than another. People think that a family is royal because it reigns : on the contrary, it reigns because it is royal." †

[130] This, therefore, is the ideal hand, for, as Herbert Spencer has said [" *Study of Sociology* " (London :

* " Mit gediegenen Kenntnissen in der Staatswissenschaft . . . verband er eine hinreichende Einsicht in die Gebiete des Kunst- und Gewerbefleiszes und der bürgerlichen Verrichtungen " *op. cit.*, p. 6.

† " Les rois abdiquent le pouvoir de juger par eux-mêmes, et les peuples, en retour, déclarent les rois infaillibles et inviolables. Rien n'arrive, rien n'existe, sans raison suffisante : une famille règne parcequ'elle a plus de vie, plus de l'esprit royale, en un mot, plus de ce qui rend une famille plus faite pour règner. On croit qu'une famille est royale parcequ'elle règne ; au contraire, elle règne parcequ'elle est royale."

¶ 142.
Musicians
hands.

Strict observance of time and measure being the necessarily precedent condition of musical rhythm, it is among subjects whose fingers are square that we find the most correct and thorough musicians ;—instrumentation is the especial forte of spatulate fingers, and melody is the peculiar province of fingers which are pointed.

¶ 143.
Scientific music.

Musicians, generally speaking, are numerous among mathematicians and algebraists ; because they of all people can mark and count their rhythm by numbers.

¶ 144.
Illustration :
M. l'Abbé Liszt.

The hands of the celebrated pianist Liszt are very large, [*i.e.*, finish in execution] ; his fingers are very prominently jointed, [*i.e.*, precision] ; his external phalanges present a highly-developed spatulation,— there we have the *power* by which he takes by storm the approbation of all who hear him.   Lean and slim, with a head which is long and severe of aspect, with a sharply-cut profile, he stands with his arms crossed, with an air which is at the same time courteous and cavalier, shaking back his long lank hair, which reminds one of Buonaparte the First Consul, and indeed he is perfectly willing to be placed in the same category of individualities.   He seats himself and the concert commences : a concert without any instrument but his, and without any performer but himself.   His fingers fly over the keyboard, and one thinks involuntarily of the tramp of an army ; one remembers Attila,[13] and one imagines

1884), p. 322] :—"The analytical habit of mind has to be supplemented by the synthetical habit of mind. Seen in its proper place, analysis has for its chief function to prepare the way for synthesis, and to keep a due mental balance, there must be not only a recognition of the truth that synthesis is the end to which analysis is the means, but there must also be a practice of synthesis along with the practice of analysis."

[13] ATTILA, a celebrated King of the Huns, who invaded the Roman Empire in the reign of Valentinian

that the *Scourge of God* is sent upon us ; or again, it seems as if a tempest howled across the desert whilst his fingers thrash the ivory keys like a downpour of living hail. We realise then that he has not over-rated his powers of entrancing us, for his fingers have the powers of a whole orchestra ; but, 'ardent and impetuous as he is, he never loses his self-possession, for his hand is not only that of an instrumentalist, it is the hand of a mathematician, of a mechanician, and, by a natural development, that of a metaphysician, *i.e.*, of a man whose genius is more pre-arranged than spon-taneous in its exhibition, a man more clever than passion-ate, and gifted with more intelligence than soul.[139]

¶ 145.
Critical genius.

Genius which is subtle and critically disposed, a strong love of polemic discussions, and an instinct of controversy, often gather themselves together in the individuality of the man whose large hand is fur-nished with square fingers of which the joints are prominent.

¶ 146.
Decline and fall of the Greek Empire.

When, shorn of all its most active and most power-

with an army of 500,000 men, and laid waste the provinces. He took the town of Aquileia, and marched against Rome, but his retreat and peace were purchased with a large sum of money by the feeble emperor. Attila, who boasted in the appellation of the " *Scourge of God*," died A.D. 453. (*Lemprière*). For an account of this monarch and of his operations upon Rome, see Gibbon's " *Decline and Fall of the Roman Empire*," chap. xxiv.

[132] It is difficult, if not impossible, to annotate a pas-sage like the above, dealing, as it does, with the name of a man now living, who even as I write is creating a stir in our very midst. His compositions and his biographies are innumerable. Mr. Arthur Pougin, in his " *Supplé-ment et Complément*" (Paris: 1881) to the " *Bio-graphie Universelle des Musiciens et Bibliographie Générale de la Musique*," by M. F. J. Fétis (Paris: 1860-65), gives a list of ten biographical works dealing with this artist--known to the world as M. l'Abbé Liszt, --the only ones of which that I know and can re-commend being J. Schuberth's " *Franz Liszts*

ful colonies, left like a head without a body, the Greek Empire, reduced to a single city, became at length extinguished, engulphed in the vortex of an absolutely abnormal state of government, it was ruled no doubt by hands such as I have just described. Until the last moment even,[133] when the scimitar of the second Muhammad hung over their devoted heads, its citizens were involved in the most incomprehensible quarrels, in abstractions, in undifferentiated distinctions, in theological disputes, to the exclusion of the duties which they owed to their mother country, not from want of courage [with which, of a sort, they were well provided], but from stupidity.

¶ 147.
Small hands.

To small and finely narrowed hands belongs the faculty of synthesis.

¶ 148.
Modern French literature.

The taste, so prevalent in France in the present day, for historic and literary works which abound in details, is a proof of the intellectual advancement of the democracy, for democracy is the laborious pro-

Biographie" (Leipzig: 1871), and "*L'Abbé Liszt*" (Paris: 1871), which are as satisfactory as biographies of living celebrities can be. M. Pougin, in the work above cited, says of him :—"Cet artiste prodigieux, fantasque, mais d'une trempe intellectuelle singulièrement vigoureuse, n'a cessé, depuis plus d'un demi siècle, d'occuper le monde de sa personne, de ses travaux, et de ses excentricités. . . . Dans ces dernières années, ayant presque épuisé tous les moyens ordinaires, il n'en a pas trouvé de meilleur que de faire croire qu'il entrait en religion. . . . Tout porte à croire pourtant qu'il n'en est rien, et que les pratiques de dévotion qu'on a remarquées chez le grand artiste ne sont encore de sa part qu'une nouvelle occasion de *réclame* et un désir toujours plus intense de faire parler de lui. . . . M. Liszt est un type à part dans l'histoire musicale du 19ᵉ siècle, et si l'on peut regretter ses défauts artistiques et intellectuels, on n'en doit pas moins apprécier ses étonnantes qualités et les facultés admirables quoique mal équilibrées, qui constituent sa personnalité."

[133] In the fifteenth century.

genitor of large hands, just as aristocracy is the indolent fountain head of small ones. Thus, a little while before the Revolution, hardly any literature was produced save for the aristocracy, which naturally preferred books which were synthetical to books which were analytic.

¶ 149.
Our taste in literature.

Our admiration for the works of artists or of authors is in a direct ratio to the sympathy which exists to a greater or lesser degree between our physical organisation and theirs.

¶ 150.
Harmony between fingers and joints.

To spatulated, and even to square hands, prominent joints are an additional beauty, seeing that they are by nature destined to the cultivation of the useful arts, which are those of combination and of calculation; but to pointed or to conic hands developed joints would be a deformity, seeing that they are destined to the prosecution of the liberal arts, which are those of intuition and inspiration.

¶ 151.
Fusion of types.

At the same time, predominance of the intuitive faculties does not necessarily presuppose an entire absence of all the talents which depend upon the faculty of combination, any more than the predominance of the talent of combination necessarily implies complete absence of all inspiration.

¶ 152.
Alexander the Great and Cæsar

Alexander proceeded, as Bossuet has remarked, by great and impetuous sallies.[134] He favoured poets,

[134] This reference is made, I presume, to the passage in Bossuet's "*Discours sur l'Histoire Universelle*" (Paris: 1786), where, after comparing Alexander with Darius, the author describes his entry into Babylon, saying:—"And after having with incredible rapidity subjugated the whole of Persia, to secure the safety of his new empire on all sides, or rather to gratify his ambition and render his name still more famous than that of Bacchus, he entered India, where he pushed his conquests even further than those of that celebrated conqueror. He returned to Babylon feared and respected, not like a conqueror, but like a god."—Vol. ii., p. 277.

[¶ 152]

whilst he merely esteemed philosophers.[135] Cæsar,
on the other hand, regarded philosophers with favour,
whilst he looked upon poets merely with cold approbation. Both of them reached the zenith of glory,
the one by inspiration supported by combination,
the other by combination supported by inspiration.
Alexander was a man with a great soul, Cæsar was
a man with a great mind.

¶ 153.
Sense of touch
in the finger-tips.

Regard being had to the fact that the sense of
external touch is most highly developed at the tips
of the fingers,[136] and that man is naturally prone to

---

[135] *Vide* Plutarch's "*Life of Alexander*." "He loved
polite learning too, and his natural thirst of knowledge
made him a man of extensive reading. The '*Iliad*,' he
thought, as well as called, a portable treasure of military
knowledge ; and he had a copy corrected by Aristotle,
which is called *the casket copy*.* Onesicritus informs
us that he used to lay it under his pillow with his sword.
As he could not find many books in the upper provinces
of Asia, he wrote to Harpalus for a supply, who sent
him the works of Philistus, most of the tragedies of
Euripides, Sophocles, and Æschylus, and the Dithy-
rambics of Telestus and Philoxenus." So much for his
love of poets ; as to his mere *esteem* of philosophers, in
the same life we find :—"Aristotle was the man he admired in his younger years, and, as he said himself, he
had no less affection for him than for his own father.
. . . But afterwards he looked upon him with the eye of
suspicion. He never indeed did the philosopher any
harm ; but the testimonies of his regard being neither
so extraordinary nor so endearing as before, he discovered
something of a coldness. However, his love of philosophy . . . never quitted his soul, as appears from the
honours he paid Anaxarchus, the fifty talents he sent
to Xenocrates, and his attentions to Dandamis and
Calanus" (*Langhorne*).

[136] *Vide* Julius Bernstein's physiology of the sense of
touch : "*The Five Senses of Man*" (London : 1883,
4th edit., p. 17), "The nerves of the skin which ter-

---

* He used to keep it in a rich casket found among the spoils of
Darius. "Darius," said he, "used to keep his ointments in this
casket ; but I, who have no time to anoint myself, will convert
it to a nobler use."

exercise that sense which, by the accuracy of its perceptions, he feels to be the most perfect and complete,
it is obvious that the desire for those employments
in which the physical sense is more utilised than the
moral, will be the more pronounced in proportion as
the spatule is the more developed in hands of that
type.

And in like manner, the more the conic phalanx is ¶ 154.
drawn out in artistic or psychic hands, the more Conic and pointed hands.
unpractical and unworldly will be the peculiar bent
of the genius. Lord Byron's hands were remarkable Lord Byron.
for extremely pointed fingers,[137] and in the same way

minate in single fibres extend only to the dermis, and
here they are observed to end in a peculiar manner in
the papillæ. Many of them contain, for instance, an
egg-shaped particle, which a nerve-fibre enters, and in
which it is lost after several convolutions round it.
They are called *tactile corpuscles*, and there can be no
doubt that they act as the instruments of the sensation
of touch. . . . They are extraordinarily numerous at the
tips of the fingers, where in the space of a square line
about a hundred can be counted " [*vide* ¶ 77]. And
the same thing is laid down by Jan. E. Purkinye in his
"*Commentatio de Examine Physiologico*" (Leipsic :
1830), and by Arthur Kollmann in his recent work, "*Der
Tast-apparat der Hand der Menschlichen Rassen und
der Affen in seiner Entwickelung und Gliederung*"
(Hamburg und Leipsic: 1883). For a complete physiology of the sense of touch as regards the hand
vide "*A Manual of Cheirosophy*" (London: 1883),
§§ 39-51.

[137] Leigh Hunt, in his bitter and ungrateful volume
upon his best friend, "*Lord Byron and some of his
Contemporaries, etc.*" (London: 1828, p. 91), says :—
" He had a delicate white hand, of which he was proud,
and he used to call attention to it by rings. He thought
a hand of this description almost the only mark remaining nowadays of a gentleman." In another place the
same writer says :—" My friend George Bustle used to
lament that, in consequence of the advancement of
knowledge and politeness, there was no longer any distinguishing mark of gentility but a white hand." Sir
Cosmo Gordon tells us that all his life Byron was dis-

[¶ 154]
Moreau.

the hand of Hegesippe Moreau was beautifully modelled.[138]

¶ 155.
The exterior phalanges.

The exterior phalanges are the *eyes* of the hand.

¶ 106.
Invariable forms.

Each type has certain formations, which an enforced labour, a labour utterly at variance with the genius of which the hand is the born instrument, can very appreciably modify; but which it cannot transform to such an extent as to render them unrecognisable.

¶ 157.
Modifications of a leading type.

One finds convincing examples of this in villages shut in on all sides by forests, and peopled exclusively by charcoal burners [for instance]; or in the hamlets perched upon the rocks of little barren islands where fishing is the sole industry. Unless the population of these places has derived its ancestry from a common source, all the types of hands will be represented with all their varieties, and the continued pursuit of an occupation which is imposed upon them, rather than chosen by them, will never change a conic finger into a spatulated one. The hand may swell, may thicken, and may lose its suppleness and its elasticity, but the innate formation will remain, just as does the instinct which is inseparable from it. To speak truly, the poet or the logician, in these hands thus altered and in these instincts thus combated and falsified, is hidden nearly as completely as is the

tinguished by an intense sensitiveness,* which showed itself particularly concerning his deformed foot, which he was continually striving to conceal. †

[138] The same sensitiveness has been recorded of the ill-starred young poet Moreau, who has been compared by Ste. Marie Marcotte to our poet Chatterton (in the 1851 edition of his poems). Sainte-Beuve in the edition of 1860, and Louis Ratisbonne in the edition of 1861, both call attention to his extremely unpractical and unworldly characteristics.

* SIR COSMO GORDON, "*Life and Genius of Lord Byron*" (London : 1824).
† *Vide* J. GALT'S "*Life of Lord Byron*" (London : 1830), p. 24.

oak in the acorn, as is the butterfly in the chrysalis, or, as is the delicately-moulded goddess in the roughly chiselled-down block of marble; but a chance phrase in a homely conversation, or a chance opinion expressed in the rudimentary debates of the conclave of village worthies, will reveal the hidden potentialities to an observing and penetrating intelligence.[159]

Finally, if, in practising this science a lingering doubt remains in your mind as to the effect of the habitual labour upon a hand submitted for your examination, you must either relinquish your task, or base your judgment on the forms which the hand originally displayed, carefully described by its owner.

¶ 158.
Effacement of characteristics.

"But," you say, "these signs which you have just described to us; are they in fact infallible indices of our intellectual tendencies? in other words, does the standard always declare the nationality? is the voice of the oracle always that of God?"

¶ 159.
Certainty of the indications.

In my opinion—Yes. But do not take my word for it; let your convictions result from your personal observations; only, do not allow yourself to be prejudiced beforehand, and do not allow a few mixed hands which are difficult to decipher, because they have momentarily confused the pilot, cause you to deny the accuracy of the compass.

¶ 160.
Confirmation by observation.

Let us now, forsaking cities and their inhabitants, follow this company of surveyors and engineers, these representatives of a class of men who are anything but poetic, and who worship God in the form of a triangle; there they go, backwards and forwards across the country, armed with poles, measuring-planks, and chains; one can tell by the joyous activity with which they pursue their task that they are exercising a pursuit of their own choice and entirely to their own taste, and that their souls, like the birds

¶ 161.
The practical hand.

, [159] *Vide "A Manual of Cheirosophy,"* ¶ 70

[¶ 161]

among the boughs overhead, or like the gazelle upon the sea-shore, lose themselves in the enjoyment of these trapezoids and squares which they map out with such ease and dexterity.

¶ 162.
Its shape.

Their hands are square or spatulated.

Whether spatulate or square, they have knotty fingers, jewelled as if with rings, with the equations which daring modern science attaches boldly to the blazing locks of asteroids and comets.

¶ 163.
Artistico-energetic hands.

Now let us penetrate into the workshops of the artillery schools and of the sciences of military engineering, into the circus and the hippodrome, theatres for the prowess of the loud-voiced descendants of Alcmene and of Leda;[140] into the gymnasia of the acrobat, the fencer, and the equilibrist; into the haunts of the poacher, of the jockey, and of the horse-trainer. Here we find hands which terminate in a spatule, and also large conic hands which are very hard: these latter combine a vague sentiment of grace with their feats of strength.

¶ 164.
Illustration:
Le Vicomte
d'Aure.

The most expert horse-trainer of to-day, the cleverest, the most progressive, and the most elegant, M. le Vicomte d'Aure, author of several excellent works upon horsemanship and horse-training, has a hand which is decidedly spatulate, but extremely supple.[141]

---

[140] Alcmene was the daughter of Electryon, King of Argos, and Anaxo [called *Lysidice* by Plutarch, "*De Rebus Græcorum*," and *Eurymede* by Diodorus (i., c. 2)], and was the mother of Hercules, by Jupiter. Leda was the mother of Castor and Pollux, who were also sons of Jupiter.

[141] M. le Vicomte d'Aure, one of the leading authorities on horsemanship, at the date at which the above was written, was the author of a "*Traité d'Equitation*" (Paris: 1834). *Vide* also M. C. Raabe's "*Examen du Cours d'Equitation de M. d'Aure, etc.*" (Paris: 1854).

If now, leaving the crowd and bustle of our fellow-men, we go and, pursue our investigations in the crystallising solitude of large libraries, beneath the inspiring tiles of aërial garrets, in the narcotising atmosphere of laboratories, within the bare walls of the schools where ushers and pedants stalk up and down among the scholars,—if, I say, we go into these different places and examine the hands of the philosophers, the artists, the poets, the mathematicians, the professors, all of whom an irresistible vocation has forced to follow the pursuits implied by the titles of these professions, we shall find them to present the appearances I have described ; that is to say :—

¶ 165.
The scholarly hand.

Those of lyric poets and of romance writers who aim at ideality, such as Georges Sand, Leconte de l'Isle, Chateaubriand, Hugo, De Vigny, Lamartine, etc., will be more or less conic as to their tips.

¶ 166.
The poetic hand.

Those of grammarians, critics, didactic, analytic, and dramatic poets, those of doctors, lawyers, geometricians, artists of the rule and line school, will have hands whose fingers are square or even spatulate.

¶ 167.
The methodical hand.

As for the polytechnic schools ; if you find in those of dynamics, mechanics, and applied science, a hand which is finely moulded or pointed, pity the ill-luck of an unfortunate poet who has strayed from his rightful vocation of a sun-worshipper or follower of Astarte, constrained to offer sacrifice to the Cyclops and the Gnome.

¶ 168.
Hands out of place.

In a word, what more can I say ? Without fatiguing you with a study whose elements are to be found on all sides, cast your eyes on your own immediate surroundings. Observe the hands of your friends, of your neighbours, and of your relations : This one devotes his whole attention to intellectual pursuits—his characteristics are poetry of soul far

ILLUSTRATIONS.
¶ 169.
Artistic type.

more than analysis of mind; he has an absorbing
passion for pictures, music, monuments, statuary,
and poetry; he prefers that things should be beautiful
rather than that they should be useful: he is easily
exalted, there is in his expression, in his gestures, in
his language, and in his dress an undefined element
of strangeness and of inspiration; he can do without
necessaries, but not without luxuries; his purse, open
to all who require assistance, is hermetically sealed
to his creditors.   At an age when most men have
left behind them the illusions of youth and have got a
grip of real life, giving themselves up with ardour to
the fruitful pursuits of a working existence, his heart,
ever young, ever accessible to the most exalted and
impractical ambitions, remains dominated by ideas of
romance and of Utopianism; he regards the world by
the light of the antique hypotheses of spiritualism, and
is profoundly ignorant of the real meanings and values
of the things of this life.   For him, the mountain tops
shower down holy thoughts from their beetling crests,
he loves high-flown language and beautiful sentiments; he prefers charm to intellect, and he prefers
grace to beauty.   He see a poetic sentiment everywhere—in the raindrops which streak the heavens
with innumerable sparkling arrows, in the window
panes which weep with the sobs of the tempest,
in the hoarse cry of the weathercock, in the figures
of light like snowy doves, which the sun traces
upon the green sward through the leaves of the
forest when they are kissed into motion by the
winds.   By night, when the moon has extinguished
her feeble beams below the watery horizon, he loves
to wander, his heart filled with a voluntary melancholy, along the moist sands of the deserted shore.
He is credulous, fond of the unforeseen, and his soul
is like the spark which, if it is not allowed to burst
into flame, dies away.

Very well ; this friend, this relation, this neighbour, has *inevitably* fingers which are conic or pointed, and a small thumb.

¶ 170.
Energetic type

Here is another subject. This one likes to work with his hands ; he digs, he hews, he prunes. He lives standing on his feet as it were, always on the move, always with a knife, a hammer, or a gun in his hand. He despises those visionaries who, continually wandering in dreamland, drag out their existence watching the stream flowing to the sea, the clouds passing overhead, and the trees swaying and whispering to one another. He likes the noise of the hunting horn and the yelping of the pack ; he is passionately fond of horses, his courtyards abound with dogs, with peacocks, with poultry, with magnificently coloured cocks strutting hither and thither with their scarlet combs jerked to one side. He is an early riser, he is a hunter, he is an angler. He can tell you, without a moment's consideration, all about everything within ten leagues of his habitation in the way of lakes abounding with fish, or heaths and moors abounding with game. He loves the sight of the restless sea ; he loves all that assists locomotion or produces activity. He likes physics and mechanics, and the turmoil of the timberyard and of the workshop.

¶ 171.
The same continued

Don't bother him about gardens aromatic with the perfumes of mystic verse, retreats heavy with silence and shadows, whose ornaments are perchance a saintly statue, a sculptured well and fresh-leaved avenues, where the laurel and the cypress embrace one another, and the dragon-fly and the dove disport themselves. He would much rather see fruit-laden orchards and wide kitchen gardens, fringed with walls on which innumerable espaliers are trained symmetrically upon the green trellis-work. Here, beneath glass, ripen the pine-apple and the cantaloup,

and here in its stony bed the limpid streamlet pursues
its even way, keeping up a running accompaniment
to the songs of the bullfinches.   Arbours, rustic seats,
shutters, swings, and what not ? all that furnishes
his bowers, his terraces, his arbours, and his house
has received from his industrious and natty fingers
its form or its finishing touches.   He is not superstitious, he describes himself as living for the present
day and for his own country, and he shakes his head
at the words of apostles of religion and travellers
alike ; he requires comfort, and looks for things that
are useful, of good quality, and sound construction.
Veterinary surgeons, masters of the noble arts of
self-defence, horse dealers, iron-masters, turners,
and huntsmen, *et hoc genus omne*, find in him an
adept, a patron, and a friend.   His manners are
frank and open, he is gifted with the qualities of
power, rectitude, and sincerity, and he is governed
by his affections rather than by his judgment.

Need I say that such a man will have hands which
terminate in a spatule, with a firm palm and a large
thumb ?

¶ 172.
Elementary
spatulate type.

Listen now to the lucubrations of yonder parvenu :
" He has been in his time a cowherd, a porter, and a
smuggler, and he is proud of it," says he with a
swagger ; "he could live on ortolans if he wanted
to, he is rich enough, goodness knows, but he
prefers pork ; let each man please himself."   His
clothes are always too big for him, and he has his
hair cropped short like that of a labourer.   Of his
three sons his favourite is the one who blacks his
own boots, and saddles and grooms his own horse.
"That's a *man*," says he ; "he could carry an ox ; the
others read and think and fiddle, but they can't even
make wine ! "   He will marry his sons, if possible, to
women who like to cook and to do their own washing,
and who would despise the luxury of a parasol.

None of your mincing, attitudinising, dancing, singing, dolly-faced misses for him ! Music ? Bah ! it sends him to sleep. These people who are profuse in their salutations and civilities, and full of the minor courtesies of life—the very sight of them, like that of rats or of custom-house officials, horrifies and irritates him. He likes to feed in his shirt sleeves, and with his dress in disorder ; he admires big women and big dogs. In days gone by, when he frequented fairs and markets, he was concerned in every quarrel, and was to be found in every gang of roysterers; he is so far a philosopher as not to believe in the "mummeries of religion." He is no connoisseur of pictures or of statues, they are all nonsense, but he has an unerring judgment in beasts and farm produce. Sciences and arts ! fine things indeed, but they have no value on 'Change or in the market. In his garden you find squares of cabbages and ranks of sunflowers. He does his own marketing, hews his own wood, κ.τ.λ.

A large, thick hand with a hard palm, spatulate fingers, and a large thumb.

But here is another subject whose dress and whose deportment argue a wholly different class of mind. **¶ 173.** The useful type He possesses to the highest degree the sentiment of respect of persons; he has a pompous manner, highly-starched linen, and spectacles. He inhabits a little town, destitute alike of commerce and of population, where one's footsteps echo on the deserted street, where the country squire is lord of all he surveys, and sacristans abound. In speaking of his patrons and of his superiors his voice becomes grave and subdued. · He knows Latin, geometry, natural history, botany, geography, archæology, a little medicine, a little jurisprudence—a little, in fact, of everything which is capable of being learnt, but hardly anything which requires to be instinctively acquired. He is not in the habit of joking, and his sparks of wit, clogged by a viscid protoplasm of pedantry, have no spontaneity, even when they are just and to the point. To words capable of the widest interpretation as regards their intellectual meanings, such as liberty, order, poetry, and so on, he insists upon giving their strictly literal and material significations. Constantly arranging, brushing, and dusting, he piles his own linen in its place after having carefully verified the marking of it, and ever since attaining his majority he has kept all his receipted bills carefully locked up. He is exact, formal, methodical, punctual; a martyr to regulations, and submissive to generally accepted usages, who regulates his life by these qualities, who is annoyed and disconcerted by any form of innovation, and whose thoughts wander at ease merely within the narrow limits of vulgar common sense. He consults his mind more than his heart, and he denies that beauty can exist in a thing which cannot be reduced to a definite theory. He likes gardens with box-bordered cross paths, which one can take in at a glance, whose trees, annually trimmed by the gardener's art, have no movement and make no rustle, and round which thickly-planted hedges, like heavy folding-screens, extend themselves in a rectangular figure.

¶ 174.
The same continued.

Such is the "proprietor" who is predestined to the honours of the municipal scarf of office, the churchwarden who is conscientiously regular in his attendance at the board; where there is legality he recognises equity, where there is diplomacy he sees science. He reveres equally the rules of syntax and the penal code. Such is the provincial academician, with his barometer, his thermometer, his telescope, his sun-dial, and his bottled monsters; at the same time lazy and fussy, instructed and shallow-minded. Such a man do we find among people of second-rate minds, a man whose fingers are square, with developed joints and a large thumb.

And if these indications are to be found, perfectly according with the tastes and pursuits of the subject, and centred in the person of a man who, being independent, or even rich, would not be forced to these habits of life, · were it not that they entirely suit his inclinations, what stronger proof can you require of the truth and trustworthiness of this system ?

¶ 175.
Application of
the system.

AUTHOR'S NOTE.—Out of four characters I have described, I have emphasised the bad qualities in particular of two, and the good qualities in particular of the other two ; but as there is no type which exhibits only the good or the evil tendencies of its nature and instincts, it will be easy for the reader to rectify any obvious bias in the above descriptions, whether for good or evil, and to complete the portraits accordingly.

¶ 176.
Note on the
above illustra-
tions.

¶ 177.
Newton.

THE Thumb, on account of the clearness and the importance of the indications which it presents, deserves to be made the subject of a distinct chapter.

"In default of any other proofs," said Newton, "the thumb would convince me of the existence of God."[112]

¶ 178.
The thumb emblematic of moral force.

Just as, without the thumb, the hand would be defective and incomplete, so, without moral force, logic, and decision [faculties of which the thumb in

---

[112] This paragraph, which has been freely quoted from M. d'Arpentigny, occurs thus in the original :—"'A defaut d'autres preuves,' disait Newton, 'le pouce me convaincrait de l'existence de Dieu'" [edition 1865, p. 59]; and speaking from what I think I may say is a fairly intimate acquaintance with the works of Newton, I have very little hesitation in recording that Sir Isaac never said it at all ! Certainly, it does not occur in his "*Four Letters to Dr. Bentley, etc.*" (London : 1756), or in his "*Treatise of the System of the World*" (London : 1731), and I have searched the complete edition of his works published in 1779 (London) in vain. He states his convictions of the existence of God, and the reasons for those convictions, at the end of "*The Mathematical Principles of Natural Philosophy*," translated by Andrew Motte (New York : 1848, p. 504-556), in a passage beginning and ending :—"This most beautiful system of the sun, planets, and comets could only proceed from the counsel and dominion of an intelligent

different degrees affords the indications] the most fertile and brilliant spirit would be a gift entirely without value.

¶ 179.
Instinct and reason.

In the same manner as the inferior animals, we have an instinctive will, an instinctive logic, and an instinctive decision; but the thumb represents only the will of reason, the logic of reason, and the decision of reason.

¶ 180.
The distinctive thumb.

The superior animal is signalised by the possession of the *hand*, the *man* is signalised by his possessing a *thumb*.[143]

¶ 181.
Pollice truncatus

From the words *pollice truncatus* (= a man with his thumb cut off), which the ancient Romans applied to the cowardly citizen who cut off his thumb to obviate the possibility of his being sent to the wars,[144] we have derived our word *poltroon*.[145]

and powerful Being. . . . We know Him only by His most wise and excellent contrivances of things and final causes, and we admire Him for His perfections," etc.* It is possible that our author made his quotation from a *coup d'œil* of this *Scholium*, in the manner in which he "closed up" the dialogue recorded in note [110], p. 107.

[143] This is a further development of the dicta of Aristotle, Anaxagoras, and Galen, discussed in note[101], p. 94.

[144] This is recorded of a Roman citizen of the time of Augustus Cæsar, who cut off the thumbs of his two sons, so as to keep them at home.

[145] As a punishment the amputation of the hand has been practised by a great many nations [*vide* "*Manual*," ¶ 22), especially among the ancients; in more modern times it has been principally in the East that this punishment has been put in force; whereupon George Sale, in his "Preliminary Discourse" to his translation of the Qur'an, says:—"Theft is ordered to

* SIR ISAAC NEWTON, "*Philosophiæ Naturalis Principia Mathematica*," editio tertia (London: 1726). "De Mundi Systemate"—*Scholium Generale*:—Elegantissimæ hæc solis, planetarum, et cometarum compages, non nisi consilio et dominio Entis intelligentis et potentis oriri potuit. . . . Hunc cognoscimus solummodo per proprietates ejus et per attributa et per sapientissimas et optimas rerum structuras et causas finales, et admiramur ob perfectiones."

¶ 182.
Thumb of the monkey.

The thumb of the monkey, which is barely flexible at all, and for this reason scarcely able to be opposed —*i.e.*, be made to act in conjunction with any of the other fingers—is looked upon by many naturalists as nothing more than a movable nail.[146]

¶ 183.
Symbolism.

Whereas, on the contrary, the human thumb is so situated and organised as to be able always to act in an opposite direction to the other fingers, and it is by this power that it symbolises the inner or moral sense which we oppose to our will, and through it to the temptations of our instincts and our senses.[147] Proofs of this assertion abound ; thus, for instance,

Idiotr.

idiots who are idiotic from birth, come into the world either without thumbs, or with thumbs which are

---

be punished by cutting off the offending part, the hand [Qur'an, chap. v.], which at first sight seemed just enough ; but the law of Justinian, forbidding a thief to be maimed [Novell., 134, cap. 13], is more reasonable ; because stealing being generally the effect of indigence, to cut off that limb would be to deprive him of the means of getting his livelihood in an honest manner." *Vide* also Puffendorf, "*De Jure Nat. et Gent.*," lib viii., cap. 3, sec. 26.

[146] Professor Sir Richard Owen, in his monograph "*On the Nature of Limbs*" (London : 1849, p. 36), calls attention to the fact that the "thumb," which is the least important and constant digit of the anterior extremity in the rest of the mammals, "becomes in man the most important element of the terminal segment, and that which makes it a 'hand' properly so called." *Vide* "*A Manual of Cheirosophy*," ¶¶ 31, 35.

[147] This important peculiarity of the thumb is duly emphasised by Galen in the passage which I have quoted at length in ¶ 35 of "*A Manual, etc.*"*

* "Τί γὰρ εἰ μηδεὶς ἀντετέτακτο τοῖς τέτταρσιν, ὥσπερ νῦν, ἀλλ' ἑξῆς ἅπαντες ἐπὶ μιᾶς εὐθείας ἐπεφύκεσαν οἱ πέντε ; ἆρ' οὐ πρόδηλον, ὡς, ἄχρηστον αὐτῶν ἐγίγνετο τὸ πλῆθος ; δεῖται γὰρ τὸ λαμβανόμενον ἀσφαλῶς ἢ πανταχύθεν κατὰ κύκλον, ἢ, πάντως γ' ἐξ ἐναντίων δυοῖν τόπων διαλαμβάνεσθαι. Τοῦτ' οὖν ἀπώλετ' ἂν εἴπερ ἐπὶ μιᾶς εὐθείας ἑξῆς ἅπαντες ἐπεφύκεσαν σώζεται δ' ἀκριβῶς νῦν, ἑνὸς τοῖς ἄλλοις ἀντιταχθέντος, οὕτω γὰρ ἔχει θέσεώς τε καὶ κινήσεως ὁ εἷς οὗτος ὥστ' ἐπιστρεφόμενος βραχείας παντελῶς ἐπιστροφὰς μεθ' ἑκάστου τῶν τεττάρων ἀντιτεταγμένου ἐνεργεῖν."

powerless and atrophied; and this is perfectly logical, for where the essence is absent, its symbol also must be wanting.

Infants up to the time when their intelligence begins to be developed, keep their hands continually closed, folding their fingers over the thumb; but in proportion as with the body the mind becomes developed, the thumb in turn folds itself over the fingers.[148]

¶ 184.
The human infant.

Epileptic patients in their fits fold their thumbs before the rest of their hands, which shows that this evil, which is instinctively apprehended before it is actually felt, affects the organisation by which one perceives and *knows*, before it affects the organisation by which one merely *feels*.[149]

¶ 185.
Epileptic patients.

At the approach of death, the thumbs of the dying, struck as it were with the vague terror of approaching dissolution, fold themselves beneath the fingers,— sure sign of the nearness of the final struggle.[150]

¶ 186.
Moribunds.

[148] " By a contraction of the flexor, and non-development of the extensor muscles, the human infant hides its thumbs in the palms of its hands until its *will* shall have developed itself and put itself into exercise ; and a test of complex *will* organisation is to be found in an early and complete development of extensor action." *Prof. Victor Horsley.*

[149] Mr. W. R. Gowers, in his exhaustive and minute work "*Epilepsy and other Chronic Convulsive Diseases: their Causes, Symptoms, and Treatment*" (London : 1881), makes frequent references to cases in which the precursory and early symptoms of an epileptic fit have appeared in the hands. Out of forty cases, in which the attacks made their first appearance in the upper limb, about thirty commenced in the hand (p. 46). In one case the premonitory symptom was a twitching of the thumb and forefinger (p. 47), in another the whole hand closed up (p. 48). During the fits, says he, " the fingers are flexed at the metacarpo-phalangeal joints, and extended at the others, *the thumb being adducted into the palm*, or pressed against the first finger (p. 74).

[150] The author is not quite right in his data here : it is true that when *rigor mortis* sets in, the thumb

¶ 187.
Man's knowledge of death.

Man alone, by reason of his having a thumb,—*i.e.*, reasoning faculties,—*knows*, *i.e.*, is familiar with, the idea of death.

¶ 188.
The thumb the seat of will

At the root of the thumb is situated the indication of the reasoning will, a power of whose intensity one forms an idea by the observation of the length and thickness of this root. It displays also, say the Cheiromants [for which reason they have called this locality the Mount of Venus], the greater or less intensity of the capacity to love.[151] And in point of fact, love is but *will* [desire].

¶ 189.
The two joints.

In the first [or lower] phalanx reside the indications of the logic, *i.e.*, of the perceptions of the judgment, and of the reasoning faculties; and in the second [or upper] phalanx, reside the indications of the invention, of the decision and strength of will, and of the initiative power.

¶ 190.
The thumb in gladiatorial shows.

The Romans, in their gladiatorial displays, raised their thumbs, and the combatant stricken down received his life. If they lowered them, he died.[152]

naturally drops towards the centre of the palm, the power of the flexor muscles being always [even at birth, *vide* note [118]] superior to that of the extensors; but this ensues some time *after* death, and not before it, unless the final struggle has been accompanied by convulsions of any kind, in which case this symptom may be observed.

[151] *Vide* "*A Manual of Cheirosophy*," pp. 224-225.

[152] "As the gladiators in the gladiatorial displays had all previously sworn to fight till death . . . the fight was bloody and obstinate, and when one signified his submission by surrendering his arms, the victor was not permitted to grant him his life without the leave and approbation of the multitude. This was done by clenching the fingers of both hands between each other, and holding the thumbs upright close together, or by bending back their thumbs. The first of these was called *pollicem premere*, and signified the wish of the people to save the life of the conquered. The other sign, called *pollicem vertere*, signified their disapprobation, and ordered the victor to put his antagonist to death."—*Lemprière*.

Strange instinctive appreciation of the initiative reposing in the development of this digit.[153]

If your thumb is narrow, lean, thin, and short in this phalanx, you are troubled with a complete absence of decision ; you are prone to accept generally-received opinions and the ideas of other people, and you are subject to doubt and endless incertitude, culminating in moral indifference.

¶ 191.
Weak thumbs

Of this continually uncertain state of your mind, of this inability to form and adhere to an opinion of your own, you will always be able to give a logical and comprehensible explanation, if your first phalanx [that of logic] is developed. If, on the contrary, your second phalanx is long and well-formed, whilst the first [the lower one] is thin and short, you will have decided opinions, strong and lasting convictions, a quick mind, and a decisive and initiative spirit; at the same time you will probably be a poor arguer, or a man more gifted with passionate conviction than with judgment.

¶ 192.
Development of the joints.

Generally speaking, a thumb which is small, mean, and poorly formed, announces an irresolute mind, and a wavering disposition in those things which are usually the result of reasoning power, and not of

¶ 193.
A small thumb.

---

[153] There are many expressions of this description in use among the French, even to-day, many of which are given by Littré in his dictionary. Of these the most interesting are :—" METTRE LES POUCES, or COUCHER LES POUCES," *i.e.*, to give oneself up, to yield, after a longer or shorter resistance,—a saying which seems to have arisen from the fact that the thumb falls into the palm of the hand when the resistance is exhausted [*vide* note ¹⁵⁰, p. 141], " Et faisant unebelle révérence se retira luy estant tombé le poulce dans la main," Carl I., 29, *Mercure.* SERRER LES POUCES, *i.e.*, to torture ; SE MORDRE LES POUCES, *i.e.*, to repent [*vide* note, ¶ 16]; DONNER UN COUP DE POUCE, *i.e.*, to strangle; and MALADE DU POUCE, *i.e.*, stingy, derived from the expression JOUER DU POUCE, *i.e.*, to pay out money at a game.

[¶ 193]

sentiment or of instinctive knowledge. Such thumbs are the inheritance of the descendants of the Foolish Virgins, a class of people who are impressionable, sensual, and dominated by their innate tendencies, but who are at the same time impartial, tolerant, naturally amiable, and able to accommodate themselves to all characters with whom they are thrown together.

¶ 194.
Its effect.

People with small thumbs are governed by their hearts [source of all tolerant feelings], and are more at ease in an atmosphere of sentiments than in one of ideas; they appreciate things more at a rapid survey of them than on reflection.

¶ 195.
Large thumbs.

People with large thumbs are governed by their heads [source of all feelings of exclusiveness], and are more at ease in an atmosphere of ideas, than in one of sentiments. They judge things better by reflection than on the spur of the moment.

¶ 196.
Illustrations.
A. Durer.

Homer
Shakespeare.

Albert Dürer, who was so powerful in his art simply by reason of the exquisite *naïveté* of his ideas, and so weak beneath the detestable tyranny of his wife;[154] Homer and Shakespeare, those impartial

---

[154] Albert Dürer's dearest friend, Wilibald Pirkheimer, has left us a sad account of Dürer's wife, who would seem to have been to him very much what Xantippe was to Socrates,* excepting that poor Albert Dürer, not being a philosopher, and not having married his wife *on account* of her ill-temper, died [as we are told by Pirkheimer] beneath the infliction. Pirkheimer in a letter [in fac-simile] to Tzerte, the emperor's architect at Vienna, published in "*Reliquien von Albrecht Dürer*" (Nürnberg: 1828) says:—"That he should have died of such a hard death [as God willed it] I can attribute to none other but his wife, who gnawed into his heart, and tormented him to such a degree as to hasten his death: for he was wasted to a skeleton, lost all his courage, and could no longer go among his friends. Also this wicked woman, though she had never known want, was continually at him day and night to

* Vide ÆLIAN, ΠΟΙΚΙΛΗ ῙΣΤΟΡΙΑ, Βιβλ., vii., κεφ. 1C, β:βλ. x., κεφ. 7, *et passim.*

reflectors of the human heart ; Montaigne, whose device was "What do I know?"[165] and who preferred supporting an opinion to stating it ; La Fontaine, who hesitated between royalism and social-

apply himself rigidly to work, in order that he should earn money and leave it to her. I have often chidden her myself for her wickedness, and warned her, and also predicted what the end of it would be."* *Vide* also "*The Artist's Married Life*," translated from the German of Leopold Schäfer by Mrs. Stodart (New York : 1862), supposed to have been written by Pirkheimer.

[165] Blaise Pascal tells us that Montaigne, "not wishing to say 'I know not,' said 'What do I know?' of which saying he made himself a device," etc. † *Vide* also Bigorie de Leschamps, "*Michel Montaigne, sa Vie, etc.*" (Paris: 1860, p. 2). On p. 493 [Appendix] of this volume, de Leschamps, speaking of the inscriptions which Montaigne had carved everywhere upon the lintels of his library, called attention to the fact that "he had particularly caused to be there inscribed the maxim Γνῶθι σεαυτόν, which antique wisdom had inscribed upon the porch of the temple at Delphi" (*vide* note [97], p. 87).

---

* "*Reliquien von Albrecht Dürer.*" (Nürnberg, 1828), Doktor Friedrich Campe. "Eigene Handschrift *Wilibald Pirkheimers.*" Auszug eines Briefes an Tzerte, Kaiserl. Baumeister zu Wien . . . "dasz er (A. D.) so eines hartseligen Todes verstorben ist, welchen ich, nach dem Verhängniss Gottes, niemanden denn seiner Hausfrauen zusagen kann, die ihm sein Herz eingenagen und dermassen gepeinigt hat, dasz er sich des schneller von hinnen gemacht hat, denn er war ausgedorrt wie eine Schraub, durft nirgend keinen guten Muth mehr suchen oder zu den Leuten gehen, also hatte das böse Weib seine Sorge, das ihr doch warlich nicht Noth gethan hat, zu dem hat sie ihn Tag und Nacht angelegen, zu der Arbeit härtiglich gedrungen, allein darum, dasz er Geld verdiente und ihr das liese . . . Ich habe sie selbst oft für ihr argwöhnisch sträflich Wesen gebeten, und sie gewarnt, auch ihr vorgesagt was das Ende hiervon seyn würde," etc. etc.

† Ne voulant pas dire "*Je ne sais,*" il dit " *Que sais-je ?*" De quoi il fait sa devise en la mettant sous des balances. qui, pesant les.contradictions, se trouvent en parfait équilibre . . . . Sur ce principe roulent tous ses Discours et tous ses Essais, et c'est la seule chose qu'il prétende bien établir." "*Œuvres Complètes de Blaise Pascal*" (Paris: 1858), vol. i., p. 426. *Entretien avec M. de Saci sur Epictète et Montaigne.*

[¶ 196]

Louis XVI

ism ; [56] and Louis XVI., who owed all his troubles to his indecision of mind,[157] must certainly have had small thumbs.

¶ 197.
Large thumbs.

With a large and well-formed thumb, however, you are your own master, and such being the case, your master is often a fool, as Henri IV. used to say.[158] Your principles are your laws, but you are inclined to despotism. You are truthful, but you are not innocent

---

[156] Martin [*"Histoire de France"* (Paris: 1879), vol. xiv., p. 252] tells us of La Fontaine :—" Il n'avait des opinions serieusement négatives que sur quelques points de Théologie et de politique où il repoussait, soit par raison, soit par sentiment, les doctrines officielles. On se rapelle ses disputes avec Racine sur la monarchie absolue." He showed his loyalty by writing several "éloges" upon Louis XIV.,° and his indifference by remaining friends with the Prince de Conti, who was disgraced in consequence of the discovery of some of his letters reflecting on the government.† There, is a letter written by La Fontaine to Conti during the latter's exile, in the " *Œuvres Complètes de La Fontaine*" (Paris: 1837, epître xx., p. 552). The league above referred to is of course that of Augsbourg, of which La Fontaine speaks in an epître, according to Walcknaer ["*Vie de La F.," op. cit.*, vol. ii., bk. v., p. 102]. "Non la ligue d'Augsbourg, que je sais moins encore," etc.

[157] Martin [*op. cit.*, vol. xvi., liv. ciii., p. 313] tells us :—" Louis XVI. is everything which is contrary to what he would himself wish to be, *i.e.*, the very embodiment of indecision. Later the vacillations of feebleness will be looked upon in his case as treacherous combinations which will land him upon the scaffold ; " and Carlyle, in his " *History of the French Revolution*" (London: 1837), makes continual allusion to this monarch's fatal indecision of mind. *Vide, e.g.*, vol. i., p. 28, vol. ii., pp. 127 and 152 :—" Louis . . . if thy heart ever formed, since it began beating under the name of heart, any resolution at all, be it now, then, or never in this world," etc.

[158] " Henri Quatre rencontra un jour dans les apparte-

---

* *Fables de La Fontaine*, liv. vii. and xi., epître xviii. and xix.
† *Vide* hereon, C. WALCKNAER, "*Histoire de la Vie de La Fontaine*" (Paris : 1858), vol. ii., chap. v., p. 188.

in the intellectual acceptation of the term. Your
power does not lie in your capacity to charm, for
grace can only exist in that which is pliant.

Souvaroff, renowned for the intensity of his will;[159]
Danton, the magnanimous soul who underwent the
opprobrium of a crime in the hope of saving his
country;[160] Galileo, Descartes, Newton, Leibnitz,
Saint-Simon the reformer, Charles Fourier, Robert

¶ 198.
Illustrations.
Danton, etc.,
etc.

mens du Louvre un homme qui lui étoît inconnu, et dont
l'extérieur n'annonçoît rien de fort distingué. Il lui
demanda, ' À qu'il appartenait ? ' le croyant de la suite
de quelque seigneur. ' J 'appartiens à moi même,' lui
dit ce personnage d'un ton fier et peu respectueux.
' Mon ami,' reprit le roi, en lui tournant le dos, ' vous
avez un sot maître.' "—" *L 'Esprit de Henri IV.*"
(Paris : 1775), p. 271.

[159] L. M. P. Tranchant de Laverme, in his "*Histoire
de Souvarof*" (Paris : 1809), lays great stress upon
Souvaroff's indomitable will. "He had a decision of
will, that no obstacle could cause to swerve" (p. 455).
Compare also the passages dealing with the same
characteristic upon pp. 438-499, and 453 (chap. vii.).
Mr. E. N. Macready also, in his work "*A Sketch of
Suwarow*" (London : 1851, p. 28), says :—"With him
. . . the simple term 'duty' was equivalent to, and
exacted, the utmost devotion of which a man was
capable."

[160] This is a somewhat uncommon view of the character
of Danton ; I conclude M. le Capitaine d'Arpentigny
has been led away by the grand rhetoric of a speech of
Danton's, made on March 10th, 1793, recorded in
MM. E. Buchez and P. Roux's "*Histoire Parlementaire de la Révolution Française. Journal des
Assemblées Nationales depuis 1789 jusqu'en 1815*"
(Paris : 1836, 40 vols.), in which he cries out :—"What
to me is my reputation ? so long as France is free, let
my name be wounded. What do I care about being
called ' drinker of blood ' ! Good ! let us drink the blood
of the enemies of the human race," etc.°

---

* " Eh ! que m'importe ma réputation ! Que la France soit
libre, et que mon nom soit flétri ! Que m'importe d'être appelé
buveur de sang ! Eh bien ! buvons le sang des enemis de
l'humanité s'il le faut ; combattons, conquerons la liberté," etc.

[¶ 198]

Owen, all of them profound reasoners and bold inventors, had infallibly very large thumbs.

¶ 199.
Voltaire.

Voltaire, who of all men who ever lived subordinated the promptings of his heart to those of his head, had, as his statue in the Théâtre Français bears witness, enormous thumbs. Certainly, the sculptor Houdon, an artist of the keenest and most delicate taste, cannot have given such thumbs to this marble statue, had not the well-known hands of his model imposed upon him the obligation to do so.[151]

¶ 200.
Thibetan customs.

The shaven-headed lama, clad in the robe of yellow wool, which enhances the brilliancy of his scarlet stole and violet dalmatic, benignly salutes his superiors after the manner of the Thibetans, *i.e.*, by putting out his tongue, and scratching his ear. Beneath those eyebrows, arched like the leaf of the peach-tree, his little beady eyes sparkle with pure contentment, by reason of the fact that he has successfully ejaculated before the assembled multitudes that sacred phrase, deeply fraught with mystic profundity, "Oh, Buddha, jewel of the lotus! Oh, Buddha, jewel of the lotus!"

¶ 201.
The same.

His science also is on a par with his devotion, and if he knows that he is forbidden to lay hold of a cow's tail to help him to ford a river which is deep and rapid, none the less is he aware of the healing and preservative powers of the flesh of the griffin and the horn of the winged unicorn; besides this, Buddha has appeared to him in a dream, and he knows that after his death he will not be thrown to the sturgeons of the Yellow River, neither will his body be exposed upon a mountain, nor burnt, nor eaten by Thibetan worms; but that he will be cut into pieces and given to dogs to eat—an apotheosis only

---

[151] There exists of this statue a well-known print, which I have before me. In it the hands are certainly thoroughly philosophic, and the thumbs, as M. d'Arpentigny says, are very large.

vouchsafed to those people whose high moral qualities are evidenced by a red bead on the top of the hat, or a peacock feather swaying in the breeze ! Hugging himself in the certainty of so glorious a fate, his heart expands, his pride increases, he compares himself with the kings of the earth, and to give himself a correct and striking and withal a dignified sense of his own merits, he raises proudly on high his right-hand thumb and exclaims, "Thus am I."[162]

The Corsicans, a stubborn race, obstinate by respect for tradition, and not, as with our Breton folk, by reason of an obstinate instinct, all have very large thumbs.[163]

¶ 202.
Corsican and
Breton
obstinacy

[162] M. d'Arpentigny has evidently taken his information concerning Thibetan customs from works like those of M. Huc [" *Souvenirs d'un Voyage dans la Tartarie, le Thibet, et la Chine*" (Paris: 1850),] and the " *Corréspondance de Victor Jacquemont avec sa Famille, etc.*, *pendant son Voyage dans L'Inde*" (Paris: 1846, 4th edn., 1869, 2 vols.),—works, which though interesting and valuable in themselves, rather come under the category of the books of travel described by Colonel Tcheng-ki-Tong in his recent work " *The Chinese, painted by Themselves*" (London : translated by James Millington, 1886), in which he says :—"Nothing is more imperfect than a note-book of travels : the first fool one meets gives a physiognomy to the whole nation whose customs are to be described. The recorded conversation with an outcast may perhaps be considered a valuable document by a traveller. . . . The fact is, the book is often written before the travels are undertaken, for the simple reason that the aim of the journey is the book to be published," etc. The sketch of Buddhism and of the Thibetan worship, which is more properly Lamaism, which is given above is highly coloured by our author from an already highly-coloured original account. Those who are interested in the subject should read the works of Stanislas Julien and of Barthélémy St. Hilaire, and Rhys David's " *Hibbert Lectures*" 1881, or his Manual " *Buddhism*" (London : 1882).

[163] The Corsicans and the Bretons have always had the reputation of obstinacy. As regards the former,

¶ 203.
Vendean super-
stition.

In La Vendée a large thumb is looked upon as a sure sign of a remarkable aptitude for the occult sciences. According to the peasants of the Bocage, no sorcerer can fail to be gifted with a rolling eye and large thumbs; the rolling eye by. reason of the malicious mobility of his spirit, and large thumbs because it is on his thumbs that the full weight of the upper half of his body is supported when, after having transformed himself into a were-wolf,* he goes at dead of night to howl and to gambol at the cross roads!

¶ 204.
Artistic instinct.

With a small thumb and smooth fingers, *whatever may be the form of the exterior phalanges [nota bene!]* one will have within oneself, not necessarily the actual talents of poetry or art, but most certainly the germs of these faculties. Only, naturally attracted as they are towards the ideal, conic phalanges will

this is said of them by the author of "*A General Account and Description of Corsica*" (London: 1739), who describes them as having "always had the character of a clownish, rough, stubborn people;" and Strabo gives a vivid account of their obstinacy and stupidity in his ΓΕΩΓΡΑΦΙΚΟΝ (book v., chap. 2.)° As to the Bretons, A. de Courson, in his "*Histoire des peuples Bretonnes*" (Paris: 1846), speaks of "the obstinacy which distinguishes them . . . whence come the extraordinary contrasts of the national character," in describing the contradictions to be found in their religious observances and superstitions. The poet Briseux also speaks of the Bretons as :—

"La race, sur le dos portant de longs cheveux,
    Que rien ne peut dompter quand elle dit ' je veux.' "

The same characteristics are called attention to by M. Daru, in his "*Histoire de Bretagne*" (Paris: 1826), p. 356.

---

* "Quo fit ut montana colentes, qui latrociniis vitam sustenant, ipsis si ut inhumaniores bestias. Itaque, quum Romani duces in insulam hanc incursionem faciunt, ac munitiones adorti, magnum mancipiorum numerum ceperunt, videre Romæ cum admiratione licet, quantum in eis feritatis ac indolis plane sit bellissimæ ac stupiditate dominos obtendunt, ut impensere pœnitet etiam si quis minimo emerit." Strabo, *loc. cit., Didot's translation.*

incline to a mode of expression, or, if you prefer the term, a manifestation of the talent, which is more spiritualistic than spiritual. As for instance Rafaëlle, Correggio, Perugino, and so on ; and among writers Tasso, Georges Sand, and others. And as to the others, I mean those whose fingers terminate in a spatule or in a square, seeing that they are attracted by what is true and real, *i.e.*, towards commonplace in the world of *things*, and towards custom in the world of ideas, *they* will incline to a mode of expression which is more spiritual than spiritualistic, such as, for instance, Teniers or Callot, Scarron, Regnard, Lesage, Béranger, and so on, whose arts lie more in the expression of real life than in the interpretation of what is really beautiful. They interest the mind and sometimes the heart, but never the soul. One appreciates them and likes them, but one does not really admire them.

Illustrations

Hands which are conic or pointed, with a large thumb, proceed in art by method, by logic, and by deduction, almost after the manner of persons with square hands and a small thumb. Such were David [the painter], Voltaire, Fontenelle, and others, all people distinguished by little or no *naïveté*.

¶ 205.
Methodical art.

If, therefore, you remember what I said about the indications found in the joints and upon the outer phalanges, you will remember that he who has conic phalanges, smooth fingers, and a small thumb, is trebly predestined to a poetic existence, whilst trebly predestined to scientific pursuits is he who, to square or slightly spatulated finger-tips, joins knotty fingers and a large thumb.

¶ 206.
Confirmations of tendencies.

It is more easy for large-thumbed subjects [by reason of the strength of will with which they are gifted] to overcome the tendencies of their natures than for people with small thumbs.[161] Again, many

¶ 207.
Modification of tendencies.

[161] Mark S. T. Coleridge's words [" *Table Talk*," Sept.

philosophers and learned professors have formulated their systems in verse of a higher or lower quality of inspiration.   But there has never lived an eminent poet who has excelled in abstract sciences.

28th, 1830] on this subject :—" Why need we talk of a fiery hell ? If the will, which is the law of our nature, were withdrawn from our memory, fancy, understanding, and reason, no other hell could equal for a spiritual being what we should then feel from the anarchy of our powers.   It would be conscious madness!"

SUB-SECTION V.

You have before you two hands of the same thickness, the same size, the same development of parts, and terminating similarly [for instance] in spatulate exterior phalanges ; only there is this difference between them : one is supple even to flabbiness, whilst the other is so firm as to be absolutely hard.

¶ 208.
Variations of consistency.

You must observe that the difference lies in the temperament and manner of life, and that, though the intellectual tendencies of these two subjects may be the same [by reason of their similarly spatulated finger-tips], their aptitudes and their moral natures will nevertheless be different, for, as Fontenelle has said, from a basis of resemblance may rise infinite differences.[165] In the love of action, of

¶ 209.
Differentiated similarities.

[165] This remark of Fontenelle's occurs in his "*Entretiens sur la Pluralité des Mondes*" ["*Œuvres Complètes de M. de Fontenelle*" (Paris : 1766), vol. ii., p. 146], "Ne faut-il pas pourtant que les mondes, malgré cette égalité, diffèrent en mille choses ? Car un fond de ressemblances ne laisse pas de porter des differences infinies." These "*Entretiens*" were published at Amsterdam in 1701, and an English translation was subsequently made by William Gardiner, entitled, "*A Week's Conversation on the Plurality of Worlds*" (London : 3rd ed., 1737).

movement, which is common to them, the soft-handed subject will seek the fulfilment of its· desire in moderate and subdued action, whilst the hard-handed subject will develop a love of energetic movement and violent locomotion. The latter will rise with the lark, whilst the former will appreciate the charms of a comfortable bed until the risen sun suggests the necessity of exertion ; as in their pleasures, the influence of their organisation will make itself felt in the choice of their studies and of their professions.

¶ 210.
Hard and soft-handed artists. The ideas of hard-handed artists are based upon sound facts, and their works have more virile strength than those of artists whose hands are soft. These latter, strongly acted upon by outward influences, are governed by ideas which are essentially those of the surface, but their works are characterised by more shades of sentiment, more diversity and more subtlety, than those of artists whose hands are hard. The

The finger-tips. little fleshy lumps which are found on the faces of the finger-tips are generally more pronounced and more delicately formed upon their hands than on those of the generality of people. Well, it is in this little fleshy protuberance that the sense of touch is the most highly developed [166]—the sense of touch, which is the sense of discernment, and, consequently, the outward symbol of moral tact.

¶ 211.
The Picardian temperament. Paris obtains from the province of Picardy, handsome men-servants, fair haired, and ruddy of tint, young apprentices, with low foreheads, who, at the same time credulous and defiant, headstrong and shallow-minded, perform their duties in accordance with the promptings of their instincts, either inertly or obstinately. Vulgarity, which is the leading feature of the physiognomy of the province, is written on every feature of their faces. Born beneath the

[166] *Vide* hereon ¶ 153 and note [136], p. 126.

thatched roofs of muddy and dilapidated hamlets, wherein ghostly visitors hold nightly revels in all the panoply of shrouds and chains, they are principally noted for want of manual dexterity, for a highly developed vanity of mind which is at once shamefaced and sullen, and by that hare-brained freedom peculiar to this class of people which is at the same time the result of maliciousness and folly. Their hands are large, red, and very hard.

In the immense forests which clothe the banks of the Dnieper, here and there are found little villages, built of wood, of the most squalid appearance, inhabited by low-class Jews and rough cowherds. Their staple industry is the rearing of huge packs of gigantic dogs, which they let loose by night for protection against wolves.

¶ 212.
Russian element-
ary hands.

With the exception of those of the Jews [who are a race peculiarly endowed with the talents of commerce], all the hands one meets with in these localities are extremely hard.

¶ 213.
Exception of
the Jews.

In the reign of Hé-Sou, Emperor of China, men lived at peace with one another, not bothering themselves about what they did or where they went. They wandered about in a satisfied sort of way, patting their chests as if they had been drums; and more or less always eating, they were supremely contented: they were ignorant of even the most elementary principles of good and evil. Hands of the soft type.[167]

¶ 214.
Chinese hands.

[167] I am not acquainted with the monarch recorded by the author, and he is not [as far as I can see] mentioned by Giles,* or by any of the standard authors upon ancient Chinese history. He may be one of the fifteen sages who formed the privy council of the Emperor Fou-Hi, who was one of the rulers of the Celestial Empire in "fabulous" times. M. J. A. M. de Moyriac de Mailla, in his "*Histoire Générale de la Chine*," traduite du

* H. A. GILES, "*Gems of Chinese Literature*" (London and Shanghai : 1884).

¶ 215.
The Hottentots.   The expiring races whom, with their red-chalked
and soot-streaked faces, Levaillant lauds and reveres
for their stagnating indolence ;[168] the fattened monks,

Ecclesiastics   corpulent caryatids of the holy kitchens of the Roman
Church, whose love of the plethoric delights of an
ever-somnolent laziness has obtained for them such
celebrity at the hands of Erasmus and Rabelais ;[169] the
flabby-visaged gate-keepers who, ensconced behind a
glazed window sash, go through their innocent lives
like oysters, perpetually opening and shutting a door,
—all these have naturally soft hands.

¶ 216.
The Guachos.   The chase of wild cattle amid the jungles of the
savannahs of La Plata is the sole occupation, the

Tong Kien Kang Mon (Paris: 1777, 13 vols.), tells us,
after giving a list of emperors who reigned B.C. 2953—
2689, the first of whom were Fou-Hi and Ching-Nong,
that the historian Ouai-Ki places between them the
names of fifteen rulers, who reigned for 17,798 years (!);
but he informs us that these fifteen were councillors and
officers of Fou-Hi [who was a most exemplary monarch],
and in the list which he gives of them appears the name
of Hé-Sou.*

[168] It is fortunate for Le Valliant that they had some
redeeming points in his eyes, for he says of these
savages : "Constantly seeing the Hottentots has never
been able to accustom me to their habit of painting
their faces with a thousand different designs, which I
find hideous and repulsive." Le Valliant, " *Voyage
dans l'Intérieur de l'Afrique*" (Paris: 1790), vol. ii.,
p. 44.

[169] It is not surprising that Rabelais was acrid on the
subject of monks, seeing that he began life himself
as a Cordelier, from which order he was expelled for
personating the image of St. Francis in his niche on a
saint's day, and playing various uncanonical tricks to
the stupefaction of the assembled worshippers. By a
bull of Pope Clement VII. he afterwards became a
Benedictine, but soon abandoned a profession for which
he was pre-eminently unsuited.   Erasmus fulminated

---

* "Alors Fou-hi composa son conseil de quinze d'entr'eux
qu'il jugea les plus sages et les mieux instruits," etc.—Vol. i.,
p. 5, note.

unique industry, of the Gauchos, who, leathern lasso in hand, and with their heels armed with huge spurs, devote themselves entirely to this form of exercise, mounted on superb wild horses. They are a race by nature nervous, agile, irritable, and prompt in expedients where prompt action is necessary, beneath an exterior which is phlegmatic ; a race which is consumed by a thirst for strife and for action, and a love of boundless horizons and unchecked liberty. A horse- or buffalo-skin forms their bed, and analogously the desiccated heads of horses or buffaloes form their chairs and tables ; in fact, their domestic furniture is constructed almost exclusively of buffalo and horse bones.[170] Their hands are very hard.

You have probably remarked ere now that a taste for agriculture and horticulture gains upon us as we grow old. This taste, feebly developed as it is at first,—potently warred against to the last by the smiles which shine upon us from gentle lips, and by the soft hands which by their touch inspire us with patriotic, poetic, and scientific enthusiasm,—grows gradually stronger and stronger, and develops its

¶ 217.
Development of tastes for manual labour.

against monks more seriously and to better purpose than Rabelais. Among the Elzevirs on my bookshelves I find a little work, entitled " *Desid. Erasmi Roterdami Colloquia*" (Amsterdam : *Elzevir*, 1677), on p. 238 of which, during a colloquy between some Franciscans and one PANDOCHEUS, an inn-keeper, the latter refusing to take them in, exclaims, " Quia ad edendum et bibendum plus quam viri estis, ad laborandum nec manus habetis nec pedes," etc. Any one who is sufficiently interested to care to know the exact opinions of Erasmus on the holy Fathers of the Church, had better read from pp. 156—173 of his "ΜΩΡΙΑΣ ΕΓΚΩΜΙΟΝ: *Stultitiæ Laus. Rot. declamatio*" (Basle : 1676).

[170] Captain Head has recorded this in his " *Journeys to South America*" (London : 1829, p. 111), where describing the family meal of buffalo-flesh, he says:— "The family and guests sit round it on the skeletons of horses' heads, handle their long knives, and cut away."

[¶ 217]

intensity in proportion as the faculties of our imaginations grow weaker; and it is when our hands, stiffened and bony and bereft of their delicate tactile organisation, offer a faithful reflex of our impoverished imaginations, that this inclination to garden, to labour with our hands, acquires more and more dominion over us.[171]

¶ 218.
Philosophy in age.

In like manner we become more steady, less credulous, and more logical in proportion as the joints of our fingers emphasise themselves by their development, and become more prominently visible.[172]

¶ 219.
Love of hard and soft hands.

Though not dead to love, hands which are very hard are seldom capable of much tenderness; and by contrast, soft hands are generally more capable of tenderness than of love.

¶ 220.
Callosity.

Callosity in a hand seems always to cast a shadow upon the mind.

¶ 221.
Advantages of a medium consistency.

For my part let me see hands which are firm without being hard, and elastic without being flabby. They indicate an intelligence which is pre-eminently wide and active in its scope, having at the same time the faculties of theory and of action, and in addition to this, whatever may be the material occupations with which they may be employed, they only harden very gradually; whilst, on the contrary, hands which are naturally very firm rapidly become extremely hard.

[171] *Vide* "*A Manual of Cheirosophy*," ¶ 210.
[172] "M. d'Arpentigny holds a theory," says Adrien Desbarrolles ["*Mystères de la Main*" (Paris: 15th edn., *n.d.*), p. 241], "that the joints develop upon hands, or may *tend* to diminish, and he gives as an example the hand of Madame Sand, whose fingers, once very smooth, have developed the lower joints since she has taken to philosophy and serious literature. Very well then. If the hands which contain the destiny can modify themselves by the direction of the will, *Destiny* can also modify itself, *and is not irrevocably fixed*." Compare "*A Manual, etc.*," ¶¶ 89–91

The same remarks apply to the skins and coats of highly-bred horses, which although much finer than those of commonly-bred horses, are always less subject to variations of texture, or to disease.

¶ 222.
Illustration.
Horses.

How admirable is the foresight of Nature, which, in proportion as the aims of created beings are higher and purer, supplies them with instruments and weapons of a keener edge and finer temper.[173]

¶ 223.
Adaptation of nature.

According, therefore, to the development of their intelligences, the individuals of a particular type divide themselves into classes, and adopt for themselves the sciences and labours which are most adapted to capacities of their particular class.

¶ 224.
Classification of intelligence.

Thus, for instance, though all are equally well constituted for the race, the finely-moulded horses of the high table-lands of Central Arabia [El Nejed] are not all equally well-gifted with swiftness of foot;[174] thus, although very widely differing in physical power, the common house-cat and the royal tiger are none the less members of the same family; in like manner it is from the same learned hands of scientific appreciation that Lavoisier,[175] and Jean Maria Farina, of

¶ 225
Differentiated similarities.

---

[173] *Vide* ¶ 86.

[174] "Nejedean horses are to Arab horses in general what Arab horses are to those of other countries."— W. G. PALGRAVE, "*A Journey through Central and Eastern Arabia*" (London: 1865), chap. ix., p. 432.

[175] Antoine Laurent Lavoisier, the eminent chemist, was born in Paris in 1743. At the age of twenty-three he received the gold medal of the Académie for a new system of town illumination, and at twenty-five was admitted a member of that august body. A most interesting biography of Lavoisier was written by A. F. Fourcroy, after his death, entitled, "*Notice sur la Vie et les Travaux de Lavoisier*" (Paris: 1796); in which his biographer tells us that "all branches of mathematical and physical science had their places in the studies of his waking hours." He was made a Fermier-Général by the Government, a post which he accepted, as it gave him leisure in which to pursue his studies. Having been

[¶ 225]

Cologne, have received the laurel-wreath, highest mark of academic distinction; in fact, it comes to this, that, as Sganarelle remarks, "there are faggots *and* faggots."[176]

¶ 226.
Rarity of versatile talents.

You must not, however, conclude that because you have a large thumb and jointed fingers terminating in spatule, you are necessarily gifted with the capacity of excelling in all practical sciences and occupations; nor that because you have a small thumb and smooth fingers you are necessarily gifted with pre-eminent talent in every branch of the fine arts; on the contrary, the pursuit of a single science, or of a limited number of sciences [to an extent limited by the scope of the faculties of each individual] absorbs, as a rule, the whole of the stock of genius with which God has endowed the generality

arrested with the rest of the Fermiers-Généraux in 1794, the most strenuous efforts were made to save his life, to all of which the judge replied, "The country has no longer any need of *savants*." [*Vide* ED. FLEURY, *"Dupin de l'Aisne"* (Laon: 1852), pp. 22-3.] He was guillotined on the 5th May, 1794, having laboured night and day in prison at the completion of his discoveries in chemistry, until summoned by the executioner. His most celebrated works are "*Mémoires de la Chimie* (Paris: 1805), and a "*Traité Elémentaire de la Chimie*" (Paris: 1789); of which an English translation appeared at Edinburgh in 1790. M. d'Arpentigny's classification of his eminent countryman with Johann Maria Farina, who received the same distinction as Lavoisier from the Academy for his Eau de Cologne, is truly an exquisite piece of sarcasm.*

[176] "Il y a fagots et fagots; vous en pourrez trouver autre part à moins, mais pour çeux que je fais——" *Valère.*—"Eh! monsieur, laissons là çc discours."— MOLIÈRE, "*Le Médecin Malgré Lui*," act i., sc. 6.

* "Comme Lavoisier. c'est des doctes mains de la science que Jean Marie Farina, de Cologne, a reçu son ridicule laurier (*voir ses Prospectus*)."—D'ARPENTIGNY.

[¶ 226]

of men. One has, indeed, known of individuals like Cæsar, Napoleon, Michael Angelo, Humboldt, Voltaire, Cuvier, Leibnitz, and others, whose colossal intellects have embraced within their comprehension nearly all the talents specially adapted to their types of character, but these examples are rare—very rare.

¶ 227.
Lower middle-class hands in France.

The large, fat, soft, spatulate hand among the middle classes in France, untroubled by the moral excitements of high-class education, finds pleasure after its own heart in the hum of conversation which hovers over the crowded cafés, and the subdued gesticulations of the lower class of clubs. Driving in a nail here, strengthening the treacherous leg of a dilapidated table there, or again, drumming listlessly on the window-panes,—it is in such phlegmatic employments that they are content to pass their irresponsible days, and pursue the even tenor of their ways. The dull and stupid contentment of insignificant towns is less irksome to them than it is to people whose hands are hard. Such subjects find a pleasure in the noise and turmoil of fairs and markets; you may see them any day, marching along, erect of gait and stick in hand, keeping military step with the evening fanfaronade of the garrison trumpeters. You may see them calmly enjoying themselves, engrossed in the dissipations of draughts, of backgammon, and of bagatelle, relinquishing to their harder-handed neighbours all wearying exercises and laborious pleasures. They do not do much themselves, but they like to see others working hard; they like [remaining quiescent themselves] to watch the spectacle of action; they do not travel themselves, but they like to read of voyages, traversing the habitable globe, riding, as one might say, on the shoulders of the energetic and actual traveller. Like D'Anville, who went everywhere without once

[¶ 227]

leaving his study,[177] such subjects find the energy to travel in the activity of their brains.[178]

¶ 228.
Instinctive development of negative characteristics.

Among subjects whose intelligences are wholly inferior, the types of nature, as a rule, hardly exhibit their *positive* characteristics at all; it is not so, however, with their *negative* peculiarities. The eagle and the ostrich have alike wings and legs, but in escaping from the hunter, or in attacking their prey, the eagle knows instinctively that he can only rely upon his wings, and the ostrich that he must trust only to his legs. Every animal is automatically conscious of the portions of their organisation whose powers are the most highly developed, whereas - many human beings are absolutely ignorant [from the moral point of view] of the particular direction in which lies their strength.

It is, therefore, the office of education to enlighten them hereon.

¶¶ 229–233.
An illustration from Chinese progressive philosophy.

" In this world," says Tseu-sse, a commentator of Confucius,[179] " man alone of all created beings is

[177] At the time when the above was written, Jean Baptiste Bourguignon d'Anville was perhaps the most celebrated geographer whom France had ever seen; he was, besides, an eminent antiquary. He is chiefly known to-day by his " *Antiquité Géographique de l'Inde et de plusieurs autres Contrées de la Haute Asie*" (Paris: 1757); " *L'Euphrate et le Tigre* " (Paris: 1779); " *Déscription de la Gaule*" (Paris: 1760); and his " *Géographie Anciénne* " (Paris: 1768). *Vide* " *Notice des Ouvrages de M. D'Anville précédée par son Eloge,*" by MM. J. D. Barbié and B. J. Dacier (Paris: 1802).

[178] " The poet has his Rome, his Florence, his whole glowing Italy within the four walls of his library. He has in his books the ruins of an antique world, and the glories of a modern one."—LONGFELLOW, " *Hyperion* " (Boston: 1881, p. 82), bk. i., ch. viii.

[179] The passage here quoted, which is a grand exemplar of the subtle casuistry and logic of the Chinese philosophers, comes from the 大學 (" *Ta Hio* "), of which perhaps the best translation that exists is G. Pauthier's " *Le Ta Hio, ou la Grande Etude. Ouvrage de Kung-*

[¶ 229.]

possessed of a sovereign intelligence which is capable of fully comprehending his own peculiar nature, the laws by which his life is governed, and the duties he owes to society.

"Gifted with this knowledge of his own nature, and of the reciprocal duties which he is bound to perform, he can, by this very fact, thoroughly comprehend his fellow-men, and the laws by which they in turn are regulated, and can thus instruct them in the duties which it is necessary for them to perform in order to carry out the mandates of the Most High.

¶ 230.

"Again, being gifted with this same knowledge of his fellow-men, and being able therefore in these matters and in this manner to instruct them, he can by analogy arrive at an understanding of all other living things, whether animal or vegetable, and can help them to carry out the mandates of the laws under

¶ 231.

fu-tzu (CONFUCIUS) *et de son disciple Thseng-tzu,"* avec un commentaire par Tchou Hi (Paris: 1837). The passage quoted above is from chap. iv., "Sur le Devoir de connoître et de distinguer les Causes et les Effets." "To know the *root* or the *cause*, that is the perfection of knowledge." That is all that remains of the fifth chapter of the commentary. It explained what one must understand by "to perfect one's moral knowledge, by penetrating the principles of actions." It is now lost. To perfect one's moral knowledge consists in penetrating the principle and the nature of actions. We must devote ourselves to a profound investigation of actions, and examine to their foundations their principles or causes. . . . . Only, these principles, these causes, these *raisons d'être* have not yet been submitted to sufficiently profound investigations ; that is why the sciences of men are not complete, absolute [*vide* note [100], p. 91]; it is for that reason also that the "*Ta Hio*" commences by teaching men that those among them who study moral philosophy must submit all the objects of nature and human actions to a long and profound examination, to the end that, starting from what they know already of the principles of actions, they can

[¶ 231]

which they live, in accordance with the requirements of their particular natures.

¶ 232. "Yet again, being able thus by knowledge to direct the existences of all things, animal and vegetable, as aforesaid, he is able by this very knowledge, and by means of his superior intelligence, to assist Providence itself in its direction of the evolution and bringing together of *beings*, to the end that they may reach their fullest and highest developments.

¶ 233. "And finally, being able to assist heaven and earth in the evolution of the laws of natural causation, he is able by this power alone to stand up in the character of a third force or potentiality, ranking equally with heaven and earth."

¶ 234.
A contrast.
It is not necessary for me to call attention to the circumstance that Tseu-sse did not live under the rule

increase their knowledge, and penetrate into their inmost natures. "In applying oneself thus to exercise for a long time all one's energy, all one's intellectual faculties, one arrives at last at the possession of a knowledge, of an intimate comprehension of the true principles of actions. Then the intrinsic and extrinsic natures of all human actions, their most subtle essentials, as well as their most gross particles, are penetrated : and for our intelligences thus exercised and applied by a sustained effort, all the principles of actions become clear and manifest." Another commentator, Ho Kiang, says on this passage of the "*Ta Hio*" :—"It is not said that it is necessary to seek to know, to scrutinise profoundly principles and causes ; but it *is* said that it is necessary to seek to understand perfectly human actions. In saying that one must seek to know, to scrutinise profoundly, principles and causes, one may easily draw the mind into a chaos of inextricable incertitudes. But in saying that one must seek to understand perfectly human actions, one leads the mind to a search after truth." I have amplified this note with comparatively full translations of the Commentaries of Tchou Hi and of Ho Kiang, as they afford an extremely characteristic and interesting example of the systems of Chinese philosophy.

[¶ 234]

of He Sou of whom we spoke a short while since. In my opinion neither Condorcet nor Saint-Simon, nor Hegel nor Chas. Fourier, have ever excelled him in the enunciation of this definition of the illimitable perfectibility of humanity and of all nature.

Illustrations.

A Few Miscellaneous Observations.

The portrait of
a strange Radish,
which grew at
Haarlem,
& was painted from
the life
by Jacob Penoy
.1672.

SUB-SECTION VI.

A FEW WORDS UPON THE SCIENCE OF CHEIROMANCY.

CHEIROMANCY, looked down upon as it is by the present generation, was in former days studied by philosophers and scholiasts of eminent celebrity and worth. Among them we may mention the names of such men as Plato,[180] Aristotle,[181] Galen,[182] Albertus

¶ 235.
The older
Cheiromants.

Aristotle.

[180] I do not know of any direct mention of the hand in any of the dialogues of Plato. Our author may have considered that the remarks which Socrates makes to Alcibiades in the "*First Alcibiades*," or the dialogue "*Upon the Nature of Man*," upon the use of the hands, and the custom of wearing rings, are sufficient basis upon which to claim Plato as a cheirosophist.

[181] Aristotle certainly seems to have taken a considerable amount of notice of the science. *Vide*, for instance, the passages in which he calls attention to the fact that length of life is indicated by the lines of the hands, which occur notably in the "*History of Animals*,"* and in his "*Problems*."† I have paid much attention to his science of the hand in "*A Manual of Cheirosophy*," in the first index to which complete references may be found *sub* ARISTOTLE.

[182] In the same way Galen has remarked at much

* ΠΕΡΙ ΤΑ ΖΩΑ 'ΙΣΤΟΡΙΟΝ, Βιβλ. Α.', κεφ. ιέ.
† "Διὰ τί ὅσοι τὴν διὰ χειρὸς τομὴν ἔχουσι δι᾿ ὅλης, μακρο-βιώτατοι; 'Η ὅλοτι τὰ ἄναρθρα βραχύβια καὶ ἀσθενῆ."—ΠΡΟ-ΒΛΗΜΑΤΩΝ ΛΔ'., ι.

[¶ 235]

Magnus,[133] Ptolemy,[134] Avicenna,[135] Averroës,[136] Antiochus Tibertus,[137] Tricasso,[138] Taisnier,[139] Bélot,[190]

length upon the hand, but more, I think, upon its importance as a member than upon the indications afforded by it. *Vide* notes [102], [101], and [117], pp. 92 and 140, and "*A Manual of Cheirosophy*," index 1, *sub* GALEN.

[183] Albertus Magnus does not seem to have paid more attention to the hand than would ordinarily be required of a mediæval sorcerer or alchemist. Godwin does not tell us anything about it in his "*Lives of the Necromancers*" (London: 1834, p. 260); nor does Naudé in his "*Apologie pour tous les Grands Personnages qui ont été faussement soupçonnés de Magie*" (Paris: 1625, chap. 18), though both of them give accounts of Albertus Magnus. If anywhere it would be in the "*Parva Naturalia*,"—" De Motibus Animalium," lib. ii., tract. i., c. v., and "De Unitate Intellectus," cap. v. *Vide* also "*Albertus Magnus, Geheimer Chiromant, etc.*" (Leipzig: 1807).

[184] Many of the Ptolemies were celebrated for their literary tastes and studies of esoteric philosophy. The Ptolemy alluded to above would probably be Ptolemæus Lagus, the founder of the Alexandrine library.

[185] Husain ibn Ab'd'allah (ابو على الحُسَين بن عبد الله) commonly called Ibn Sina, or Avicenna, one of the best known of the Arabian writers of medicine, physics, and metaphysics, lays great stress upon the importance of the hand in his numerous commentaries and treatises upon the works of Galen, of Aristotle, and of Hippocrates. Oriental scholars who feel interested in this subject may find the animadversions to which I allude in his ـكتاب القانون فى الطب لابو على الشيخ الرييس ابن سينا الخ

("*The Medical Canons* (القانون [!]) *of Avicenna, with treatises on Logic, Physics, and Metaphysics*"), which were published at Rome in 1593.

[186] Muhammad ibn Ahmad ibn Rushd (بن احمد بن رشد جُزّ) commonly called Ibn Rushd, or Averroës, like Avicenna a commentator of Aristotle, was an Arabian philosopher, who lived at Corduba in the twelfth century. His chief work ابو الوليد محمّد &c., published in Rome in 1562, contains very little on the passages of Aristotle to which I have called attention in "*A Manual, etc.*"

[187] Antiochus Tibertus was the pseudonym of one of

Frœtichius, De Peruchio,[191] κ.τ.λ., all of whom have
handed down to us reflections, and, in some cases,
long treatises on the art of divination by the observation of the lines traced upon the palms of the
hands,—treatises which amply prove the high esteem
with which they regarded the science. We are told
that Aristotle, having found upon an altar dedicated
to Hermes a treatise on this subject, engraved in
letters of gold, made a great point of transmitting
it to Alexander, as a study worthy the attention of

the earliest of the cheiromants who have left behind
them works on the subject. The principal of these are
"*Ad Illustrem Principem Octavianum Ubaldinum
Merchatelli Comitem A. Tyberti Epistola*" (Bononiæ:
1494), and "*Antiochi Tyberti de Cheiromantiâ
Libri III., denuo recogniti. Ejus idem Argumenti
Cheiromantiæ, etc.*" (Moguntiæ: 1541).

[198] Tricasso, commonly known as Patritio Tricasso da
Cerasari, was one of the most celebrated cheiromants
that the world has known. His principal works were
"*Chyromantia de Tricasso de Cerasari . . . . nuova-
mente revista*" (Venice: 1534), and "*Enarratio
Pulcherrima Principiorum Chyromantiæ, etc.*"
(Noribergæ: 1560). For a fuller catalogue of his works
vide in "*Bibliographiâ Cheirosophicâ,*" p. 421.

[188] The most celebrated work of Taisnier now extant
is his "*Opus Mathematicum. Octo libros . . . .
quorum sex priores libri absolutissimæ Cheiromantiæ
theoricam . . . . continent, etc., etc.*" (Coloniæ
Agrippinæ: 1562), [*vide* Appendix: *Bibliographia
Cheirosophica*],—a work which has been freely epito-
mised and translated from by authors of the seventeenth century.

[190] VIDE "*Les Œuvres de M. Jean Bélot, Curé
de Milmonts, Professeur aux Sciences Divines et
Célestes, contenant la Chiromence, etc.*" (Lyon:
1654), pp. 118; (Rouen: 1669), pp. 480; and (Liége:
1704), pp. 528.

[191] Le Sieur de Peruchio was the author of one of the
most leading works on cheiromancy proper that has
reached us to-day; it is called "*La Chiromence, la
Physionomie, et la Géomence, etc.*" (Paris: 1656, 1657,
1663). I have quoted some of his aphorisms in "*A
Manual of Cheirosophy,*" p. 116.

[¶ 235]

some highly-cultivated and developed intelligence. This treatise, written originally in Arabic, has been translated into Latin by Hispanus.[192]

¶ 236.
False arguments
from a correct
basis.

Starting, however, from a few readily admissible principia,—principia admitted in fact by physicians of note to be incontestable, the Cheiromants have deduced arguments so utterly absurd, that they have ended by causing themselves to be discredited even by the most ignorant and the most credulous. At the same time, however, one finds here and there among their mummeries decisive indications resulting from

[192] I do not know where this statement originated; probably among the vaticinations and literary irresponsibilities of some of the older cheiromants; certainly for many years every writer on the subject has reproduced the statement which I have myself recorded in "*A Manual of Cheirosophy*," ¶ 58. The "*Ciromancia Aristotelis*" there referred to, as also the MS. [Brit. Mus. Eg., 547], is recognised as a *supposititious* work only, having probably been compiled by some student or commentator from Aristotle's multiplied references to the hand [*vide* note [181], p. 179]. I presume that the account arose in this way : There is no doubt that when Aristotle—subsidised by Alexander—made his expedition into Asia for the purpose of compiling his "*History of Animals*," he was in the habit of sending the results of his investigations to Alexander, as they were completed, and in the course of this journey visiting Ægypt, he there picked up a quantity of the occult knowledge of the Ægyptian magi—cheiromancy among them. I have called attention in another place [*Manual*, ¶ 58] to the fact that Aristotle's works, ΠΕΡΙ ΖΩΩΝ ΜΟΡΙΩΝ and ΠΕΡΙ ΤΑ ΖΩΑ ΙΣΤΟΠΙΟΝ, teem with references to the hand and to the art of cheiromancy. Now, as to the connection of Hermes, two hypotheses present themselves to my mind. First, the worship of Hermes [under the name Teti, Thoth, or Taut] occurs earlier in Ægyptian records than in any others ; he occurs as early as the eleventh dynasty, and being regarded as the inventor of hieroglyphics, *all literary compositions were dedicated to him.* Hermetic philosophy is said to have originated with him, and Clement of Alexandria mentions 42, Iamblichus mentions 20,000, and Manetho mentions 36,525 books devoted to this particular cult.

repeated observation which it is convenient to admit.[193] Such as, for example, the following :—

Persons whose fingers are supple, and have a tendency to turn back, are gifted with sagacity, curiosity, and address. [*Manual*, ¶ 151.]

¶ 237.
Fingers turned back.

Persons whose fingers seem clumsily set upon their hands, and whose fingers all differ as to their terminal phalanges, are wanting in strength of mind. The cheiromants condemn them to misery and to intellectual ineptitude. [*Manual*, ¶ 149.]

¶ 238.
Mixed hands.

If your hand held before a candle shows no chinks or crannies, *i.e.*, your fleshy fingers adhere to one another parallel throughout their length, it is a sign of avarice. [*Manual*, ¶ 152.]

¶ 239.
Close-fitting fingers.

Very short and very thick fingers are a sign of cruelty. [*Manual*, ¶ 133.][194]

¶ 240.
Short, thick fingers.

Astrology and medicine were particularly in his line, and undoubtedly any *papyrus* or *tablet* dealing with cheiromancy would have been dedicated to him. It is more than probable that the *altar* was merely a votive tablet or a papyrus hanging in one of the temples of Ibis [*Hermes*], and that this is what has given rise to the statement which is under discussion. The second hypothesis yields, I think, to this one in the matter of probability,—viz., Aristotle was a great friend of Hermias, the tyrant of Atarnea, and spent some time with him on this same journey [*vide* Diogenes Laertius]. It is possible that it was here that he learnt what he did of cheiromancy, but I think the hypothesis given above is the more probably correct one. *Vide* on this point E. A. W. Græfenham, "*Aristoteles Poeta ; sive Aristotelis Scolion in Hermiam*" (Mulhusæ: 1831), a most fascinatingly interesting and scholarly little work.

[193] It must be remembered that at the time when the above was written the only works on the science of Cheiromancy were those of such authors as are cited in the immediately preceding notes. M d'Arpentigny had not seen the works of Desbarrolles [with whose labours he was then unacquainted], which were not published until some years afterwards.

[194] Such a hand as this was the hand of Marchandon,

¶ 241.
Long, thin fingers.

Long and thin fingers are usually those of diplomatists, of deceivers, of card-sharpers, and of pickpockets. [*Manual*, ¶ 135—141.]

¶ 242.
Flat fingers.

A tendency to theft is indicated by a flattened condition of the outer or nailed phalanges.[195]

¶ 243.
Smooth fingers.

Curiosity and indiscretion are the leading characteristics of persons whose fingers are smooth and transparent. [*Manual*, ¶ 153.]

¶ 244.
Smooth and conic.

Smooth and conic fingers are an indication of loquacity and levity of mind.

¶ 245.
Strong fingers.

Strong and large-jointed fingers are a sign of prudence and of ability. [*Manual*, ¶ 163.]

¶ 246.
Gesticulation.

To move the arms about violently with the fists clenched whilst walking is a mark of promptitude and of impetuosity.   The habit of keeping the thumb hidden beneath the other fingers indicates a sordid and avaricious mind. [*Manual*, ¶ 248.] [196]

the murderer, whose atrocious crime struck Europe with horror in July 1885.   A cast of the hand is preserved in the museum of anthropology in Paris, and a description of it, with some cheirosophic notes, appeared in *La République Française* for 15th August, 1885.

[195] I do not know where M. d'Arpentigny can have found this interpretation of a spatulate (?) finger-tip. Compare "*Manual, etc.*," ¶ 169.

[196] The mobility of the hand is not its least expressive property.   It is, of all the parts of the body, the most movable and the most rich in articulations ; over twenty joints concur in the production of this multiplicity which furnishes its physiognomical character.   It cannot help denoting the character of the *body* to which it is so nearly attached, of the *temperament*, and consequently of the *heart* and of the *mind*."   GASPARD LAVATER, "*L'Art de connaître les Hommes, etc.*" (Paris : 1806), vol. iii., p. 1.

A PARTICULAR system of education, applied exclusively to a particular class of mind, often results in the perversion and abnegation of that intellect by its possessor himself. How fortunate, therefore, are those whose intellectual aptitudes, having been appreciated and understood early in life, have served as a basis upon which their early training has been founded and built up. They become at once a happy fusion of the man who has learnt [*i.e.*, the man of education], and the man of inborn faculties [*i.e.*, the man of innate talent]; thus they have two impulses to direct their lives which are practically one; they come upon the stage of life armed with ideas which they have acquired, and supported by an intelligence, by instincts which harmonise with those ideas; and, whilst those whose talents have been stunted by an illogical form of education are constantly retarded and embarrassed by doubt, the former subjects attain practically without effort the front rank in any profession they may take up.

¶ 247.
Mistaken
training for
after-life.

But how few young people there are who are sufficiently fortunate to have been *understood* when they were young enough to be guided in the path most advantageous for them, and how few teachers of youth there are who are ready to abandon all stereotyped

¶ 248.
Appreciation of
talents in youth

rules and methods, and to adopt a separate system for each individual genius.   It would not be too much to expect this from a parent, but it is obvious that such an effort, generous as it must necessarily be, must always be beyond the venal solicitude of a stranger.

¶ 249.
Early recognition
of genius.

If this volume has any value, it must lie, as I have before remarked, in the fact of its furnishing the means of recognising the physical signs [signs which I think I have described with sufficient clearness] of the special bent of every man's individual intelligence. At eight years old, or even at six years old, a child's hand is sufficiently developed to render practicable the interpretation of its particular aptitudes and faculties ; whether he will be a man of contemplation or a man of action, whether he will affect the study of ideas, or the practice of actual things.   I trust that my observations have placed me in the path of truth, and that the primary cause of unnatural and improper educations will now disappear.

¶ 250.
Ex pede
Herculem,

By the formation of a dog's foot you can tell for what particular kind of chase he is most fitted ; by the shape of a horse's hoof you can tell what is his breed, and what qualities particularly distinguish him. In the same way, by examining our hands with care, we cannot help recognising the fact that they sum up, as it were, the whole of our minds, and that the tracing, the diagram, as it were, they afford us of our intelligence, cannot fail to be an interpretation which is at the same time profound and true.[197]   It is in this sense, and not in the sense which is given to

[197] " Every hand, in its natural state—*i.e.*, without taking into consideration unforeseen accidents,—is in perfect analogy with the body of which it is a member ; . . . the same blood circulates in the heart, in the brain, and in the hand ! "   GASPARD LAVATER, " *L'Art de connaître les Hommes par la Physionomie* " (Paris : 1806), vol. iii., p. 1.

it by the Cheiromants, that we must interpret the
celebrated passage from the Book of Job xxxvii. 7 :—
"In manu omnium Deus signa posuit ut noverint
singuli opera sua." [108]

¶ 251.
Imitative hands

Nature, in endowing the whole monkey-tribe with
identical instincts, has equally endowed them with
identical hands ; and whilst mentioning these animal
imitators, I may add that jugglers, conjurers, mimics,
and actors have nearly all of them, like the monkey,
spatulate fingers.

¶¶ 252—261.
Symbolical
expressions.

There is a saying to the effect that a man has
"hidden his thumbs," which signifies that he has
abdicated all strength of will to act for himself.

¶ 253.
Giving the hand.

A young girl, in "giving her hand," yields up her
liberty ; the man in the marriage ceremony does not

---

[108] I have considered and discussed this passage at
much length on pp. 55-58 of "*A Manual of Cheirosophy.*" The quotation which I have left in Latin, as
quoted in the text by M. d'Arpentigny, translates :—
"In the hands of all men God *has placed signs*, that
each man may know *His* (God's) works," and is a
most unfortunate misquotation of the text as it appears
in the Vulgate, "In manu omnium hominum *signat*,
ut noverint singuli opera sua," which translates :—
"who signs [or *seals*] the hand of every man," etc.
In our Authorised Version the verse is rendered :—
"He sealeth up the hand of every man ; that all men
may know His work," or, as the Revised Version hath
it, "That all men whom He hath made may know [*it*]."
Anyone who will read the preceding and following
verses (6 and 8) will see that no idea of cheirosophy
or divination of any kind was in the writer's mind. I
have cited the leading commentators on the verse in
"*A Manual, etc.*" It has always surprised me that
so exact and learned a writer as d'Arpentigny should
have fallen even half-way—as it seems he has—into this
hackneyed error. It is, of course, one of those cases in
which, as Sir Walter Scott says ["*Letters on Demonology and Witchcraft,*" vi.], people claim the support
of Scripture for their own theories without regarding in
any way the niceties of translation which should be
examined in such matters.

[¶ 253]

give his hand; thus he does not swear obedience, but undertakes to protect.

¶ 254.
A blow.

Almost any verbal slander can be overlooked, but once a man has raised his hand against another the insult is past forgiveness. It is true that neither Diogenes nor Christ preached a doctrine such as this, but man is governed by rules which are other than those established by cynic or Divine utterances.[199]

¶ 255.
Hiding the hands.

The ancient Persians, as a sign of absolute submission, kept their hands constantly hidden in the folds of their robes when in the presence of the king.[200] [201]

---

[199] "And unto him that smiteth thee on the one cheek, offer also the other."—Luke vi. 29. "For ye bear with a man, if he smiteth you on the face."—2 Cor. xi. 20. The passage of Diogenes to which I conclude our author refers, is the account of his celebrated remark to Antisthenes, which is thus rendered by M. C. Zevort in his "*Vies et Doctrines des Philosophes, Diogène de Laërte*" (Paris: 1847, vol. ii., p. 11, book iv., c. 2). "One day Antisthenes threatening him with his stick, he (Diogenes Laërtius) stretched out his head, saying, 'Strike on! you will not find a stick hard enough to drive me away from you when you are speaking.'"* But Diogenes did not always show the same meekness to his assailants: on p. 35 of M. Zevort's book we find, "A man having jostled him with a beam and cried 'Take care!' Diogenes struck him with his stick, crying in turn 'Take care!'" So that one of the author's illustrations is at least doubtful.

[200] We find a passage which tells us of this custom in the "*Cyropædia*," where we are told that Cyrus in the procession [alone] kept his hands outside his robes,† and the custom of concealing the hands in the presence of superiors obtains even to this day. Sir John Malcom, in his "*History of Persia*" (London:

---

* ΔΙΟΓΕΝΟΤΣ ΛΑΕΡΤΙΟΥ ΒΙΩΝ ΚΑΙ ΓΝΩΜΩΝ ΤΩΝ ΕΝ ΦΙΛΟΣΟΦΙΑ, etc., Βιβλ. Γ'., β'., Διογενής.—*Didot's Edition*, p. 138.

† ΧΕΝΟΡΗΟΝ, ΚΥΡΟΤ ΠΑΙΔΕΙΑΣ Βιβλ. Η'., κεφ. γ.—"Καὶ οἱ ἱππεῖς δὲ πάντες παρῆσαν καταβεβηκότες ἀπὸ τῶν ἵππων, καὶ διειρηκότες τὰς χεῖρας διὰ τῶν πανδύων, ὥσπερ καὶ νῦν ἔτι διειρουσιν, ὅταν ὁρᾷ βασιλεύς . . . Ἐπὶ δὲ τούτοις, ἤδη αὐτὸς ἐκτῶν πυλῶν πρου φαίνετο ὁ Κῖρος . . . τὰς δὲ χεῖοας ἔξω τῶν χειρίδων εἶχε."

As a mark of abnegation, to express the knowledge which we instinctively possess of our weakness and of our insignificance, we clasp our hands in praying to God.[202] For, after all, what is a man if he be without hands ?[203]

¶ 256.
Clasping the
hands in prayer.

Very little, according at any rate to the opinion of Lysander, who put to death the Athenian prisoners captured at Ægos-Potami, because they had decreed that they would cut off the thumbs of all prisoners of war who should fall into their hands in the victory which they regarded as a certainty.[204]

¶ 257.
Cutting off the
hand.

1829, vol. ii., ch. xxiii., p. 399), says, "Looks, words, the motions of the body, are all regulated by the strictest forms. When the king is seated in public, his sons, ministers, and courtiers stand erect with their hands crossed." The custom is not mentioned by Brechillet Jourdain ["*La Perse*" (Paris: 1814)], but there are many of my readers who must have observed it among the members of the *suite* during the visit of the Shah to Europe in 1873-4.

[201] Among the Arabs the posture of the profoundest submission and respect is standing with the hands behind the back, the right grasping the left. Sir R. F. BURTON, "*Arabian Nights*," vol. iii., 218.

[202] It has often occurred to me that the Moslem attitude, *i.e.*, standing with the palms of the hands turned upwards, is the more appropriate and symbolical. When reciting the Fatihah [the opening chapter of the Qur'an], the hands are held in this position as if to receive a blessing falling from heaven ; after which both palms are passed down the face to distribute it over the eyes and other organs of sense [BURTON, "*Arabian Nights*," vol. v., p. 80].

[203] Compare "*A Manual of Cheirosophy*," ¶ 5.

[204] M. d'Arpentigny quotes Thucydides as his authority for this passage ; as a matter of fact, Thucydides never mentioned either Lysander or the battle of Ægos-Potami in his "*History of the Peloponnesian War*." It is Xenophon who gives us the account of Lysander having made this slaughter of Athenian prisoners, because they had decreed that, should they win, all the prisoners they took should have their right hands cut off, Adimantos only being spared, because he had opposed

¶ 258.
Raising the
hand in the law
courts.

It is the right, and not the left hand, which is raised in taking the oath in the law-court, because the right hand, being the one of which we make the most use, it affords for this reason a more perfect representation of our physical, intellectual, and moral worth, than the left.[205]

¶ 259.
A parallel.

In like manner the foreman of the works, who over-looks the whole of the process of construction, is a better personification and representative of the absent master-builder than the labourer, who is employed only upon some secondary but exclusive task.

¶ 260.
Hand in the
Royal Arms
of France.

The hand of Justice, which figures among the insignia of our [French] royal families, is always a *right* hand.[206]

this horrible decree in the Assembly.—*Vide "A Manual of Cheirosophy,"* ¶ 22.*

[205] This formality will have doubtless been observed by many of my readers who have frequented the courts of the "Palais de Justice." Until comparatively recently the custom obtained in this country, and in Scotland it is the practice to this day, to raise the right hand in taking the oath in court. "In taking a *great* oath . . . the gods used to lift up their hands, as *Apollo* in the poet bids *Lachesis* χεῖρας ανατεῖναι. Little thought he how the Scripture makes the like action of the true God in several places. Men, when they swore a *great* oath, laid downe their handes upon the altar as we do upon the New Testament, whereas in a lesse, or in a private oath, made to such or such man, according to the Roman fashion, they laid their hand upon the hand of the party to whom they swore. This ceremony, I remember, *Menelaus* in *Euripides* demanded of *Helen* besides the words of her oath" [*Helen*, v., 834]. F. ROUS, *"Arch. Att.,"* p. 278.

[206] This emblazonment of the House of Orleans was unknown to me. I have, however, received a document on the subject from M. A. Daubrée, Ex-President of

* XENOPHON, ᾽ΕΛΛΗΝΙΚΟΝ, Βιβλ. Β᾽., κεφ. ά (31):—"᾽Εν-ταῦθα δὴ κατηγορίαι ἐγίγνοντο πολλαὶ τῶν ᾽Αθηναίων ἅ τε ἤδη παρανενομήκεσαν καὶ ἃ ἐψηφίσμενοι ἦσαν ποιεῖν, εἰ κρατήσειαν τῇ ναυμαχίᾳ, τὴν δεξίαν χεῖρα ἀποκόπτειν τῶν ζωγρηθέντων πάντων."

When one feels a presentiment of the wrath of God, and of the approach of the chastisement by which it will be manifested, one says, He is going to stretch forth His right hand, *i.e.*, He is going to strike with intelligence and discernment.[207]

¶ 261.
The Hand of God.

We kiss the hand of a prince in token of our submission,[208] that of a father or of a protector in token of respect and gratitude, those of holy men in token of veneration, those of fair women in token of adoration;—and all this because royalty, paternity, sanctity, and beauty are real powers, and all real power has the attribute of enchaining and of subjecting individuals.

¶ 262.
Kissing hands

L'Institut de France, of which the following is a transcript and translation : " From the moment when the Duc d'Orleans ascended the throne in August 1830, he retained for a few days the arms of his family, that is to say, three fleurs de lys. He soon considered that he ought to adopt others,—the tables of the law upon an escutcheon, on either side of which are banners, and on one side the royal sceptre, and on the other *the Hand of Justice*. It is this latter which is *a right hand upraised*. It cannot be said that it forms part of the arms of France ; it is an *accessory* of the escutcheon, according to the Marquis de Flert."—A. DAUBRÉE. This information was obtained from the Marquis de Flert, at present the head of the House of Orleans.

[207] " The saving strength of His right hand."—Psalm xx. 6. " Why drawest Thou back Thy right hand?"—Psalm lxxiv. 11. " The right hand of the Lord doeth valiantly."—Psalm cxviii. 16. " When Thou stretchest forth Thy hand to heal."—Acts iv. 30. " Behold therefore I have *stretched forth My hand over thee and have diminished thy food*"—Ezek. xvi. 27. " Therefore is the anger of the Lord kindled against His people, and *He hath stretched forth His hand against them*, and hath smitten them."—Isa. v. 25, κ.τ.λ. A vast collection of phrases in which the word " MANUS " figures symbolically may be found in Ducange's " *Glossarium Mediæ et Infimæ Latinitatis* " (Niort and London, *reprint:* 1885), vol. v.

[208] One of the oldest tributes of respect and submission. *Vide* several instances recorded by Xenophon in

¶ 263.
Symbol of power,
and universal
study of the
hand.
From which data I deduce this maxim : that the hand is the symbol of all *power*. The study of the hand has at some time or another engrossed the attention of every race of living men.

¶ 264.
The cheirosophy
of Manou.
"Creatures which are passive," says Manou, "are the natural prey of those who are active; creatures without teeth are the natural prey of those which have those weapons; creatures who have no hands are the natural prey of those who have those members;"[269] and he continues to the effect that the part of the hand situated at the root of the thumb [which, as I have said, is the seat of reasoning will],[210] is consecrated to the Vedas; that the part consecrated to the Creator is at the root of the little finger [which being the finger of the heart is always pointed, because the heart is always more or less poetic, and consequently credulous]; and that the part devoted to the lesser gods [probably looked upon as the symbols of action, as manifested in the arts, sciences, and liberal professions], is to be found at the tips of

the seventh book of the Cyropædia.[*] F. Rous also, in his "*Archeologicæ Atticæ*" (London: 1685, p. 278), says, "It was either this kisse, or a kisse of their owne hande which they anciently termed *labratum*. I have read of a kisse of the hande when they did the reverence to the gods, with putting the forefinger over the thumb [perhaps upon the middle joynt], and then giving a turn on the right hande, as it is in Plautus [*In Curculi*]:—
    "Quo me vortam nescio, si deos salutas vorsam censeo."
Compare with this note "*A Manual, etc.*," ¶¶ 18 and 19, and notes thereto.
[209] This passage, which is implied by the order of creation and superiority laid down in the first lecture of Manu, may be found in the edition of "*The Ordinances of Manu*" cited in note [235], p. 202.
[210] *Vide* ¶ 188.

---

[*] Κεφ. έ :—"Καὶ ἡ γυνὴ δὲ ἀνωδύρατο, καὶ δεξαμένη, τὴν χεῖρα παρὰ τοῦ Κύρου ἐφιλήσετε, καὶ πάλιν ὡς οἷόν τ᾽ἦν προσήρμοσε ; " and a little further on we are told of Gadatas and Gobryas, " ἔπειτα δὲ Κύρου κατεφίλουν καὶ χεῖρας καὶ πόδας, πολλὰ δακρύοντες," etc. etc., and so on, *passim*.

the fingers.[211] I did not know this explanation [in which I find the germs of my system] when I established the bases upon which I started to write this book.

Abd-el-Kader bears on his banner a red hand blazoned upon a blue field.[212]

In Tripoli they hang a little metal hand upon all objects, such as temples, houses, or palaces, which they wish to preserve against the effects and influence of the evil eye.[213]

¶ 265.
Arms of Abd-el Kader.
¶ 266.
The evil eye.

[211] The best translation of this passage to cite as a commentary to the above is to be found in "*Manava-Dharma-Shastra, or, the Institutes of Manu,*" by Sir Wm. Jones (Madras: 4th edition, 1880), lecture ii., vers. 58 and 59. "Let a Brahman at all times perform the ablution with the pure part of his hand denominated from the Veda, or with the part sacred to the lord of creatures, or with that dedicated to the gods; but never with the part named from the Pitris. The pure part [of the hand] under the root of the thumb is called *Brahma*, that at the root of the little finger *Kaya*, that at the tips of the fingers *Daiva*, and the part between the thumb and index *Pitrya*.

[212] A red hand is a very favourite emblem of conquest in the East. A legend connected with this is a great favourite in Constantinople, and is thus referred to by Théophile Gautier in his "*Constantinople of To-day*" (London: translated by R. H. Gould: 1854):—"I sought in vain in St. Sophia for the imprint of the bloody hand, which Mahomet II., dashing on horseback into the sanctuary, imprinted upon the wall, in sign of taking possession as conqueror, while the women and maidens were crowded round the altar as a last refuge from the besieging army, and expecting rescue by a miracle, which did not occur. This bloody imprint of the conqueror's hand—is it an historical fact or only an idle legend?" [p. 281.]

[213] A superstition which is not by any means confined to Tripoli. It is generally supposed to have had its origin in Naples, where, as a preservative against the *jettatura*, a little coral or metal hand is in great request and favour, being either cast in the position known as the devil's horns, or in the position prescribed for the invocation of the episcopal blessing [*vide* ¶ 13

¶ 267.
Turkish hand
symbolisms

The Turks, a nation essentially contemplative and inert, have only been able to find in the hand a kind of rosary, of which the fourteen joints constitute the beads; God is represented by the entirety of the hand, and each finger represents one of the cardinal maxims :—*e.g.* : Belief in Allah and his Prophet, Prayer, Alms, Observation of the Ramadan, and the Hadj pilgrimage.[211]

¶ 268
Wizard hands.

Hands which are large, dry, wrinkled, very knotty, and pointed-fingered, irresistibly suggest ideas of uncanniness and isolation, which are to the last degree unattractive/ When hailstones, like an insurgent mob, or like a tribe of gipsies striking camp, roll and dance about on the pavement in the streets, and on the sounding tiles of the roof, it is hands of this type which the benumbed sorceress strives vainly to warm, muttering beneath her cloak of owl feathers.

¶ 269.
Cheirosophy and
Physiognomy
compared.

For the hand has its physiognomy like the face, with this difference, that as it reflects only the immutable bases of the intelligence, it has all the permanence

ante and "*A Manual, etc.*," ¶¶ 23 and 24]. The hand-charm most in use along the north coast of Africa has all its fingers extended; a description thereof may be found in Chas. Holme's recent "ODD VOLUME MISCELLANY," No. 15 (London: 1886).

[211] Compare Al Qur'ân of Muhammad, chap. ii., in which these directions are all specified for the guidance of true believers. George Sale, in the Preliminary Discourse which precedes his translation of the Qur'ân (London: with a Memoir of the Translator: 1865), says in the 4th section :—" The Muhammadans divide their religion, which they call *Islam*, into two distinct parts: *Iman, i.e., faith*, or theory; and *Din, i.e., religion*, or practice; and teach that it is built upon five fundamental points, one belonging to faith, and the other four to practice. The first is the confession of faith, that 'There is no God but the true God; and that Muhammad is His prophet' . . . The four points relating to practice are: 1. Prayer; 2. Alms; 3. Fasting; 4. the Pilgrimage to Mecca."

of a material symbol.[215] Mirror as it is of the soul, of
the heart, of the mind, and of the spirit, the physiognomy of the face is endued with all the charms of
variety, but as it is, to a certain extent, subject to the
dictates of our will, the accuracy of its indications
cannot be guaranteed; whereas the physiognomy of
the hand always bears the stamp, whatever it may be,
of our genius.[216] [217]

There are hands which naturally attract us, and
there are hands which excite in us repulsion. I have
seen hands which seemed covered with *eyes*, so
sagacious and so penetrating was their appearance.
Some, like those of the sphinx, suggest an idea of
mystery; some betray folly and strength combined
with activity of body; others again indicate laziness,
joined to feebleness and cunning.[218]

¶ 270.
Characteristic
hands.

There are people who fancy they are serious,
because they are of a lugubrious and miserable state
of mind; there are others who, like the Abbé Galiani,
resembling clocks which keep good time, but whose

¶ 271.
Peculiar varieti
of mind.
Abbé Galiani.

---

[215] *Vide " A Manual, etc.,"* ¶ 78.

[216] " Particular hands *can only* belong to particular
bodies. This is easy enough to verify; choose a hand
for an example, compare it with a thousand others, and
amid this number there will not be a single one that
could be substituted for the first. . . . It is—just as
much as the other parts of the body—an object for the
attention of physiognomy; an object so much the more
significant and striking from the facts that the hand
cannot dissemble, and that its mobility betrays it at
every moment. I say that it cannot dissemble, because
the most accomplished hypocrite, the most experienced
deceiver, could not alter the forms of his hand, nor its
outlines, proportions, or muscles, nor even of a part of
it; he could not protect it from the gaze of the observer,
save by hiding it altogether." GASPARD LAVATER,
" *L'Art de connaître les Hommes, etc.* " (Paris: 1806),
vol. iii., p. 1.

[217] Compare also Desbarrolles " *Les Mystères de la
Main,*" 15th edn., pp. 419—421.

[218] " Whether in motion or in a state of repose the

striking apparatus is out of order, contradict the
wisdom of their behaviour by the folly of their conversation.[219]   Joseph de Maistre fancies he is an ardent
admirer of truth, whilst in reality it is power of
which he is so assiduous a votary.   His mind is
like the tower at Pisa, grandly proportioned, solid,
and *crooked*.[220]   Again, there are people who are

expression of the hand cannot be misunderstood.   Its
most tranquil condition indicates our natural propen-
sities ; its flexions, our actions and passions.   In all its
movements it follows the impulse given to it by the rest
of the body.   It attests, therefore, the nobility and the
superiority of the man ; it is at the same time the inter-
preter and the instrument of our faculties."—LAVATER,
loc. cit.

[219] In this respect the celebrated Abbé Galiani was a
most extraordinary contradiction.   Eugène Asse, in his
"*Lettres de l'Abbé Galiani avec une Notice Biographique*" (Paris : 1881, vol. ii., p. xxxviii.), "If
Galiani was often a regular Neapolitan clown, he was
also, and most often, a true *savant*, a profound thinker,"
etc.; and in another place he says (p. xxv.): " L'Abbé
Galiani entra, et avec le gentil abbé, la gaieté, la folie,
la plaisanterie, et tout ce qui fait oublier les peines de
la vie."   His greatest friend, Grimm, said of him : " It
is Plato with the whims and the gestures of Harlequin "
Corréspondence Littéraire " (Nov[bre] 15, 1764), tome vi.,
p. 116.

[220] Joseph de Maistre was the Piedmontese ambassador
to the Court of St. Petersburg.   Finding that the
philosophers of his day were attacking the Catholic
religion, he entered the lists with enthusiasm as the
defender and apologist of the See of St. Peter, and wrote
in its defence his two books, " *Du Pape* " (Lyon : 1836),
and "*De l'Eglise Gallicane dans son Rapport avec
le Souverain Pontif*" (Lyon : 1837).   Immediately
he was violently attacked on all sides by the progressive writers of the French philosophical schools, and as
many works were written in his defence as were written
against him.   M. du Lac, in his preface to R. de Sezeval's
" *Joseph de Maistre, ses Détracteurs et son Génie* "
(Paris : 1865), commences with the words : " Pendant
de longues années les écrivains du liberalisme se sont
plu à répandre sur le nom de M. de Maistre la haine,
le ridicule, et le mépris, ... sous quelques ineptes

moral, not for the good of their souls, but for the pleasure of talking about it; others there are who make the foibles of great men their particular study,[221] all that they know of Vincent de Paul being that he cheated at cards.[222] These, like Balzac, because they are of a subtle mind, fancy themselves spiritualists;[223] those affect concealment that they may preserve an in-

Honoré de Balzac

sarcasmes n'ont-ils pas tenté d'étouffer sa parole et sa gloire?" M. d'Arpentigny, whose whole work teems with liberalism of the most pronounced description, would naturally follow in the wake of these detractors of a writer who wrote in defence of the most bigoted conservatism which it is possible to imagine.

[221] "Note, Sancho," said Don Quixote, "that wherever virtue may be in a high degree, there it is hunted down. Few or none of the past but were calumniated of malice; for examples: Julius Cæsar, most courageous, most prudent," etc., etc.—"*Don Quixote*," pt. ii., ch. ii.

[222] I cannot conceive what our author is hinting at here; St. Vincent de Paul has been cited by numberless authors as a model of all that is pure and good. Louis Abelli, Bishop of Rodez, cites many instances of his having been calumniated, and of his imperiousness under the circumstances, but does not mention this particular slander,* and it is difficult to imagine a thing of the sort of a man, one of whose most celebrated dicta was, "Gentlemen, let us pay as much attention to the interests of our fellow-men, as to our own; let us act loyally and equitably, let us be straightforward" (Abelli, book iii., c. 17, p. 260). M. Capefigue, also, in his work "*St. Vincent de Paul*" (Paris: 1865, p. 9), says:—"What motive caused his sea voyage to Marseilles? Some say that he had taken flight before an abominable calumny (the most holy men have always been calumnied, etc.);" —this may be the matter referred to in the text.

[223] There is no doubt that Balzac was endowed with one of the most subtle mental organisations that France has ever produced; but throughout his life he fostered the idea that it was in spiritual, intuitive writing that he excelled, and thought his "*Comédie Humaine*" and his "*Cousin Pons*" vastly superior to such wonderful works as his "*Illusions Perdues*" which give us per

---

* "*Vie du Vénérable Serviteur de Dieu, Vincent de Paul*, par Louis Abelli, Evesque de Rodez (Paris: 1664), liv. iii., p. 168.

[¶ 271]

cognito, and disguise themselves that their faces may not become familiar.   The hand will not reveal such shades of character as these.

¶ 272.
Greek hands as found in sculptures.

Of all the antique statues which are to be found in European museums, two only have reached us with their hands remaining, or rather with one hand remaining, on each.   Without these precious relics we should be absolutely ignorant of the Greek standard of beauty as applied to the hand.   As it is, we know that they required that it should be large, with strong smooth fingers, a large thumb, a medium palm, and square finger-tips.   Such, at all events, is the single hand of the exquisite statue of the son of Niobe, which one sees at Florence.[224]

¶ 273.
Greek appreciation of physical strength.

The Greeks, surrounded as they were by barbarous nations, and continually in danger of seeing their fragile and lightly-constructed republic overwhelmed by war, held physical strength in high esteem ; and with good reason.   In the course of their education wrestling, racing, boxing, fencing, and swimming were held to be of equal, if not of more, importance than the training of the mind.[225]   This being the case, their

haps the best possible illustration of the subtlety of his imagination.   *Vide* also Théophile Gautier's " *Honoré de Balzac* " (Paris : 1859).

[224] It must be borne in mind that the above was written before 1843, and that since then excavations have brought to light many beautiful specimens of Greek statuary with their hands in their original conditions. At the moment, indeed, that I write, the labours of Count Charles Lançkoroñski are drawing to a close ; and the illustrations to his work, describing his discoveries of Greek statuary in Asia Minor, at Idalia, and in Rhodes, will show us many representations of Greek hands.   For the rest, many such may be seen in the British Museum ; the statue cited above by M. d'Arpentigny certainly presents an exquisitely-modelled hand, but I do not think it is necessary to go to the Uffizii Palace for an unique illustration.

[225] Perhaps the most complete and curious, and at the same time pedantic and diffuse epitome of the educa-

idea of beauty was naturally different to what ours is, seeing that we are not threatened by the same dangers as they were, that we make use almost exclusively of projectile weapons, which are easy of manipulation, and that, brought up in the focus as it were of spiritual Christianity, we are surrounded by Christian and civilised nations like ourselves.

Forgetfulness of self, and calmness in danger, are to-day more necessary to our soldiers even than physical strength and bravery.

¶ 274. Necessity of calmness in action.

Large hands, particularly if they be hard, are a sign of physical strength, and as the Greeks could not

¶ 275. Greek love of a large hand.

'tion of Greek youths is Francis Rous' *"Archeologiæ Atticæ Libri Septem"* (Oxford : 1658). The training of children was, of course, very different in different states of Greece : thus, for instance, in Sparta babies were not wrapped in swaddling clothes, the prime object being to harden the body of the young warrior against the influences of pain and exertion. Among the Doric tribes it was, therefore, customary to expose weakly children in some open place, when, if they were strong enough to bear the test, they were brought up either by their parents, or by anyone who would rescue them from the exposure. This subject is treated of at much length by Rous [cap. viii., lib. v., "De Expositione Infantum"]; and it is to this custom, of course, that Creusa refers in the *Ion* of Euripides :—" The son she bore, she also did cast forth;" and again, " She came where she expos'd and found him not."* Among the Ionian tribes, on the other hand, the graceful and harmonious development of the body and mind, ease of bearing and demeanour, were particularly aimed at. This was especially the case at Athens, where children were always accompanied by an older companion who taught them good behaviour, and respect to their elders. Their education consisted principally of letters, music, and gymnastics. Lucian, in his apology of gymnastics [Luciani *" De Saltatione Dialogus"*], lays great stress upon the value of gymnastics in the training of youths, as a preventive against laziness and its accompanying vices. *Vide* hereon E. Guhl and W. Köner's work, *"The Life of the Greeks and Romans,"* translated by F. Hueffer (London : 1873).

* Wodhull's "Euripides," " *The Greek Tragic Theatre,*" (London : 1809), vol. v., p. 109.

13

conceive beauty without strength, a large hand was
to them a great beauty, following the same analogy
which with us accounts for the fact, that to us a
small hand is the most beautiful by reason of the
greater delicacy of mind which it reveals. It is, of
course, a matter of common knowledge that the
Greeks, whatever their station in life, never rode
when they might walk, did their own cooking, and
otherwise performed a great deal of manual labour
which would to-day be looked upon with aversion
and contempt. They performed these labours not
only without repugnance, but even with pleasure, by
reason of the inherent love of detail which, as I have
said, is the special attribute of large hands.[226]   From
all of which facts I gather that in Greece, not only at
the time when princes tended the flocks, and princesses
washed their own linen, and when the clergy excelled
in the professions of butcher and baker, but even in
the time of Pericles, large hands were abundant.

¶ 376.
Love of finish.

Large hands whose palms are of a medium development prefer that which is finished and exquisite, to
that which is grand and large. The Greeks only
founded small states, and erected no monuments of
remarkable size.[227]

[226] The principal occupations of the men being, as I
have indicated in the last note, of the more active and
virile description, it is in the manners of the women that
we notice principally the simplicity of which our author
speaks.   "The chief occupations of women, beyond the
preparing of meals, consisted in spinning and weaving.
In Homer we see the wives of the nobles occupied ·in
this way; and the custom of the women making the
necessary articles of dress continued to prevail, even
when the luxury of later times, together with the
degeneracy of the women themselves, had made the
establishment of workshops and places of manufacture
for this purpose necessary.   Antique art has frequently
treated of these domestic occupations."—E. GUHL and
W. KÖNER, *op. cit.*, p. 186.

[227] It is difficult to say whether in this passage

In Paris, notwithstanding their enormous hands, the Flemish journeymen-tailors are immensely sought after, on account of the fineness of their work.

¶ 277.
Flemish tailor hands.

Redouté, our celebrated flower painter [a school of painting naturally minute] had great big hands like a bricklayer. He used to laugh at the innocence of provincial poets and journalists, who, arguing by the delicacy of his work, used to compare his fingers to those of Aurora, "scattering roses as he went."[228]

¶ 278.
Redouté, the flower painter

Now, little hands, on the contrary, affect not only the large, but the colossal; in fact, one is inclined to come to the conclusion that everything must be ordered in obedience to the laws of contrast. It is towards the dwarf that the giant is irresistibly attracted, and in like manner it is by the giant that the dwarf is invariably fascinated. The Pyramids, the temples of Upper Egypt, and of India, have all been built up by people whose staple comestibles have been rice, gourds, and onions, that is to say, by the people who are the most delicate, and whose hands are the most delicate in the world. These hands were small and narrow, spatulated and smooth, as is evidenced by the representations of them which we find in the contemporary bas-reliefs with which these structures are ornamented.

¶ 279.
Vast works of small hands.

The Pyramids, etc., etc.

d'Arpentigny refers to the fact that the Greek nation consisted of many small states, each with its separate government, and so on, but united by community of race and religion ; or that the Greek colonies consisted only of towns round the shores of the Mediterranean, with but small territorial possessions attached to them. Again, the author must have overlooked the Parthenon and the other majestic buildings of the Acropolis.

[228] Joseph Redouté, born in 1759, was a celebrated flower painter attached to the court of Louis XVI. He was known as "le Raphael des Fleurs," and died at Paris in 1840. *Vide* Ch. Dezobry and Th. Bachelet's "*Dictionnaire de Biographie et d'Histoire*" (Paris: 1857), and "*Annales de la Société*" (tome xiii.), "Notice sur J. Redouté."

¶ 280.
Préault,
the sculptor.

The sculptor Préault, having a small thumb and smooth fingers which are delicately spatulated, proceeds entirely by enthusiasm and inspiration; and as his hands are very small [for a sculptor], ample proportion, power, and energetic treatment are more important to him than exact measurements and grace;[229] his sculptured horse on the bridge known as the "Pont de Jéna" seems from his springing position to carry away with him the whole block from which he was carved; it is not so much a prancing horse as a rock.

¶ 281.
Balzac: Love
of minutiæ.

Balzac, with his large conic hands, liked to count the fruit on the espalier, the leaves on the hedge, the separate hairs in his beard; he took a delight in physiological details, and might have invented the microscope had it not been invented before he was born.[230]

¶ 283.
Georges Sand's
breadth of treat-
ment.

Madame Sand, whose hands are very small, excels especially in psychological developments: her very

[229] Auguste Préault, the sculptor [b. 8th October, 1809; d. 11th January, 1879], of whom a most minute and interesting account may be found in Ernest Chesnau's "*Peintres et Statuaires Romantiques*" (Paris: 1880, p. 119, "Auguste Préault"), has been described as the Doré of sculpture. Weird, morbid, fantastic, and enormous, his work seems to have been the reflection of his intellectual organisation, which his biographer describes as "nervous as a woman, sensitive, and impressionable." On p. 119 he says, "There exists a precious cast of the right hand of Préault. It is remarkable for its smallness and the elegance of its proportions. By the absence of developed joints, and the fineness of the phalanges, Desbarrolles, the cheiromant, would recognise rapid intuition; the short, thick, spatulate thumb is that of a man of action and of stubborn will; in the confused and multiplied lines of the palm we recognise the fatal imprint of a destiny doomed to the agitation of a continual struggle ["*Manual*," ¶ 421]; the line of art deeply traced, and the Mount of Jupiter high in the hand" ["*Manual*," ¶ ¶ 429, etc., and 614, etc.].

[230] Honoré de Balzac was very proud of his hands.

details are immense; in like manner she might have invented the telescope.

There are laws which seem to be equitable, but which are not so in reality; the law of conscription is one of these. The duties, the necessities imposed by it, light and easy for spatulate and large-palmed hands, are overwhelming for conic and pointed hands, more especially so if they are also soft. What does it matter to hands which have a large hard palm, that the barrack rooms are hideously bleak and bare; that the life of the camp is brutalizing in its monotonous slothfulness; what to them is the coarseness and the insipidity of the food, the passive obedience and the automatic life? What do they matter to hands that are spatulate, with a large thumb, these eternal exercises, the monotonous activity of the work of mines, tunnels, and trenches, and the everlasting agitation of the 'tween-decks? But this same noise, these scenes, these labours are inexhaustible sources of moral and physical anguish for souls whose outward and visible signs are hands which are narrow and pointed.

¶ 234.
Unnaturalness of universal conscription.

And what shall we say of the Indian laws which compel a son to take up and follow the same handicraft as his father? Is it not obvious that the legislature would do better to order that men whose hands—that is to say, whose implements of labour—are identical, should adopt identical pursuits? [231]

¶ 285.
Indian laws of succession.

Théophile Gautier, in his "*Honoré de Balzac*" (Paris: 1859), tells us [p.11]: "We remarked his hands, which were of a rare beauty, regularly *clerical* hands, white, with medium and dimpled fingers, with brilliant rosy nails. He was proud of them, and smiled with pleasure when any one noticed them. He invested them with a symbolism of aristocracy. . . . He had even a kind of prejudice against those whose hands and feet were wanting in delicacy." Compare note :::, p. 187.
[231] *Vide* note [218], p. 185.

¶ 286.
Injustice of
property
franchise.

But, tyrannous and unnatural as is such a law as this, it is not more so than the one that in our own country makes property the sole qualification for electoral franchise.   It is a matter of common knowledge that fortune is more quickly and surely acquired by the exercise of manual skill, and of *physical* forces and activity, than by the pursuit of science, and the practice of *moral* forces and activity ; and the result is, that this franchise is nothing more than the oft-told story of the predominance of interest over principles, of industry over art, science, and philosophy ; the superiority of working over thinking hands.

¶ 287.
Illustration.

For many years past the university of Caen, an institution notable for its eminent professors, has harboured among all the persons it employs one single elector,—the gate-keeper !

¶ 288.
Necessity for
combination.

It is not right that it should be thus ; nor would it be right if the contrary were the case ; for universal life is not to be governed simply by exalted and philosophic ideas ; it has to be directed also by the common and vulgar ideas of which big spatulate hands have a far clearer perception [even though their intelligence is limited to this alone] than those more finely cut types among which all kinds of high-flown ideas and colossal schemes abound.   We must hear what each has to say ; so-called " representative governments," in which every primordial instinct requires—for its development and for the defence of the interests of which it is the fundamental principle— to satisfy conditions which are unnatural to it, are " representative " in name only.   Man is a creature of mixed composition ; he has a soul and he has a body, and both the agencies which support his body and those which support his soul must be equally attended to.   There can be no valid reason for debarring either of them from taking part in the mental debates which have for their object the physical and moral advantages and improvement of the man. Certain things there are which can only be well done by hands, and with instruments, which are homely and common ; and again, there are certain others that can only be perfected with the most finished and delicate tools. One can cut paper better with a wooden paper knife than with a blade of gold ; one can only engrave fine stones with highly tempered implements of steel.[232]

In the United States, where they appreciate the value of wooden paper-knives [*i.e.*, of ordinary minds], and of steel instruments [*i.e.*, of subtle intelligences], both are equally called upon to direct the conduct of public affairs ; and what has been the result of this co-operation within the space of a half century even ? The well-being of the individual and the prosperity of the community, moral greatness, and material power, —invincible proofs that as regards government, and the due appreciation of human faculties, the people of the United States are on the right road.

¶ 289.
Universal franchise in the United States.

There are truths which apply equally to all the types of humanity, there are others again which appeal only to particular classes and communities. The first unite mankind upon a common ground, the second divide them into separate classes ; from which we deduce the necessity of toleration, and the duty of not looking askant upon the good fortune of others. We must endeavour to appreciate the good points even of those whom we endeavour in vain to understand ; even if merely from motives of curiosity, for appreciation will often lead to an understanding.

¶ 290.
Necessity of toleration.

I have as yet made but little progress with my subject, and I have already repeated myself many times ; but I have done so intentionally. One has to

¶ 291.
Reiteration of these principles

[232] A commentary to the French and Italian proverbs : " À gens de village trompette de bois," and " A villano dà dono da villano."

[¶ 291]

impress a new idea in much the same manner as one teaches a foreign language ; the words, the idioms, and the principles of grammar must be frequently remembered and repeated, so that the ear may become accustomed to them, and the mind may become familiar with them.[233]

I pass now to the description of the various types.

[233] " *Iterations* are commonly loss of time.   But there is no such Gaine of Time, as to *iterate* often the *State* of the *Question :* For it chaseth away many a frivolous Speech as it is coming forth."—F. BACON, " *Of Dispatch*," 628.

The Elementary Type,

At first the infant
Mewling & puking in the nurse's arms

SUB-SECTION VIII.

ELEMENTARY HANDS. [Plate IV.]

ELEMENTARY
TYPE.

FINGERS big and wanting in suppleness, the thumb short and turned back [as a rule], the palm extremely big, thick, and hard [this last is their most prominent and characteristic peculiarity].

¶ 292.
Characteristics.

In Europe they undertake manual labour, the care of stables and the long programme of coarse work, which may be carried out by the dim flickerings of the light of instinct,—to them belongs war, when there is no personal prowess called into requisition ; to them belongs colonisation, when it is merely a matter of mechanically watering the soil of a foreign land with the sweat of the labouring brow.   Shut in on all sides by material instincts, they attach no importance to political unity, save from the physical point of view.   Their convictions are formed in a groove, which is inaccessible to reason, and their virtues are generally those of a negative description. Governed as they are by routine, they proceed more by custom than in answer to their passions.

¶ 293.
Capabilities of
the type.

In those of our provinces in which these hands abound, as for instance in Brittany and in La Vendée, instinct and custom, which are the bases and mainsprings of genius in the country, preponderate over reason and the spirit of progress, which latter are

¶ 294.
Superiority of
manual labour
in the provinces.

the bases and mainsprings of genius in towns. In the provinces manual labour is more honourable than professional skill.

¶ 295.
Romantic hands.
Gypsies.

Beware of seeking such climes as these, O you who love the ornamental sides of life, silken shoon, and the tinkle of the guitar at night beneath the flower-laden balcony; the races with sallow complexions and flowing locks, with melancholy faces which peer from beneath huge drooping hats, who have left the print of their footsteps in this gloomy country, upon commons decked with the soft green tufts of the broom, and have left behind them nothing more valuable than the Homeric luxury of the cabin of Eumæus.[233a]

¶ 296.
Dulness of the
type.

Strangers to anything like enthusiasm, elementary hands indicate feelings which are heavy and sluggish in rousing themselves, a dull imagination, an inert soul, and a profound indifference. They were much more common among the people of Gaul when the

---

[233a] "But Ulysses found Eumæus sitting in the portico of his lofty dwelling, which was built in a beautiful and spacious position for the accommodation of his swine, out of stones which he had carried thither, and he had crowned it with sloe bushes. And outside he drove in stakes at frequent intervals, and inside he made twelve styes for the swine close to one another. . . . And having brought him (*Ulysses*) in, "he made him rest upon a great thick couch of rushes and wild goat's skin."*

    * " Τὸν δ' ἄρ' ἐνὶ προδόμῳ εὗρ' ἥμενον, ἔνθα οἱ αὐλὴ
    'Υψηλὴ δέδμητο, περισκέπτῳ ἐνὶ χώρῳ
    Καλή τε, μεγάλη τε, περίδρομος· ἥν ῥα συβώτης
    Αὐτὸς δείμαθ' ὕεσσιν, ἀποιχομένοιο ἄνακτος
    Νόσφιν δεσποίνης, καὶ Λαέρταο γέροντος,
    'Ρυτοῖσιν λάεσσι, etc., etc.

    *   *   *   *

    'Ως εἰπὼν ἀνόρουσε· τί θει δ'ἄρα οἱ πυρὸς ἐγγυς
    Εὐνὴν, ἐν δ' οἰῶν τε καὶ αἰγῶν δέρματ' ἔβαλλεν
    'Ενθ' 'Οδυσσεὺς κατέλεκτ'· ἐπὶ δὲ χλαῖναν βάλεν αὐτῷ
    Πυκνὴν καὶ μεγάλην, ἥ οἱ παρελίσκετ' ἀμοιβὰς,
    'Εννυσθαι, ὅτε τις χειμὼν ἔκπαγλος ὄροιτο."

    HOMER, "*Odyssey*," bk. xiv., ll. 5-10; 518-522.

reindeer and the beaver found in it an atmosphere congenial to their organisations, than they are in the present day.

Among the Laps they are in an immense majority, for they escape the inherent evils of the polar latitudes by their absolute inertness.

¶ 297.
Among the Laps.

Organs which are practically insensible can only convey imperfect ideas to the brain. The outer, is merely the reflection of the inner man. As is the body, so is the mind, and *vice versâ*.

¶ 298.
Hands and minds.

In the East Indian Empire, the country of gold and of silk, a blessed clime in which the earth, bathed in the rays of the tropical sun, bursts every year into plentiful harvests, the legislature, in the interests of a community composed of dreamers, of poets, and of enthusiasts, has been obliged to counteract the complete absence of the elementary type,—of the type by which in the north the trades of street porter, butcher, skinner, tanner, roof cleaner, and scavenger are without the least repugnance generally followed.   The Pariah who with us is a natural institution, is in Bengal a legal one, *i.e.*, is the artificial product of a political arrangement.   There is no doubt that without the moral degradation, systematically encouraged by the law in a considerable class of the population, the trades which I have just enumerated, abject but useful as they are, would in India find no hands to take them upon themselves.[234]

¶ 299.
Importance of elementary hands in India.

[234] Grellmann is of opinion that from the Pariah race in India sprang the gypsies which overrun the three continents,* and Philippus Baldæus, in his " *Wahrhaftige Ausführliche Beschreibung der Ost-Indiaschen Küsten, Malabar, etc.*" (Amsterdam : 1672, p. 410), speaks of them as "a wicked crew, who in winter steal much cattle." They belong to what is called a "right-hand caste," having a rather higher position than the lowest "left-hand castes."   M. A. Sherring, in his

* H. M. GRELLMANN, " *Dissertation on the Gypsies*," translated by Th. Raper (London : 1787), chap. vi., p. 168.

¶¶ 300—308.   "The name Soudra" (or Pariah), say the laws of
Manou on the   Manou, " signifies by the first of the two words of which
Pariah.   it is composed *abject servility*, and by the second
dependence.[235]

¶ 301.   " Pork by the smell which it exhales, the dog by
his look, and the Soudra by his touch annuls the merit
of the most holy. act. [Lect. iii., vers. 178-9.]

¶ 302.   "Though a Soudra may be freed by his owner, he
is not delivered from a state of servitude, because
servitude being his natural condition, who can exempt
him from it ? [Lect. viii., ver. 414.]

¶ 303.   " By the law a Soudra cannot hold possession of any
property in his own right ; everything which is his belongs by right to his owner. [Lect. x., ver. 124, *et seq.*]

¶ 304.   " If any one attempts to instruct a Soudra in Holy
Writ he is without doubt everlastingly damned.   The
pulpit, the sword, the palette, commerce, and agriculture are all alike forbidden ground to the Soudra.
[Lect. iv., vers. 80-1.]

¶ 305.   "When a king permits a Soudra to pronounce a
judgment before his face, his kingdom is thrown into
as dire a state of distress as a cow in a morass. [Lect.
viii., v. 21.]

recent volume, " *Hindu Tribes and Castes*" (Calcutta :
1881, vol. iii., chap. iv., p. 130), says : "They are
regarded by the Brahmins as defiling their presence,
and are not allowed to dwell in villages inhabited by
Hindus, but live in their outskirts. . . .   They are a
dark-skinned race . . . intensely ignorant and de-
based. . . .   The Madras Presidency contains nearly
five millions."   Under British rule in India the lower
castes are for the most part free to act as they choose,
so the Pariahs perform many of the sordid occupations,
which no high caste Hindu could undertake.
[235] Lect. ii., vers. 31-32.   M. d'Arpentigny has quoted,
I find, from the " *Manava-Dharma-Sastra ; Lois de
Manou, traduites du Sanskrite*" par A. Loiseleur
Deslongchamps (Paris: 1833); a much better translation
is, to my mind, that of Dr. A. C. Burnell, " *The Ordinances of Manu* " (London : 1884).   *Vide* note [200], p. 182.

"If a Soudra dare to place himself beside a ¶ 306. Brahmin he is branded upon the hips, and forthwith banished, or at least the king must order his flesh to be gashed upon the hips.

"The king must sentence a Soudra who has dared ¶ 307. to advise a Brahmin on a point of duty, to have boiling oil poured into his mouth and ears.

"With whatever member a Soudra strikes a ¶ 308. superior, that member must be mutilated. Thus he will have his tongue cut out for slander, etc."[236]

And so on *ad infinitum.*

Excepting in polar latitudes, real elementary hands ¶ 309. are no longer to be found, save among nations of Tartar or Sclavonic origin. Among the latter races, however, they exist in huge quantities, and in some localities they are to be found without any admixture of the more nobly endowed types. I have lived in the reeking huts of these peoples, devoted, like the solid-ungulous animals, to an eternal serfdom,[237] and I have found them to be as dead alike to all condition of happiness or of misery as the lower animals, with which they share their squalid habitations. In war they are signalised by a brutal and ferocious intrepidity.

Elementary hands in Europe.

[236] Concerning the condition of the Soudra [or Çudra] Madame de Staël has written :—" I read continually a few pages of a book, entitled '*La Chaumière Indienne.*' I know of no more profound study in practical morality than the picture of the condition of the Paria,—of this man of an accursed race, abandoned by the whole universe, crying by night among the tombs; causing horror to his fellow-men without having by any fault deserved this fate, the refuse of this world into which he has been cast by the gift of life. There we have an instance of a man literally thrown upon his own resources, no living creature interests himself in his existence ; there is nothing left for him but the contemplation of nature ; and she satisfies him."—" De l'Influence des Passions." " *Œuvres Complètes de Madame de Staël*" (Paris : 1836), vol. i., p. 167.

[237] *Vide* note [131], p. 122.

¶ **310.**
Cruelty of the Huns.

Such must have been the Huns, those barbarians whom nothing could rouse from their brutal apathy, save the spectacle of great cities in flames, or that of their droves of horses galloping in hideous confusion with the bodies of their victims hanging from their necks, tied by the hands to their manes.[238]

¶ **311.**
Elementary hands in Gaul.

Such were the hands which in Gaul substituted the ordeals by fire, and by water, for the formulas of the accepted jurisprudence,—methods of investigation well adapted to their intelligence and to their physical sensibilities.[239]

¶ **312.**
The Lithuanians. Tacitus.

" As uncivilised as they were unclean, the Lithuanians, in the time of Tacitus, had neither arms nor horses, nor even huts. For sustenance they relied on the herbs of the earth ; for clothing, on the skins of animals ; and for resting-places the earth itself. Their sole means of existence were their arrows, which, in default of iron, were barbed with sharpened bones ; men and women alike followed the chase, and shared its spoils. To protect their children from wild beasts and from the inclemency of the weather, they put them to sleep among the interwoven branches of

[238] " The numbers, the strength, the rapid motions, and the implacable cruelty of the Huns were felt and dreaded and magnified by the astonished Goths, who beheld their fields and villages consumed with flames, and deluged with indiscriminate slaughter." Gibbon's *" Decline and Fall of the Roman Empire,"* chap. 26 (iii.). The whole of this section of Gibbon is full of instances illustrative of the coarseness and barbarous cruelty of the Huns.

[239] " A far more remarkable and permanent superstition was the appeal to heaven in judicial controversies, whether through the means of combat or of ordeal. . . . It does discredit to the memory of Charlemagne that he was one of its warmest advocates " (Baluzii " *Capitularia*," p. 444).—H. HALLAM, " *A View of the State of Europe during the Middle Ages* " (London : 1853), vol. iii., p. 294, chap. ix., pt. i. The recourse to the trial by ordeal was prohibited, we are told by Bouquet [tome xi., p. 430], by Louis le Débonnaire, but traces of the institution lingered as late as the eleventh century.

the trees.   This was their first resting-place in youth, and their last retreat in old age."[240]

One can imagine to oneself the sort of hands which in such a climate as theirs such habits of life would lead one to suppose they possessed.

Still, the influence of such hands was a lasting one, a fact which may be inferred from some of the laws which were still in force in the fourteenth century.   Such, for instance, as that one which ordained that slanderers should be condemned to remain on all fours and bark like a dog for the space of an hour, or that other which decreed that the man who was convicted of having feasted on a fast day should have his teeth broken.

In our country it is to quasi-elementary hands that one owes the existence of those gardens beloved by bees ; gardens filled with thyme, and with vegetables straggling among wallflowers and violets, where the watercresses lie in the running stream ; where the blackbird warbles in the hedge ; where everything flourishes and rejoices beneath the smiling sky.

To them do we owe the villages that we see, perfumed with straw, and with the aroma of the cattle-stall, where all day long one may see huge oxen wandering about the roads, where at every cross-road in front of a spangled niche, the votive lamp of some golden-shoed and scarlet-cheeked saint raises its feeble glimmer ; rustic communities whose

[240] This is a passage taken from the concluding words of the forty-sixth chapter of Tacitus' book, "*De Moribus Germaniæ*," which reads in the original : " Fennis mira feritas, fœda paupertas : non arma, non equi, non penates ; victui herba, vestitui pelles, cubile humus ; sola in sagittis spes, quas, inopia ferri, ossibus aspe-rant.   Idemque venatus viros pariter ac fœminas alit. Passim enim comitantur, partemque prædæ petunt. Nec aliud infantibus ferarum imbriumque suffugium, quam ut aliquo ramorum nexu contegantur : huc redeunt juvenes, hoc senum receptaculum."

[¶ 316]

distinguishing features are the tavern, the row of lime trees, and the dove-cot.

¶ **317.**
Manners and customs of the Turks.

The original manners and customs of the Turks, a nation of mid-Asiatic origin, have hardly received any modifications from time, and we shall presently see the reason of this. Given over entirely to fatalism, and, therefore, to absolutism, they prefer [by reason of a hankering after the liberty of the pristine savage] an arbitrary and absolute form of government, whose action may be termed intermittent, to a government which is regular, and whose action is consequently continuous.[241] They are ruled to-day, as they have been in past ages, exclusively by instinct, looking upon instinct as a gift of God, as if reason was not also a gift of God. They look upon instinct, in fact, as the only infallible guide of human conduct. They fancy, with their vacuous solemnity and gravity, that it is a fitting substitute for everything,—study, reflection, experience, and science.[242] The prime favourite

[241] "The Turks, governed by the principle of fatalism, have much of the passive immobility of the animals themselves."—THEOPHILE GAUTIER, "*Constantinople of To-day*," translated by R. H. Gould (London: 1854).
[242] "One cannot long reside in Turkey without being made aware of the fact that the entire fabric of Mussulman society is founded upon the Qur'án, which claims to be of Divine origin, and, therefore, unalterable. . . . It may well be called a body politic, constructed after the pattern of the Middle Ages, struggling for continued existence amid the blazing light of the civilisation and knowledge of the nineteenth century." — Murray's "*Handbook for Travellers in Constantinople*" (London: 1871, Introduction, p. 30). Of all writers who have discussed and criticised the Turkish character, none have given to European readers a better account than the author of "*Stamboul and the Sea of Gems*" (London: 1852); and Edmond de Amicis, in his volume "*Constantinople*" [translated by Caroline Tilton (London: 1878), p. 127], gives a graphic account of the incomparable indolence of the Turk. "They look," says he [p. 134], "like philosophers all bent on the same thesis, or somnambulists walking about unconscious of

[¶ 317]

Mahmoud II
Achmet Fevzi.

of the civilising Sultan Mahmoud,[243] Achmet Fevzi Pacha, had been in turn cobbler, coffee-house-keeper, water-carrier, boatman, and probably the intimate friend of every stray dog in the Imperial city,[244] when Mahmoud, struck by his prepossessing appearance, took it into his head to make him his pipe-bearer;

the place they are in, or the objects about them. They have a look in their eyes as if they were contemplating a distant horizon, and a vague sadness hovers round the mouth like people accustomed to live much alone and shut up within themselves. All have the same gravity, the same composed manner, the same reserve of language, the same look and gesture."

[243] Mahmoud II., who is always known as the "Reformer of Turkey," was the father of the two late Sultans, Abdul Medjid and Abdul Aziz. To him are due all the reforms which have taken place during the present century, and to him belongs the fame of the world-known extermination of the Janissaries in the Et-meidan [Meat-market] in 1826. We find on p. 434 of Jacob's [et variorum] "History of the Ottoman Empire" (London: 1854) a full catalogue of his reforms. "The reign of Sultan Mahmoud II.," say the authors, "was the most eventful in every way that has occurred in the history of the Ottoman Empire since the commencement of the present century, and that he and the more enlightened of his ministers" [notably Hussein Pacha] "communicated that impulse to the Turkish mind of which the vital proofs are being so conspicuously afforded at this hour," etc. Achmet Fevzi was not the only plebeian who was raised to high dignity by Mahmoud II. Riza Pacha, who was the favourite alike of Mahmoud, of the Sultana Walidé, and of Abdul Medjid, was originally a grocer's boy in the spice-bazaar. In theory every Moslem is equal. Compare also B. Poujoulat, "Voyage à Constantinople" (Paris: 1840), vol. i., p. 219.

[244] "Constantinople is an immense dog kennel; the dogs constitute a second population of the city, less numerous, but not less strange than the first. Everybody knows how the Turks love them and protect them. I do not know if it is because the sentiment of charity towards all creatures is recommended in the Korán, or because, like certain birds, the dogs are believed to be bringers of good fortune, or because the prophet loved them, or because the sacred books speak of them, or

from this he became overseer of the harêm, after
which he was gazetted a colonel of the guard, and sent
as ambassador to St. Petersburg.   To-day sees him
admiral-pacha of the fleet.[245]   What a skilled mariner
he must make [1838].[246]

because, as some pretend, Muhammad the Victorious
brought in his train a numerous staff of dogs, who
entered triumphantly with him through the breach in
the San Romano gate. The fact is, that they are highly
esteemed, and that many Turks leave sums for their
support in their wills."—E. de Amicis,"*Constantinople*"
(London: 1872), p. 108. Compare the passage in the
sixth chapter of the Qur'án: "There is no kind of beast
on earth, nor fowl which flieth with wings, but the same
is a people like unto you ; we have not omitted anything
in the book of our decrees: unto their Lord shall they
return."—*Sale's* translation. Compare also *Savary's*,
and note the passage in chap. viii. [" The Cafés "] of
Th. Gautier, " *Constantinople of To-day* " (London:
1854), upon the kindness of Turks to animals.

[246] Baptistin Poujoulat, in his " *Voyage à Constantinople, etc.* " (Paris: 1840, 2 vols.), gives us an account
of Achmet Fevzi [vol. i., p. 229], which shows us that
M. d'Arpentigny has been a little hard upon him : " Il
y a quinze ans," says he, " qu' Achmet Fevzi était
cafetier dans le vallon des Eaux Douces d'Europe, il
avait amassé assez de piastres pour faire l'acquisition
d'un caique. . . . Un favori du sultan, se promenant un
jour dans le Bosphore, fut frappé de la jolie figure
d'Achmet Fevzi, le prit avec lui, et lui donna la place de
Chiboukji au serail. Il devint ensuite inspecteur du
linge au Harém, puis *bachki* de Mahmoud. Puis il
entra dans le *mabein* ou personnel du grand seigneur.
Dans la mémorable journée de la chute d'Ojak, Achmet
Fevzi donna des preuves d'une grande bravoure et quand
on forma les troupes regulières, le grade de capitaine de
cavalerie de la garde compensa son courage. Achmet
ne resta pas longtemps dans ce grade, il avança successivement en dignité dans l'armée, et parvint jusqu'au
grade de *mouschkir* ou général-en-chef de la garde.
L'Ex-cafetier fut envoyé en ambassade extraordinaire
auprès de l'Empereur de Russie . . . . A son retour de
Saint Petersbourg, Achmet fut nommé grand amiral.
*La manière dont l'ancien batelier du Bosphore a fait
son chemin, est l'histoire de la plupart des hauts personnages de l'Empire Ottoman.*"

[246] A grim commentary on the concluding sentences

In 1821 the Tchobanbachi, or chief shepherd, was elevated to this dignity ; an old man whose hair had grown gray in the enviable quietude of a pumpkin in a frame, who had never done more than count heads of sheep, and who fell into the water the first time he went on board the admiral's flag-ship.[247]

¶ 318.
The Tchoban-bachi of 1821.

M. Fontanier gives the following account of his entry into Sapanja, one of the principal towns of Anatolia [the ancient Bithynia] :—

¶ 319.
Fontanier's entry into Sapanja.

"At last we entered the town, and I took up my abode in a tavern, the sole hostelry to be found in these parts. Once installed, after having arranged my carpet, and having sat myself down cross-legged, holding my pipe in one hand and the inevitable coffee in the other, I entered into conversation with mine host, who lost no time in giving me the customary welcome,[248] and plying me with a string of questions, to which I had by this time become thoroughly

The Osman welcome.

of the above paragraph is afforded by the following sentence from Samuel Jacob's "*History of the Ottoman Empire*" (London : 1854, p. 433): "A few weeks afterwards the fleet, under Achmet Capitan Pacha, was carried to Alexandria by that traitor, and delivered up to the Pacha of Egypt."

[247] From V. Fontanier's "*Voyages en Orient entre-pris par Ordre du Gouvernement Français de l'année 1821 à l'année 1829*" (Paris : 1829), p. 16, "It is needless to say that to become captain-pacha of the fleet, no knowledge of the sea is in any way necessary. To give an instance of this, in 1821 the Tchoban-bachi," etc.

[248] The Turkish salutation is as follows : The visitor bends low, extending the right hand, with which he touches his breast, lips, and forehead, saying, as he does so, "Selam aleïkūm ; " the master of the house simultaneously imitates this motion, repeating, "Ve aleïkūm selam." This is *only* among Turks. To a Christian, or Ghiaour, the ordinary nod of the head and the Turkish "Good-day" [صباحكز خير اولسون افندم], or, صباحكز خير اولسون [صباح شريفكز خير اولسون] is considered sufficient.

acclimatised, and to all of which I had the necessary answers on the tip of my tongue, thus, What is going on ?   What is *not* going on ?   Whence do you come ?   Whither do you go ?   Have you much money ?   Have you a passport ?   Are you a spy ?

"Four or five Turkish travellers, separated from me by a wooden railing which divided the raised floor of the caravanserai into various compartments, listened with indifference, and smoked with imperturbable gravity.   Then each one in turn, without making any further draft on his imagination, gravely put to me the same questions of which they had just received the answers.   On my side it was merely a question of memory, and above all of patience, for had they been twenty I should have had to repeat the same answers twenty times."[249]

¶ 320.
Lethargy of the Turks.

And throughout every class among the Turks we find this same mental lethargy ; only read their romances, only listen to the recital of their dreams, eternally filled with accounts of diamonds by the bushel, and of voluptuous houris by the troop, of hidden treasures suddenly discovered by the aid of some sorcerer whose good graces have been won by some spontaneous act of common hospitality, and you will amply realise that nothing is more repugnant to them than mental effort of any kind, which they avoid making by throwing the blame of all things upon fatality ; and manual labour, against which they protest by their fondness for, and their belief in, talismans and charms.[250]

¶ 321.
Its causes

This manner of looking at things they owe to their physical constitutions which owe their equability to their civil and religious institutions.   They feel them-

---

[249] V. Fontanier, " *Voyages en Orient, etc.*," " Deuxième Voyage en Anatolie " (Paris : 1834).
[250] *Vide* note [1], p. 183.

[¶ 321]

selves that any attempt to regenerate them as a nation would be futile, and that the tide of civilisation [as we understand the word] would be as fatal to them as the waters of the ocean would be to river fishes.[25]

In 1817 some one remarked to Faslé-Bey, Colonel of the Imperial Guard, that the reforms of Mahmoud seemed to be achieving considerable progress. "The Osmanlis," replied he, "remain buried in their prejudices; they are like madmen to whom the right road has been pointed out, but who persist in travelling in a different direction." "But still, one sees many Mussulmans adopting the European costume, and surely that proves that they seek after civilisation." "Those Mussulmans," replied Faslé-Bey, "are like men dressed as musicians, but who have no idea of music. Turkey is at the present moment in a wretched condition; she is like a cistern from which water is constantly being drawn, but into which none is ever put back." "Your opinion of the country is a very despondent one." He replied by this verse of the Qur'án, "Unto every nation is a fixed term decreed; when their term, therefore, is expired they shall not have respite for an hour, neither shall their punishment be anticipated."[252]

¶ 322.
Faslé-Bey on the conservatism of the Turks.

[251] This work bears internal evidence of having been written between the years 1835 and 1838, when the reforms of Mahmoud II. [vide note [249], p. 206] had not yet had time to make themselves felt with all their force as they are now.

[252] This occurs in the tenth chapter of the Qur'án. I do not know whence M. d'Arpentigny obtained his version, which reads : "Aucun peuple ne peut avancer ni reculer sa chute ; chaque nation a son terme fixé ; elle ne saurait ni le hâter ni le retarder d'un instant ; Dieu seul est éternel." The above conversation is condensed from B. Poujoulat's "Voyage à Constantinople" (Paris : 1840), vol. i., pp. 234-5.

¶ 323.
Christianity and
Islam.

If a man is a Christian he *hopes* [positive force]; if he is a Muhammadan he *resigns himself* [negative force].

¶ 324.
Each type is
proud of its own
characteristics.

It is part of the nature of every type to revel in its idiosyncrasies, and to despise and mistrust anything which is foreign to it. [I shall have occasion to recur to this observation.] Where we find the combination of reason and science, intuitive instinct is frightened at itself. Among nations in which the elementary type predominates, they pride themselves on not being able to read or write; and they are taught concerning a god who is the friend of ignorance and poverty of soul.

¶ 325.
Illustrations.

Thus, for instance, in Barbary, the possession of a book is looked upon as a crime,[253] and in Turkey idiotcy is looked upon with reverence, as being something holy.[254]

Among the Kalmuks each family has in its tent a machine called the Tchukor, consisting of a cylinder

[253] Wicked from the point of view of its being a strange and uncanny possession. The Rev. M. Russell, in his "*History of the Barbary States*" (Edinburgh: 1835), says of the inhabitants with every show of reason: —"Since the sun of knowledge rose again in Europe, the shades of intellectual night appear to have fallen with increased obscurity upon all the kingdoms of North Africa" [chap. iv., p. 145]. The same remark applies all over the Ottoman Empire. Dr. Shaw, in his "*Travels relating to Barbary*" [vol. xv. of "*Pinkerton's Voyages and Travels*" (London: 1814), p. 637], remarks :—"As for the Turks, they have no taste at all for learning, being wonderfully astonished at how the Christians can take delight, or spend so much time and money, in such empty amusements as study and speculation. . . . If we except the Koran, and some enthusiastic comments upon it, few books are read or inquired after by those persons of riper years, who have either time or leisure for study and contemplation."

[254] Edmond de Amicis, in his chapter "Galata" [*op. cit.*, note [241], p. 207], gives an account of the consternation produced by the continual appearance of idiots in the streets of Constantinople without any let or hindrance.

¶ 326.
Ka'm ık prayer-wheels

-covered with manuscript prayers and hymns, which is put into motion by means of a mechanical arrangement, which is wound up like a roasting-jack. This apparatus, by turning, blesses and prays to God for the whole family,—an extremely convenient manner of attaining everlasting happiness without making too great an effort over the process.[255]

[255] " Perhaps the most marvellous invention which the Thibetan has devised for drawing down blessings from the hypothetical beings with which his childish fancy has filled the heavens, are the well-known praying wheels,—those curious machines which, filled with prayers, or charms, or passages from holy books, stand in the towns in every open place, are placed beside the footpaths and the roads, revolve in every stream, and even [by the help of sails like those of windmills] are turned by every breeze which blow over the thrice sacred valleys of Thibet."—T. W. RH. DAVIDS, "*Buddhism, being a Sketch of the Life and Teachings of Gautama, the Buddha*" (London: 1882), ch. iii., p. 210. Every Kalmuk or Thibetan Buddhist has also a private prayer-wheel of his own, which, being filled with a tight roll of parchment, on which is inscribed many thousands of times the formula, " *O Buddha, Jewel of the Lotus!*" he constantly revolves from *right to left*. A friend of mine tried very hard to purchase one of these apparatus, but in vain, for the owners believe that, should anyone turn them from *left to right*, all their prayers will proportionately be cancelled! *Vide* also concerning this curious superstition Huc and Gabet's "*Voyages*" (Paris: 1850, vol. i., p. 324); General Alexander Cunningham's "*Ladak*" (London: 1854), p. 374; Davis in the *Transactions of the Royal Asiatic Society*, vol. ii., p. 494; Klaproth's "*Reise in den Kaukasus*" (Halle: 1812-14), vol. i., p. 181; and Jas. Ferguson's "*Tree and Serpent Worship*" (London: 1868). "So also," says T. W. Rh. Davids [*op. cit.*, p. 210], "these simple folk are fond of putting up what they call 'Trees of Law,' that is, lofty flagstaffs with silk flags upon them, blazoned with that mystic charm of wonder-working power, the sacred words, '*Om Mani padme hum*' ['Ah! the jewel is in the Lotus']. Whenever the flags are blown open by the wind, and 'the holy six syllables' are turned towards heaven, it counts as if a prayer were uttered, not only upon the pious devotee at whose expense it was put up, but also upon the whole country side."

¶ 327.
National
characteristics.

Some nations there are, who have left behind them a glorious reputation for superlative horsemanship, such, for instance, as the Parthians,[256] the Persians,[257] the Thessalians,[258] etc., or for having left behind them the most stupendous and indestructible monuments, like the Cyclops, Egyptians,[259] etc., or for having lived free, and valiantly maintained the democratic form of government, *e.g.*, the Athenians. Again, we know concerning the Sybarites that they dressed horses to perfection ; that theirs was a republican form of government ; that they were adepts at the precise hewing and elegantly magnificent super-imposition of huge stones. Nevertheless the word

Sybarites.

" Sybarite," classed in the present day among de-rogatory epithets, is no longer applied to a man excepting as an insult. Whence comes this state of things ? Is it because they slept upon beds of roses ? A moment's reflection will show us that, besides the fact that such beds could not be in very general use, a bed of roses could not be more com-fortable than one of straw, and would be more a

[256] The superlative horsemanship of the Parthians, and their custom of firing their arrows whilst pretending to fly [whence our term " a Parthian shot"], have been a theme to almost every poet and prose writer of anti-quity. Thus we have the passages in Horace, book i., Ode 19, v. 11, and book ii., Ode 13, v. 17 ; Herodian, " *Historiarum Romanarum*," book iii. ; Lucan, *pas-sim* ; Virgil, " *Georgic.*," iii., l. 31 ; " *Æneid*," vii., l. 606.

[257] Referring no doubt to the " *immortal* guard " of 10,000 horsemen who were attached to the person of the king of Persia.

[258] The Thessalian cavalry was, after the Parthian, the finest in the ancient world.

[259] " The most solid walls and impregnable fortresses were said, among the ancients, to be the work of the Cyclops, to render them more respectable " (*Lemprière*). The builders of the Pyramids of Thebes, of Carnac, and of Luxor, hardly require a note to illustrate their masonic capacities.

matter of experiment or of pageantry than of effemi-
nacy or sensuality.   No, the Sybarites, a rich and
a civilised nation, having been overwhelmed and
destroyed by barbarians, were slandered by their
conquerors,[260] who execrated in them all the instincts
of civilisation in the same way that the Cimbrians
and the Teutons,* who were overwhelmed by civilised conquerors, have been calumniated by their
exterminators, who loathed in them their barbaric
instincts.[261]

Cimbrians and
Teutons.

Like that of the other types, the Elementary Type,
whilst it remains irresistibly attached to the tendencies

¶ 328.
Modifications c
the type.

[260] The luxury and fastidiousness of the Sybarites have
been chanted by many a writer; by none more than by
Ælian, who, in his accounts of Smindarides ["*Var. Hist.,*"
ix. 24, and xii. 24], aims at the whole nation.   Of their
valour, however, and of their power as a nation, there
is no question.   Sybaris was finally reduced, 508 B.C.,
by the disciples of Pythagoras, after a long and vigorous
resistance against the town of Crotona.
[261] The brazen bull to whom the Cimbri and Teutons
were in the habit of making their sacrifices of human
blood is thus mentioned by Plutarch in his "*Life of
Caius Marius*":—"The barbarians now assaulted and
took the fortress on the other side of the Athesis, but
admiring the bravery of the garrison, who had behaved
in a manner suitable to the glory of Rome, they dismissed
them upon certain conditions, having first made them
swear to them upon a brazen bull.   In the battle which
followed this bull was taken among the spoils, and is
said to have been carried to Catulus' house, as the first-
fruits of the victory" [*Langhorne's Translation*].
Strabo ["*Geography,*" book vii., cap. ii. (3)] describes the
rite of the Cimbrians sacrificing their prisoners, and
catching their blood to draw auguries from, but does
not mention the brazen bull.   The best quasi-classic
account which we have of the Cimbri and Teutons is
Christophorus Cellarius' "*Dissertatio Historica de
Cimbris et Teutonis*" (Magdeburg: 1701).

---

* They used to sprinkle with human blood the altar of a
brazen bull, their principal idol, but barring this they were just,
content, brave, and devoted to their leaders and to their friends.
—AUTHOR'S NOTE.

[¶ 328]

of its nature, modifies and transforms itself according to the times and places in which it lives.

¶ 329.
Personifications of national characteristics.
—
Elementary types.

Greece, still in a condition of barbarism and governed by instinct [as is natural to every society which exists in a state of syncretism], saw the embodiment of its idiosyncrasies in the formidable features of Polyphemus ; later on she saw it adorned with the natural grace and repose of rustic moralities. Caliban in rough and foggy England, Mælibæa beneath the scented pines of the Sabine hills, Sancho amid the joyous turmoil of the Castilian hostelries, are after all merely different re-incarnations of this same idea.

¶ 330.
Illustration.
Conico-elementary.
General Rapp.

The General Rapp seems to me to have been the best expression of the conico-elementary type, as it has manifested itself among the upper classes of our society under the Empire. He was a man in appearance round, broad, highly-coloured, and of striking individuality, with manners at the same time sumptuous and rustic, theatrical and soldierly, who required *either* a luxurious bed and a delicate arrangement of furniture, *or* a truss of hay and a wooden spoon. In Dantzig, in 1812,[262] where we used to call him the Pasha on account of his pomp and peculiar order of merit [that of the sword], he liked to drive about in an open vehicle, magnificently dressed, lounging rather than sitting with his mistress, an affected German with prominent cheek bones, to whom his inferior officers used to pay court as if she had been a queen. His magnificent feasts, at which there figured daily a hypocritical and despised dish of boiled horse-flesh, were an insult to the miseries

[262] General Jean Rapp [b. 1772, d. 1821], was re-instated in the command of Dantzig in December 1812 by Napoleon. Here he sustained one of the most memorable sieges of the century [January 1813.—January 1814].—LAROUSSE "*Dictionnaire du XIXe Siècle,*" Art. "Rapp.") See also Martin's "*Histoire de France*" (Paris : 1879), vol. iv., p. 67.

of the soldiers, to whom he would habitually and willingly give money with his own hands, but whom he abandoned in his carelessness to the rapacity of writers, and of commissariat agents. At the theatre, where the subaltern epaulette could only gain admission to the pit, eight or ten boxes were ornamented with the colours of his startling and vividly-insolent livery. In the same way that his own henchmen, beneath the rustling plumes of their aigrettes of cock feathers, had his name always on their lips, the name of the Emperor, his master, was always on his; he owed his promotion, firstly, to his exalted fetish-worship of the Hero of November;[263] secondly, to his grand audacity; and, lastly, to a kind of rough flattery, seasoned with a kind of capricious good-fellowship, which he used to great advantage. Without any cultivated talents, but not without tact and subtlety, he had recourse in all circumstances of difficulty to a faculty for dissembling which he summoned to the aid of his incapacity and of his ignorance. Such, however, was his opinion of himself that he considered us well paid for our hardships, and considered himself quit of any obligation to us for our troubles, when he had said to us on parade " that he was

[263] Napoleon was known as " L'Héros de Brumaire," because it was on the 9th November, 1799, that he brought about the *coup d'état* from which he emerged First Consul, and which was the first step towards the establishment of the empire. *Vide* H. Martin's " *Hist. Franc.,*" vol. iii., pp. 80-88. Rapp, in his own Memoirs [*"Mémoires du Général Rapp"* (Londres : 1823), chap. i., p. 4], says concerning the friendship which existed between him and Napoleon :—" Zeal, frankness, and a certain aptitude in arms, gained for me his confidence. He has often said to those around him, that it would be difficult to have more natural good sense and discernment than Rapp. They repeated these praises to me, and I confess that I was flattered by them. . . . I would have died for him to prove my gratitude ; and he knew it." On pp. 217-90 he describes the siege of Dantzig.

[¶ 330]   satisfied with us." Beyond these peculiarities he was a good man, hating set speeches, serviceable, always ready with similes and parallels, and generous often to a fault.

¶ 331.
Susceptibilities of the type.

Elementary hands are; for the most part, more accessible to the charms of poetry than to those of science. It was to the lyric measures of Orpheus, and to the harmonies of the flute of Apollo, that in the old Greek world the first communities of men were formed, and the first towns were built.[261]

¶ 332.
Superstitions of the type

In the depth of the forest, or on the deserted sea-shore, by night, when the boundless ocean moans with the murmur of the tempest, hands which are elementary are the more troubled by phantoms, spectres, and pallid apparitions, in proportion as their finger-tips are more or less conical. But whatever be the form of the terminal phalanx, the type is always much influenced by superstition. Finland, Iceland,[265] and Lapland abound with wizards and sorcerers.

¶ 333.
Weakness of the type.

Elementary-handed subjects, whom neither inertia nor insensibility have been able to protect from pain and sorrow, succumb the more readily to their attacks, from the fact that they are generally entirely wanting in resources and in moral strength.

[261] In allusion, I presume, to the legends that Orpheus was one of the *Argonauts*, from whom the Lemnian and other races are said to have sprung; and that Apollo, the god of music, is said to have assisted Neptune in raising the walls of Troy.

[265] Icelandic legends are full of tales of sorcerers and elves. Several occur in Mr. C. Warnford Lock's volume, "*The Home of the Eddas*" (London: 1879), which may be taken as specimens of the kind of legend which is most popular. The curious in such matters should consult also Dr. Wagner's "*Manual of Norse Mythology. Asgard and the Gods, the Tales and Traditions of our Northern Ancestors*," adapted by M. W. Macdonall, and edited by W. S. W. Anson (London: 1884), which is a complete epitome of Norse, Finnish, and Icelandic superstitions.

The Spatulate Type.

And then the whining school...

with his satchel

And shining morning face

Creeping like a snail unwillingly to school

A B C

SUB-SECTION IX.

SPATULATE HANDS. [Plate V.]

In this chapter I propose to deal only with spatulate hands whose thumbs are large—that is to say, with those in which the instinct which is peculiar to these hands, supported by the promptings of the brain, makes itself the most clearly manifest. The intelligent reader, after what I have said anent small thumbs, will be able easily to appreciate the intricacies of the mixed types which form themselves upon the groundwork of a spatulate hand.

¶ 335. Large-thumbed subjects.

By a spatulate hand I mean one, the outer phalanges of whose fingers present the appearance of a more or less flattened-out *spatula* (to borrow a term from the dispensing chemist). [*Vide* Plate V.]

¶ 336. The spatulate hand.

The spatulate hand has undoubtedly its origin in the latitudes where the inclemency of the climate, and the comparatively sterile nature of the soil, render locomotion, action, movement, and the practice of the arts whereby the physical weakness of man is pro-

¶ 337 Its origin.

15

[¶ 337]

tected, more obligatory upon man than they are under more southern skies.

¶ 338.
Its principal
characteristics.

Resolute, rather than resigned, the spatulate hand has resources for the resistance and conquest of physical difficulties, of which the conic hand is absolutely ignorant. The latter, more contemplative than active, prefer [especially among meridional nations] the ills to which the flesh is heir, to the exertions requisite to overcome them. The self-confidence of spatulate subjects is extreme ; they aim at *abundance* and not [like the elementary subjects] merely at *sufficiency*. They possess in the highest degree the instincts of *real* life ; and by their natural intelligence they rule matters mundane, and material interests. Devoted to manual labour and to action, and consequently endowed with feelings which are more energetic than delicate, constancy in love is more easy to them than it is to minds which are poetic, and which are more attracted by the charms of youth and beauty than by a sense of duty and by ethics. All the great workers, the great navigators, the great hunters, from Nimrod to Hippolytus and Bas-de-Cuir, have all been renowned for their sobriety and continence.[266]

¶ 339.
Parallel in the
Greek mytho-
logy.

Of the goddess of the dazzling forehead, Diana, the white-footed, the finely-formed, whose immortal life devoted to the chase is spent in the liberty and activity of the woods, the Greeks made a personification of chastity.[267]

[266] I do not know upon what episodes Nimrod, "the mighty hunter before the Lord" [Gen. x. 9], and Bas-de-Cuir base their claims to sobriety and continence. The Hippolytus mentioned is, I presume, the son of Theseus and Hippolyte, the Joseph of profane literature. Compare the passages in the 3rd book of Ovid's "*Fasti*" [l. 268], and in the 7th "*Æneid*" [l. 761, etc.].

[267] According to Virgil [*loc. cit.*], Diana restored to life Hippolytus, named in the preceding note, as a reward for his exemplary chastity.

With smooth fingers, spatulate subjects like comfort as well as elegance, but their elegance is of a fashionable rather than an artistic kind.

Our colonists of the Antilles, people for the most part luxurious and sparkling, who find delight in movement, in dissipation, in dancing, and in billiard and fencing saloons, who love to struggle for the mastery with vicious horses, whose sole amusements are hunting, fishing, and conquests in love,—these colonists, I say, necessarily the descendants of adventurers distinguished by a love of hazard and of action, have, probably almost all of them, hands like those of our circus riders and of the satellites of our jockey-club stables, that is to say, smooth-fingered and spatulate.

Larger spatulate hands are much more numerous in Scotland than they are in England, in England than they are in France, and in France than they are in Spain, and in mountainous than in flat countries.

The painter Ribera, whose natural bent always led him to paint more or less ugly people, always gave the people he painted [as also did Murillo and Zurbaran] fingers which were more or less pointed ; which would certainly not have been the case had it not been that the generality of the hands that he saw around him, being thus pointed, gave him as it were a law in the matter. Big, square, and spatulate hands abound, on the contrary, in the pictures of the Dutch and German schools.

In Spain it is in Galicia and Asturia that one sees the most spatulate hands, and it is from this rocky and mountainous district that all the muleteers, and all the labouring people that one finds in the peninsula originally come.

The Kabyles inhabit, as we know, the slopes and valleys of the Atlas range ; they are the most spatulate-handed, and also the most hardworking people to

[¶ 345]

be found in Algeria. The Bedouins of the desert, an indolent and ferocious race, are hardly more than shepherds, and their hands are enormous. Superstition is the sole sentiment by which they are in any way affected.

¶ 346.
Swiss hands.

The Swiss are actuated alike by the love of labour, patience, and obstinacy. They are a race of people only very slightly poetic, upon whom God, who has placed them upon a soil which is subject to landslips and avalanches, has bestowed by way of compensation a love of mechanics and dynamics.[268]

¶ 347.
Russians and Cossacks.
Elementary and spatulate hands.

Among the Russians the elementary hand is the most common, and among the Cossacks [who are a Mongolian race], the spatulate. The Russians lead sedentary lives, and travel in carriages; the Cossacks lead an active life, and travel on horseback. The Russians are mercers, innkeepers, shopkeepers, and bankers; the Cossacks are artizans, and construct for themselves the implements and utensils of which they make use. The Russians owe their military glory to their discipline, it is a characteristic which distinguishes them as a *race;* among the Cossacks, who aim at renown as the result of personal prowess, glory is a thing which attaches only to *individuals.*

¶ 348.
Pre-eminence of the type on colonists.

For the following reasons the most stable colonies are formed by spatulate-handed people rather than

[268] At this point we find in the original a paragraph which reads as follows :—"'The wants of man,' says Lady Morgan, 'are his most powerful masters, and the means adopted for the satisfaction of his wants are infallible indices of the real position of a nation upon the ladder of civilisation ; for the highest possible point of social refinement is no more than a more perfect development of certain physical resources; and the most lofty aspiration of human knowledge is simply a more judicious application of the faculties which have been given us for the support of our existence.'" I do not know the works of the "Lady Morgan" referred to, so I have no means of identifying the passage.

by others:—(a) Almost insensible to art or to poetry, they are endowed with a very small share of the instincts which lead to moral instability. (β) They attach themselves to a country merely for the material benefits which accrue to them therefrom. (γ) Manual labour is agreeable to them rather than antipathetic; and it is the same with all kinds of active exercise. (δ) They suffer from the absence of abundance, but not from absence of the superfluous, for they are only slightly sensual. They are more greedy than epicurean, and you will find among them more faithful husbands than gallant "sigisbés,"[269] more Frères Jean than Panurges.[270] (ε) Their love of locomotion renders them comparatively insensible to the annoyances, I will not say of exile, but of expatriation. (ζ) Accustomed as they are by the multiplicity of the wants which assail man in our northern latitudes [their indigenous habitations] to rely principally on their own exertions, they have no innate objection to solitude. (η) Finally, they are apt at sciences which are merely those of physical necessity, and which in ordinary life affect only those things which are constant and immovable.[271]

France, after having populated with hands of this description, Canada, and certain districts of Louisiana,

¶ 349.
Effect of emi-
gration on
France and
Spain.

[269] This word [the Italian *cicisbeo*], derived from the old French *chiche* and *beau*, meaning the *cavaliere servente* of the Middle Ages, is rapidly becoming [socially *and* philologically] extinct. A most interesting article upon the word is to be found in Larousse's "*Dictionnaire du XIXᵉ Siècle*" (Paris : 1866-77).

[270] M. d'Arpentigny has, I think, adopted extreme illustrations in citing Frère Jean and Panurges as examples of the glutton and of the gallant. *Vide* ¶ 215, and note [169], p. 157.

[271] Compare with these paragraphs Bacon's Essay " *Of Plantations*" [1625], and the very interesting remarks on colonists, which are recorded in S. T. Coleridge's " *Table Talk*," under date 14th August, 1831.

felt [her feelings have often saved her by checking her ideas] that she could go no further in that direction without injuring herself; and Spain, having consigned nearly all her hands of this description to America, and having thus deprived herself, not only of soldiers and agriculturists, but also of that moral counterpoise which the ideas which these hands represent, afford to ideas which are mystic, sensual, artistic, and poetic, has only just stopped short of absolute extinction from physical and moral exhaustion.

¶ 350.
North Ameri an hands.

Whence comes the severely practical commonsense of the North Americans, if it is not from these "working hands," scattered over a space which they can comprehend without enervating themselves, and from resting their faith upon institutions which harmonise with their instincts ?

¶ 351.
Dutch and Spanish hands contrast.d.

Had it not been for the intervention of northern genius by the hands of the Flemish and Walloons in the affairs of Southern Europe in the sixteenth century, the glory of Charles V., deservedly great as it was, might perhaps have been nothing more than that of an ordinary victorious prince.[72]   Certainly Spain owes much of her solidity to the qualities which she found among her Flemish, and which were absolutely wanting among her Spanish subjects.   To this day these two nations are distinguished by the most startling contrasts: the Spaniards are prompt and

[72] Charles V. of Germany, being the grandson of Ferdinand and Isabella of Spain, took possession of the Spanish throne as Charles I. of Spain on the death of his grandfather in 1516, his grandmother being mentally deranged.   He ascended the throne of Germany in 1519, and was one of the most victorious and chivalrous princes that the world had seen since Alexander the Great.   It was in his reign that the celebrated diet at Worms was held [1521], at which Luther made the great declaration of his principles.   He relinquished the crown of Spain in 1556 to his son Philip II.

violent, but constitutionally indolent ; the Dutch, on the other hand, are slow and cold-blooded, but constitutionally laborious ; the Spanish are only stubborn under the influence of passion, the Dutch are so only under the influence of self-interests. Beneath an appearance of inertia which is almost stupid, the Dutch hide an extremely keen intelligence of positive facts ; the Spaniards beneath an air of phlegmatic gravity, conceal an imagination constantly running upon excitements and adventures. The Dutch can only thoroughly comprehend what appertains to real life, and they take a special pride in wanting for nothing ; the Spanish can only appreciate phases of life which are romantic and contemplative, and pride themselves on being able to do with very little. Before the gifts of hard work, such as science and wealth, which are so dear to the Flemish, the Spaniards esteem those of chance, such as beauty, valour, genius, and good birth.

It was with his large, square-handed Dutchmen that Charles V. established, turned to good account, and organised the countries which he had conquered with his thin and pointed-handed Spaniards.[73]

¶ 352.
Combination of the two.

In proportion as the love of arbitrary facts forms the basis of the instincts of every man who is fond of material power [just as *action* requires liberty for its exercise, and spatulate-handed people are always active, or at any rate restless], so is liberty, wherever they are in majority as they are in England and the United States, a political institution ; a fact which does not prevent, but rather proves, that of all people in the world the English and the Americans are the most prone to exclusiveness and individuality.

¶ 353.
Liberty of the type.

M. Dupin the elder, whose motto is, " Every man

¶ 354.
Dupin ainé.

---

[73] *Vide* the preceding note.

[¶ 354]

for himself and by himself," has large, ugly hands, which are knotty and spatulate.[274]

¶ 355.
Predilections of the type.

It is from the restless crowd of spatulate subjects that we get these eternal gaugers and everlasting measurers, whose admiration for works of architecture measures itself by the greater or less extent of the surface of those monuments; their instinct of grandeur is not in the form, it is in the number; they are governed by arithmetic. That which does not astonish them [and they are not easily surprised] does not please them, but you will always find them going into ecstasies over those colossal monoliths, whether ornamented or not, whose unearthing, whose transport, or whose erection awakens in their minds ideas of muscular effort and mechanical industry, which are pleasant to them.

¶ 356.
The artist and the artizan.

In the north, where spatulate and square hands are in a majority, the artist is swallowed up by the artizan; in Italy, in Spain, and even in France the artizan is effaced by the artist. In the north there is more opulence than luxury, in the south there is more luxury than wealth.

¶ 357.
Habitations of the type.

You are a man of cultivation, but still you do not particularly care for the beauties of architecture and

---

[274] A. M. Jean Jacques Dupin, known as Dupin Aîné, to distinguish him from his brothers Charles and Philippe, was born in February 1783, and died in November 1865. He was one of the most eminent lawyers that France has ever known, though his reputation has mainly survived as a politician of the "Vicar of Bray" school, by reason of an elastic political conscience, which gained for him the appellation of " le chamélion political." He was elected a *député* in 1827, and became a member of the council of Louis Philippe in July 1830. In 1832 we find him President of the Chamber of Deputies, and in 1848 President of the Legislative Assembly. After the *Coup d'Etat* of 1851, he disappeared for a while, re-appearing, successful and energetic as ever in a new cause, on the restoration of the house of Orleans in 1857.

[¶ 357]

of antique monumental sculpture. A town is fine in your eyes, if it is divided by long straight streets, cutting each other at right angles; if it has symmetrical squares filled with uniformly-built houses, and public gardens planted with regularly-trimmed trees. As for statues, you do not find them necessary, and you can do without marble basins, columns, and caryatids; but you like green shutters, neat pavements, and white walls, whose painted doors are ornamented with shining brass knockers. You require that the town, regular and prosperous, shall suggest to the observer respectability, felicity, and order. It has been built and decorated by people notable for good sense rather than for imagination. Nothing that is either useful or even comfortable is missing, but the "family fool" will seek there in vain for his Divine pastures—poetry. Well, these predilections announce in you a hand which is spatulate or square; it is in England, in Belgium, or in the north of France, countries where your type predominates, and where consequently its concomitant genius is alone appreciated, that you must fix your abode.

Start thither at once! and may this inexorable symmetry rest lightly upon your soul!!

SPATULATE HANDS [*continued*].

National Characteristics and Hands.

¶ 358.
Increase of the type.

WHEN society is dominated by a single idea, the men who embody and personify that idea naturally achieve power and wealth, and with power and wealth they win women of their own kind. Thus the idea becomes supported by a number of adherents who are bound to it by ties of organisation and relationship, and becomes far more powerfully-supported than it would. have been if the idea had not already become powerful. It is a well-known axiom in stud-farms that the horse generally [though not always] hands down with his physical form his mental intelligence to his progeny. To a certain extent the same remark applies to men.

¶ 359.
Effect of exclusiveness.

Left alone, men remain at a standstill, like the peoples who prevent the infusion of foreign blood into their veins by the enforced seclusion of their womankind, and by the separation of their race into *castes.*[275]

¶ 360.
Progressive and unprogressive nations.

On the one hand, nations among whom these usages do not exist have made progress by their wars [which have, by a great poet, been described as the motive-power of the human race] and by invasion ; whilst, on the other, those who have been

[275] As, for instance, the Turks [*vide* note [242], p. 206] and the Indians.

deprived of the salutary influences of the infusion of foreign blood have degenerated, instead of transforming themselves, as a natural result of the constant reiteration of the same causes. Distinguished only by a downward tendency, the people of the Indies are to-day practically what they were in the time of Alexander.[276]

¶ 361.
Permanence a sign of excellence.

And we may remark by the way, that if the force of the genius of a man may be measured by the greater or less permanence of his work, what admiration we must feel for the high and penetrating intelligence which constructed the yoke beneath which, for five thousand years, succeeding generations of Hindoos have consented to bow.[277]

[276] If proof were wanted of this statement, it might be found in the Laws of Manu, of which a scholarly English translation and edition has been made by Dr. A. C. Burnell:—"*The Ordinances of Manu; translated from the Sanskrit*" (London: 1884). These laws, by which the entire native populations of the East Indies are governed, are of incomparable antiquity, though it is probable that the first regular codification and tabulation of them took place somewhere between Anno Domini and A.D. 700. The question with all the evidence is discussed at length in Dr. Burnell's translation above cited; in India they refer their origin to 1250 B.C. and this date was accepted as correct by Sir William Jones. (Professor Monier Williams, in his "*Indian Wisdom*," London: 1875, p. 215), put the date at 500 B.C.; Johaentgen ["*Ueber das Gesetzbuch des Manu*" (Berlin: 1863), p. 950] dates them at 350 B.C.; and Schlegel at 1000 B.C. Following, however, the line of argument adopted in his *Introduction* by Dr. Burnell, his date is, I should think, approximately correct for the tabulation, but I understand from a Brahmin of very advanced Oriental and European education that their origin is probably correctly dated by Sir William Jones.

[277] *Vide* the preceding note. The origin of Buddhism is lost in the shadows of the remotest antiquity; as much as can be known on the subject may be gathered from Rhys David's "*Buddhism*" (London: 1882), or from Spence Hardy's "*Manual of Buddhism*" (London: 1853 and 1880).

¶ 362.
Changeable and
· unchangeable
laws.

In the United States, whither new people continually betake themselves from all parts of the world, the laws undergo, every year, modifications suggested by the changes which have taken place in the temperament of the nation,[278] whereas in China and Japan, empires hermetically sealed against any influences of foreign extraction, the laws [however important or un-important they may be] remain stationary, like the national wants from which they take their origin,—a state of things brought about by the unchanging nature of the national organisation.[279]

¶ 362a.
Definition of
law.

These things prove the correctness of Montesquieu's definition of law : " Law," says he, " is a necessary relation. among men, resulting from the existent nature of things."[280]

[279] It will probably occur to the reader that this might be said of any nation among whom the constitutional form of government exists.

[279] The present centralised form of government was substituted for the previous feudal form about 2,000 years ago ; Thomas Meadows, in his work " *The Chinese and their Rebellion* " (London : 1856, ch. ii., p. 22), says :— " All Chinese law is carefully codified and divided into chapters, sections, and sub-sections. Some parts of this law are as old as the Chinese administrative system itself [*v. sup.*]. One of the oldest, and by the people most venerated, of the codes is that which most nearly concerns themselves—the penal. This, commenced 2,000 years ago, has grown with the nation," etc.   Sir Rutherford Alcock, in " *The Capital of the Tycoon* " (London: 1863, ch. ii., p. 62), gives us the same informa-tion concerning the antiquity of these laws [p. 223], and makes on the subject of Japanese legislation the following remark [p. 410] :—" A land so strangely governed by unwritten laws and irresponsible rulers—I say unwritten, for though the ministers tell me a written code exists, I have been unable to obtain a copy. A country without statute law or lawyers does seem an anomaly with a civilisation so advanced."

[280] This aphorism occurs in Montesquieu's book " *De l'Esprit de la Loi*" (Paris: 1816, liv. i., ch. i.), where he says :—" Laws, in the widest acceptation of the term,

The harsh and inflexible laws of Sparta were well adapted to the spatulate descendants of the Heraclides, just as the elastic laws of the Athenians were well framed for the brilliant and mobile genius of that nation,—that nation "enslaved by the extraordinary and loathing the commonplace; who in their fondness of fine oratory, trusted their ears even more than their eyes, and who, not reasoning with ordinary perspicacity upon any of the things which concerned them nearly, always led away by brilliant speech, and thus, as one might say, drawn about by the pleasures of their ears, seemed in their assemblies to be spectators ranged in a theatre to hear a sophistical discussion, rather than citizens deliberating upon matters which concerned the state."[281]

¶ 363.
Sparta and Athens contrasted.

M. Souvestre tells us that in the district of Léon, in Brittany, there are some villages whose inhabitants live a life which is a continual round of activity, excite-

¶ 364.
Breton villages Their uniformity.

are the relations rendered necessary by the nature of things; and in this sense all living things are subject to laws."* M. d'Arpentigny has not transcribed as correctly as he might have done.

[281] The above paragraph is the substance of the 38th chapter of the 3rd book of the "*Peloponnesian War*" of Thucydides, in which Cleon tells the Athenians that their orators, who wish to stir up the question of the Mytileneans again, must either maintain a paradox to display their talents, or must be bribed to make the worse appear the better cause. He tells them that it is their own folly that gives them encouragement; their passion for novelty; their admiration of talent; tempting the orators to labour to gratify their craving for intellectual excitement rather than to propose sound sense to them in simple language.†

---

* "Les lois, dans la signification la plus étendue, sont les rapports nécessaires qui dérivent de la nature des choses; et dans ce sens tous les êtres ont leurs lois."

† Ζητοῦντές τε ἄλλο τι ὡς εἰπεῖν ἢ ἐν οἷς ζῶμεν, φρονοῦντες δὲ οὐδὲ περὶ τῶν παρόντων ἱκανῶς· ἁπλῶς τε ἀκοῆς ἡδονῇ ἡσσώμενοι καὶ σοφιστῶν θεαταῖς ἐοικότες καθημένοις μᾶλλον ἢ περὶ πόλεως βουλευμένοις."—ΘΟΥΚΥΔΙΔΟΥ ΣΥΓΓΡΑΦΗΣ Βιβλ. Γ΄ (38).

ment, and holiday-making, whilst there are others
which are entirely populated by people who are
chronically melancholy, discontented-looking, and
morose ; and he attributes this state of things very
reasonably to the religiously-followed custom of the
exclusive inter-marriage of people of the same
village.[282]

¶ 365.
Ideal
communism.

Communism, as understood and defined by certain
theoreticians of the present day, might be practicable
in a small nation whose racial idiosyncrasies are kept
pure from the influences of foreign blood. Indeed,
what was the government of Sparta at the bottom,
but a kind of wisely-organised communism—well,
the Spartans alone, of all the Greek tribes, never
admitted a stranger to the freedom of their city.[283]

[282] Emile Souvestre, "*Les Derniers Bretons*" (Paris :
1854, vol. i., p. 20), speaking of the inhabitant of
Léon, says :—" His joy is serious, and only breaks out
in flashes, as if in spite of himself. Grave and concentrated, he shows but little interest in his dealings with
the external world." And later on [p. 45], speaking of
the inhabitant of Cornonaille, he says :—" As in the rest
of Brittany, the religious taint is perceptible, it is
mingled everywhere with a light-hearted gaiety. I have
already said that it is in the festive solemnities of life,
rather than in lugubrious ceremonies, that you must
seek his character. Poetic and bright in pleasure, he is
awkward and trivial in sorrow; it would seem as if the
' Léon-ard ' and he had shared life between them, — to
the one joys, and to the other sadness. Thus, if you
go into the neighbourhood of Léon, ask to assist at a
deathbed scene or a funeral ; but if you are among the
mountains of Artres, go in for betrothals and wedding-
breakfasts." It is evidently from these passages that
M. d'Arpentigny has taken this paragraph.

[283] This is what we find in Plutarch's " *Life of
Lycurgus.*" He says :—" He would not permit all that
desired it to go abroad and see other countries, lest
they should contract foreign manners, acquire traces of a
life of little discipline, and of a different form of government. He forbade strangers, too, to resort to Sparta
who could not assign a good reason for their coming,

The strange epoch which extended from the ninth to the twelfth century, was essentially that of hard, spatulate hands. On the death of Charlemagne, who had endeavoured to reconstruct the principles of Roman civism, the hard, spatulate subjects relapsed into the *individualism* which characterises them. The state of society which they brought about, a society divided into innumerable little groups, each independent of the others, could only realise the meaning of an idea by the examination of its outward and visible form.[84] Each group had its peculiar leader, catchword, device, and standard ; every profession had clothes peculiar to itself. Without these external signs all would have been chaos, for, when all are enveloped in an equally dense haze of ignorance, order and civilisation [when they exist] are apparent in matters material rather than in matters ideal. Besides, all these hands, clothed as it were in gauntlets of brass, aspire to command ; they desire, they search after war,—war, or else its simulacra, tournaments and the chase. To them are these long

not [as Thucydides says] out of fear that they should imitate the constitution of that city, and make improvements in their own virtue, but lest they should teach his own people some evil. For along with foreigners come new subjects of discourse ; new discourse produces new opinions ; and from these things necessarily spring new passions and desires which, like discords in music, would disturb the established government. He, therefore, thought it more expedient for the city to keep out of it corrupt customs and manners, even than to prevent the introduction of a pestilence." — *Langhorne's Translation.* Compare also what Bacon says on this peculiarity of the Spartans in his Essay "*Of Greatnesse of Kingdomes and Estates*," contrasting their exclusiveness with the liberalism of the Romans.

[85] This state of things which ensued, as our author says, was entirely owing to the rise of the feudalism which Charlemagne [*v. sup.*] had always endeavoured to repress. Henri Martin in his "*Histoire*" [vol. ii.,

cavalry rides and warlike commotions of steel,
especially dear. Glory and honour to the strong,
shame and misfortune to the weak. What is overt
licence but the offspring of power? and the less it is
restrained by law or philosophy, the more fascinating
it is. Thus they attain to the enjoyment of the mere
pleasures of the senses—the only pleasures capable of
being appreciated at a time when intellectual enjoy-
ments are utterly ignored, except in the retirement of
the cloister.

¶ 367.
Characteristics
of the type.

Spatulate hands are valiant, industrious, and
active; they have the power and the genius of the
hands of Cyclops; they forge impregnable armour,
and cover the earth with battlemented castles which
rear themselves upon the crests of rocky promon-
tories, protected by deep waters and impenetrable
forests. They build huge dungeons, theatres of orgies
and terrible tragedies, haunts everlastingly ringing
with shouts, haunts which they attack, defend, and
contend for, with a ferocity which is terrific. At rare
intervals, pointed-handed subjects devoted to prayer

Contrast,

liv. xiii., p. 364] says:—"Le Génie de l'Empire
Frankain, en remontant au ciel, laissait les peuples
occidentaux à l'entrée d'une des plus longues et plus
douloureuses crises qu'ait euc à traverser l'humanité, de
la crise qui enfanta la société féodale." Eyre Evans
Crowe ["*History of France*" (London: 1858), vol. i.,
p. 38], speaking of the commencement of the ninth
century, says:—"The development of wealth and the
accumulation of money came to give society a new and
different impulse. That of the first ages of the
modern world was limited to the destruction of the
classification of society which existed in the ancient
world. In it men were slaves, citizens, functionaries,
or emperor; the modern world came forth without these.
It presented a territorial aristocracy, replacing the
functionary, and exercising his authority, nullifying the
emperor, ignoring the citizen; and with an agricultural
class of many grades, but never descending to the
abjectness of a slave."

and to celibacy, opening the gateways of the monasteries, implore the benefits of the peace of God. The
real peace of God, it has been properly observed,
belongs to the twelfth century.

If these powerful hands had not conquered all
women with whom they were thrown into contact,
their reign would have been of shorter duration.
They raised themselves, and in a manner multiplied,
by means of the axe and the sword, and it was the
axe and the sword which, precipitating them in turn,
one after another, from the heights of their rugged
deserts, put an end at last to their brutal and savage
domination.

¶ 368.
Rise and fal.
of the type.

In Russia the nobles have acquired such rights
[whether by custom or by law] over the women on
their estates, that the population hardly resents at all
the periodical enforced sale of all the young people
in their villages.[285] Well, these nobles, a proud and
rampant class of men, ostentatious and avaricious,
full of vices and of cunning, say that they are of a

¶ 369.
Russian socia
science.

[285] " It is especially in the remoter provinces that the
power of the rich nobles is unrestrained, and the
oppressed would accuse his oppressor in vain. The
master can, if the caprice so takes him, sell his serfs or
change them like any other merchandise," etc.—M.
CHOPIN, " *Russie*" (Paris : 1838), vol. i., pp. 25 and
31. The Marquis de Custine, in his work " *La Russie
en* 1839 " (Paris : 1843, vol. i., p. 331), says :—" The serf
is the chattel of his master ; enrolled from his birth
until his death in the service of the same master, his
life represents to this proprietor of his labour merely
a part of the sum necessary for his annual capricious
expenditure." Sir A. Alison, on the other hand,
approved rather than otherwise of the institution of
serfdom, on the ground that serfs were attached to the
soil, and, though they could be sold *with* it, they could
not be sold *without* it,—" a privilege of incalculable
value, for it prevents the separation of husband and
wife, parent and child."—" *History of Europe*" (London : 1854). Serfdom was abolished by Alexander II.
in 1861. Alfred Rambaud [" *History of Russia*,"

[¶ 369]

race superior to that of the people ; in this way they
are bringing about the ruin of their influence, by
multiplying among the masses the number of individuals, already not a few, to whom they have
transmitted their spirits with their blood. The in-
continence of the great is the hotbed in which are
developed the germs of the liberty of the humble.[286]

¶ 370.
The hand as an indication of race.

You can tell those people who pride themselves
upon a descent in the direct line from the hardy
braggarts of the ninth century, and who at the same
time pride themselves upon the possession of a hand
which is fine and pointed, that the two pretensions
are incompatible. Every gentleman descended from
the old fighting nobility, has necessarily a spatulate
hand. If his hand be pointed and fine, he must
search his pedigree for some infusion of gentle or
ecclesiastical blood, or else he must resign himself to
the presence of a bar sinister in his escutcheon,
whether it be properly quartered or not.[287]

¶ 371.
Breton land-owners

Encumbered with untidy servants and yelping
hounds, the habitations of the small landowners of
Brittany exhale, just as they did in the time of
Duguesclin, a continual odour of animals and litter.
As ignorant of new ideas as are Chinese artists of

translated by Leonora Lang (London: 1879), vol. ii.,
p. 392] tells us that " the new imperial commission . . .
admitted the principle that the emancipation should
not take place gradually, but that the law should ensure
the *immediate* abolition of serfdom; that the most
effectual measures should be taken to prevent the re-
establishment of the seignorial authority under other
forms. . . . From these deliberations resulted the
new law " [February 19th—March 3rd, 1861].

[286] The above was, of course, written before the
abolition of serfdom in 1861. Since he wrote, the
events foretold as above by M. d'Arpentigny have taken
place, as he points out in this paragraph.

[287] Compare on this point ¶¶ 10 and 154, and "*A
Manual of Cheirosophy,*" ¶¶ 78 and 108.

the [for them incomprehensible] rules of perspective, these families, which I believe to be extremely ancient, which drink to excess, which blow horns, which understand their broad-shouldered horses, and are connoisseurs of their short-legged terriers, and whose connoisseurship stops short at these things, have always spatulate hands.

The Tcherkesses consider the chase, pillage, and military exercises to be the most honourable occupations possible for young men. Laws and obedience are unknown to them, and they can be governed only by eloquence and by the inspiration of respect and admiration. A handsome woman, a fine horse, an illustrious ancestry, a hardy ·constitution and sparkling accoutrements, all of which inspire one with courage only to look upon them, are in the eyes of the whole nation the most precious benefits that can be conceived. There the serf waits upon the free man, who waits upon the noble, who, in his turn, waits upon the prince. The despotism which is exercised makes what they have to suffer tolerable. Spatulate hands![288]

¶ 372.
The Tcherkesses.

Each type looks for assistance in all its decisions of any importance to the resources with which it is most richly endowed. We have seen that elementary hands obtain from their physical inertia [as regards obedience to law] the wages of innocence; whilst, appealing to address and to bodily strength, hands which are spatulate and hard think that they are appealing to the judgment of God. And note this,

¶ 373.
Dependence of types upon their characteristics.

[288]. M. le Capitaine has taken his views on the Tcherkesses from H. J. von Klaproth's *"Reise in den Caucasus"* (Halle : 1812-14), of which a French epitome, entitled *"Tableau Historique, Géographique, Ethnographique, et Politique du Caucase et des Provinces Limitrophes entre la Russie et la Perse"* (Paris and Leipsic : 1827), on p. 70 of which a minute account of the Tcherkesses and their peculiar views may be found.

that, however much the salient characteristics of each type may be, in the present day, effaced by the combined effects of the crossing of races, of civilisation, and of education, if, of two persons with whom you converse, one is remarkable for a profound apathy about anything which does not immediately concern him, and the other is remarkable for a spontaneous sympathy with prowess, whether well-directed or not, and with physical strength, a highly-developed palm will infallibly be the possession of the first, and a spatulate, or at all events a square hand will characterise the second.

¶ 374.
Dominant
classes

In the same manner that if foxes and lions were members of society [as in the days of La Fontaine], the power would be in the hands of the most cunning among the former and the strongest among the latter : so each type when it is dominant, when it governs, never fails to select as its agents of all kinds, individuals in whom its genius is the most perfectly reproduced. This is what I have just now [but in different words] expounded ; but this eccentric tendency of each type of mankind, being the natural explanation of the differentiated civilisations which have governed upon earth, I have thought it necessary to emphasise my point to obtain for it the attention which it deserves. In the tenth century, which was the epoch of a civilisation directed by hard, spatulate hands, the all-powerful cohort of the ecclesiastics recruited itself from among the ranks of soldiers and

Gerbert.

mechanicians. Gerbert, who afterwards became pope [under the name of Sylvester II.], was raised from the rank of a simple monk to that of archbishop, because he had invented a clock with a balance movement.[89]

[89] This is, I think, too arbitrarily stated by our author, probably for the sake of forcible illustration. There is no doubt that, as C. F. Hock justly remarks :—" Gerbert

The twelfth and thirteenth centuries were under the rule of psychological ideas, the reins of the world were held by priests and theologians, such as Suger, St. Bernard, Abélard, and so forth.[190]

Francis I. and Leo X. are reputed "great," because their tastes, chiming in with those of the age in which they lived [which was one in which civilisation was

was far in advance of his age in the extent of his knowledge and the aptitude with which he applied it . . . . this scientific activity bore its fruits." [" *Histoire du Pape Sylvestre II.*," traduit de l'allemand par J. M. Axinger (Paris : *n. d.*), p. 385]. Gerbert was a monk in the monastery of Fleury, in Burgundy, who made a journey to Cordova, in Spain, for the purpose of learning some of their foreign arts. He is said to have been the first to have introduced Arabic numerals into north-western Europe, as also the clock with the balance action.* P. F. Lausser, in his " *Gerbert, Étude Historique sur le Xème Siècle*" (Aurillac : 1866, p. 180), tells us that he put up a clock in the cathedral of Rheims. William of Malmesbury [lib. ii., c. 10] ascribes to him various magical powers. "The first step that he made into public life consisted in his being named preceptor, first to Robert, King of France, the son of Hugh Capet, and next to Otho III., Emperor of Germany. Hugh Capet appointed him archbishop of Rheims, but that dignity being disputed with him, he retired into Germany, and becoming eminently a favourite with Otho, he was by the influence of that prince raised, first to be archbishop of Ravenna, and afterwards to the papacy by the name of Sylvester II.—WM. GODWIN, " *Lives of the Necromancers*" (London : 1834), p. 231.

[190] This statement is amply carried out by subsequent writers. A. Huguenin, in his " *Suger, et la Monarchie Française au XIIe Siècle*" (Paris : 1857), lays the greatest possible stress upon the enormous influence of Suger over Louis VI. and the country in general Of his connection with St. Bernard he says [chap. xxxvi., p. 210] :—" St. Bernard and Suger, these two lights of the Church in the twelfth century, . . . were destined to

* "On lui attribue aussi la construction de la première horologe mue par des poids."—Larousse, "*Dictionnaire du XIXe Siècle.*"

[¶ 376]

governed by artistic hands], led them to elevate above everything else beauty of external construction.[91]

¶ 377.
Colbert and
Louvois.
Order.

With square hands at the head of affairs, the genius of material order and administrative science triumphed in the seventeenth century. No longer are artists sent forth as ambassadors, and no longer do they rise to the cardinalate. Colbert and Louvois hold the reins of government, and on all sides we find etiquette, arithmetic, regularity, and tact : bow down to them,

harmonise the two great interests of which they were then the representatives. . . . Suger was the most clever and equitable mediator in the interests of Europe. . . . [p. 374] Inventor of the science of politics in an age which knew but very little of its real secrets, he applied himself to the renaissance of public justice. By the help which he gave to St. Bernard he helped ecclesiastical reforms at the same time that he put political interests in harmony with those of religion, . . . but a glory which is due to him alone is that of having established the first foundations of public administration and finance." Abélard was more of a philosopher than either Suger or St. Bernard, and it was as such that he influenced the destinies of the empire in the twelfth century. Charles du Rémusat ["*Abélard*" (Paris : 1845), vol. i., p. 270] says of him : "Voltaire alone, perhaps, and his position in the eighteenth century, would give us some idea of what the twelfth century thought about Abélard. . . . Scholarship, the philosophy of five centuries, cites no greater name than his."

[291] Concerning Francis I.'s love of art and his influence on letters in general, M. Capefigue's work, "*François I. et la Renaissance*" (Paris : 1845), gives us voluminous particulars. Martin, in his history [vol. viii., bk. xlviii., p. 125, *op. cit.*], tells us that a taste for an elegant, accomplished, picturesque, and versatile civilisation was the sole affection to which Francis always remained faithful ; he deserved the title of "Père des lettres" far more than that of "Roi Chevalier," and points out that, with all his faults, Francis never ceased to promote the interests of art. In the same way Leo X. [*vide* ¶ 10, p. 36] was a great patron of the arts. "The claims of Leo X.," says William Roscoe, in his "*Life and Pontificate of Leo X.*" (London :

O ye people! for they hold the sceptre of administration.²⁹²

To-day we elect our senators, our ministers, and our statesmen, from the ranks of lawyers, financiers, and public caterers. A very small proportion of principle is sufficient for the wants of human society; the south has discovered and proclaimed them to the world. It was its mission to do so. Now we have arrived at an age of practical actualities; it is the task of the north, sovereignly endowed with the intelligence of things real, to practise and inculcate these principles. Well, public caterers, merchants, engineers, and industrial labourers belong almost exclusively to the types which govern in the north.

In France, under the Empire, the advent of hard, spatulate hands at the head of society was a misfortune, for there is only one class of ideas which stands lower in the scale of intelligence than those of which such hands are the born instruments. Listen to the veterans of the Empire, and you will under-

¶ 378.
Materialism of to-day.
Spatulate type.

¶ 379
Rule of hard spatulate hands in France.
The Revolution

1846, vol. ii., p. 394), "to the applause and gratitude of after-times are chiefly to be sought for in the munificent encouragement afforded by him to every department of polite literature and elegant art. It is this great characteristic which, amidst 250 successive pontiffs, ... has distinguished him above all the rest, and has given him a reputation which, notwithstanding the diversity of political, religious, and even literary opinions, has been acknowledged in all countries." Compare also Erasmus' opinion of this pontiff [Lib. i., Ep. 30].

²⁹² Anyone who reads Voltaire's "*Siècle de Louis XIV.*" will appreciate the fact that the symmetry and regularity of the court of that monarch were to a very great extent due to the characters of these two statesmen. Note, for example, the passages concerning Colbert and Louvois in chap. xxix. of that work. There is no doubt, however, that the Comptrôleur-général Colbert greatly encouraged the patronage of art by his master, in whose reign the Louvre, Saint-Germain, and Versailles were mainly built.

stand with what a small amount of brains one can arrive at fame and reputation, under the sovereignty of the sword, and in the full blast of the apotheosis of physical force. And yet, O you magnanimous subjects of the philosophic hand, Joubert, Hoche, Marceau, Lafayette, Desaix, Kléber![293] you who founded in our midst both liberty and equality; you men of warm hearts and grand physiognomies, at the same time progressive and conservative, minds which were grand and simple, austere disciples and followers of those Dorians who prayed the gods to vouchsafe to them their worldly advantages in a beautiful form,— your glory, before which we prostrate ourselves, now that the sunbeams of liberty have opened our eyes, paled for an instant before that of these bureaucrats, so profound was the blind and stupid intoxication into which they had plunged us by sheer force of false splendour, of continuous action, and chronic uproar!

[293] Of the philosophic type no better examples than these six heroes of the French Revolution could have been given by the author. A characteristic story is told concerning Marceau and Kléber. The former having been appointed commander-in-chief of the army in La Vendée, and having made it a condition that Kléber should undertake the command with him, "Hold thy peace, my friend," said Kléber, "we will fight together and we will be guillotined together."—THIERS, "Histoire de la Révolution Française," chap. xvii., Décembre 1793.

SUB-SECTION XI.

Catholics and Protestants, Lyricism, Mysticism.

LOVERS of art, of poetry, of romance, and of mystery, pointed hands require a Deity such as they imagine him to be. Lovers of science and of reality, spatulate hands require a Deity fulfilling the requirements of their reason.

¶ 380.
Deities of the types.

For the first, with its festivals and its contemplation, we have Catholicism ; for the others, with its rigorous deductions and its activity, we have Protestantism.

¶ 381.
Catholicism and Protestantism.

Protestantism has increased rapidly in the north, where spatulate hands abound, and has hardly penetrated at all in the south, save in mountainous districts, where the same hands, for the same reasons, are equally abundant.[291]

¶ 382.
Protestantism in the North.

Regard being had to the softness of the climate and the relative fertility of the latitudes where they

¶ 383.
Effect of latitude upon religion.

[291] *Vide* Voltaire's " *Siècle de Louis XIV.*," ch. xxxvi., in which, speaking of the expulsion of the Vaudois, the Albigenses, and the other Protestant cults, he says, " The English, endowed by nature with the spirit of independence, adopted them, modified them, and composed of them a religion peculiar to themselves. Presbyterianism established in Scotland, in the unhappier times, a kind of republic, whose pedantry and harshness were much more intolerable than the rigour of the climate."

flourish, Catholic nations being as susceptible to love as the Protestant, I say that it can only be by reason of a chaste repugnance of the spirit, and to satisfy a craving for moral purity, which is more imperious with them than the mere pleasures of the senses, that they impose upon themselves the inconveniences and privations attendant upon the doctrines of the indissolubility of marriage, and the enforced celibacy of the clergy. There is no doubt that the nations among whom these two institutions do not exist, however ethereal in other respects may be the atmosphere in which their poets and their painters have their being, are gifted with less spirituality than those where they do exist.

¶ 384.
The Protestants
after the Edict
of Nantes.

It has been said that it was for the purpose of regaining, by the sheer influence of wealth, their vanished political power, that the Protestants in France, after their persecutions in the time of Richelieu, applied themselves with such enthusiasm to the industrial arts and to trade. I do not hold this opinion myself. The same moral impulsion which had made them embrace Protestantism, and which made them cling to it, could not fail to urge them to the study of exact sciences, and to the practice of the mechanical arts.[295] People of poetic temperament only require proofs in rare cases; scientific minds, on

[295] On this point *vide* Ch. Weiss' "*Histoire des Réfugiés Protestants de France*" (Paris, 1853, ch. ii., p. 30), of which an English translation exists, entitled "*History of the French Protestant Refugees*" (London: 1854, p. 24):—"The Edict of Nantes was for the Protestants the inauguration of a new era; . . . obliged to apply themselves to agriculture, commerce, and industry, they took every advantage of this compulsion ; . . . the Protestant middle classes in the towns devoted themselves to industries and commerce with an activity and intelligence, and at the same time with an integrity, which have never been surpassed in any country." The edict was signed in April 1598, but the parliament, fearful of

the other hand, balance everything, check everything, and cannot believe anything without proof positive. All over the world the Protestants, not by reason of their training, but on account of their organisation, surpass the Catholics in mechanical, and are surpassed by them in the liberal, arts. Being the more active workers, they are the less capable of resignation.

Effect of religion on industries.

It is not merely two different ideas which are at issue in religious warfare, it is two diversified organisations, two chosen races of men, acting in obedience to instincts which are diametrically opposed. It is that which renders these wars so bloody and so cruel.

¶ 385.
Holy Wars

I must not forget to point out the fact that our inferiority to the English in industrial arts and scientific mechanics dates only from the revocation of the Edict of Nantes.[296] Denis Papin, the inventor of the high-pressure steam engine worked with a

¶ 386.
Protestant
mechanical
industry.

Denis Papin.

its consequences, delayed for a long time its registration, to the great annoyance of Henri IV. *Vide* Voltaire's "*Histoire du Parlement de Paris*," chap. xl.

[296] "The result of this despotic act was that rather than conform to the established religion, 400,000 Protestants —among the most industrious, intelligent, and religious of the nation—quitted France, and took refuge in Great Britain, Holland, Prussia, Switzerland, and America. The loss to France was immense, the gain to other countries no less. Composed largely of merchants, manufacturers, and skilled artisans, they carried with them their knowledge, taste, and aptitude for business." —"*Chambers' Encyclopædia*." *Vide* Voltaire's "*Siècle de Louis XIV.*," ch. 36, which is perhaps the best account of the effects of this step. The revocation of the edict followed close upon the massacres and persecutions known as *Dragonnades*, of which no better account exists than a little work, entitled "*A Narrative of the Sufferings of a French Protestant Family at the Period of the Revocation of the Edict of Nantes,*" written by J. Migault, the Father (London: 1824). *Vide* also "*Bishop Burnet's History of His own Time,*" book iv., 1685. .

piston, was obliged (being a Protestant) to take refuge in England at that epoch.[297]

¶ 387.
Civil war.

Italy and Greece, countries equally mapped out in plains and high mountains, have in all ages, to an extent which has led to their comparatively utter subjection, been the theatres of intestinal quarrel and civil war ; but alike in the pursuits of thought and of action, the grandest genii who have glorified humanity have sprung from their midst.

[297] Denis Papin, the illustrious descendant of a Protestant family at Blois, was born on the 22nd of August, 1647. Having been originally intended for the medical profession, he attracted the attention of Var. Huyghens, and was by him summoned to Paris in 1671, where in 1674 he became demonstrator to the Académie. It was whilst thus employed that he invented his high-pressure engine, of which, however, Colbert, who was then comptrôleur-général, thought but little. It was probably in consequence of the scanty recognition of his talents that he left for England in 1675. " Many of his biographers," says the Baron A. A. Ernouf, in his "*Denis Papin, sa Vie et son Œuvre*" (Paris: 1874), " have thought that this expatriation was one, and not the least regrettable, of the consequences of the religious intolerance of the Government, whose watch upon 'ceux de la religion' was then taking the form of a persecution." The fact also that Huyghens, his patron, had returned to Holland probably contributed to his decision to leave Paris.* In London he suffered many hardships until he was protected by Boyle, as he had been in Paris by Huyghens, and became curator of the museum of the Royal Society. Here he invented his celebrated Digester [1681], and here he lived until 1687, when he went to Marpurg, and married one of the French Protestant colony settled there. We find continual notices of his work in the *Journal des Sçavants* until the revocation of the Edict of Nantes in 1685, after which it is diplomatically silent concerning so

* " Huyghens et Roëmer quittèrent la France lors de la révocation de l'Edit de Nantes. On proposa, dit on, à Huyghens de rester ; mais il refusa, dédaignant de profiter d'une tolérance qui n'aurait été que pour lui. La liberté de penser est un droit, et il n'en voulait pas à titre de grâce."—Note by Louis Barré to chap. xxi. of Bry's edition of Voltaire's " *Siècle de Louis XIV.* ' (Paris: 1856), p 237.

Among the ancients, the inhabitants of the moun- tains had gods of their own, different from those of the inhabitants of the plains, and these gods only became amalgamated in proportion as the two races mixed their blood by inter-marriage. To-day Europe acknowledges but One God, who is adored in cere- monies which are frigid, severe, splendid, solemn, magnificent, or passionate, according as the worship takes place in Switzerland, Scotland, France, Ger- many, Italy, or Spain.

¶ 388.
National
religious ritual

The Old Testament, whose every page (like those of the Sagas,[208] the Eddas,[299] the Havamal, and the

¶ 389.
Old and New
Testament
contrasted.

notorious a Protestant. At Cassel, whither he repaired in 1695, he completed all his noted inventions, but ended badly in consequence of the failure of his steam-cannon, and the consternation caused by the appearance of his steamboat, which was destroyed by the infuriated watermen of the Weser. After this he returned to Eng- land [1708], but encountered the hatred of Newton [who was then president of the Royal Society] on account of his friendship for Leibnitz; he returned to Germany, having been reduced to the greatest poverty in 1712, after which all trace of him is lost. He pro- bably died about 1714.

[208] "Saga" is an old Norse or Icelandic word, denot- ing any historical tradition which in old times was handed down by word of mouth. The sagas are partly historical and partly legendary, and deal principally with the old heroic exploits of the Volsungs and the Niblungs. The sagas have been collected and criticised in a most interesting volume by Bishop P. E. Muller, entitled "Saga Bibliothek" (Copenhagen: 1817-20). Dr. Wag- ner's introduction to his work "Asgard and the Gods" (London: 1884, p. 1) tells a charming story in which "Saga" is described as a goddess, "sunk in dreamy thought, while Odin's ravens fluttered around her and whispered to her of the past and future." She points to the scrolls which lie scattered round her, and says to her visitors, "Are ye come at last to seek intelligence of the wisdom and deeds of your ancestors? I have written on these scrolls all that the people of that distant land thought and believed, and that which they held to be eternal truth." [Compare also pp. 10, 48, and 49 *op. cit.*]

[299] "The primary signification of the word *Edda*," says

[¶ 389]

Niebelungenlied)[300] is ablaze with war, movement, energy, and action, is much more highly appreciated by the people of the North, and particularly by Protestant nations, than is the New Testament, which, written by a nation in a state of slavery, and, therefore, interested in the depreciation of the virtues of actual strength in favour of those of weakness and resignation, exalts above all qualities self-abnegation, servitude, repose, and peace.

¶ 390.
Protestant and Catholic ritual contrasted.

Bereft as they are by their organisations of the sentiment of the plastic arts, the Protestants, discard-

Warnford Lock, in " *The Home of the Eddas* " (London: 1879, p. 1), "is a *great-grandmother*, . . . but the noun is no longer used in that sense. It survives only as the proper name of two literary works, the finest and greatest classics of the Gothic, or, if you will, Teutonic races of men. The elder Edda, *the* Edda *par excellence*, is occupied chiefly with skald-ship figures, and - forms the *Ars Poetica* of the old Norse tongue. The younger Edda contains the heathen mythology of Scandinavia, which equals in beauty and interest, and in some respects excels, that of ancient Greece and Rome." These books are continually referred to by Dr. Wagner [" *A Manual of Norse Mythology* " (London: 1884), *passim*], and are known as the *Edda Sœmundar hins Froda* and *Snorri Sturleson's Edda*. Sœmund lived between 1054 and 1153 A.D., and Sturleson was born in 1178, and was assassinated in 1274. The best editions for the student are those of Brynjölf Svendson [A. A. Afzelius (Stockholm: 1818)] and of Munch (Christiania: 1847). *Vide* also on this word J. B. Depping's introduction to Th. Liquet's " *Histoire de Normandie* " (Rouen: 1835).

[300] The " *Nibelungenlied* " dates in its present form [from the earliest known MS.] from about 1210 A.D. It is mainly a collection of barbarous and rude recitals of the warlike exploits of the old German heroes Gunther, Siegfried, Haco [or Hunding], and of Brynhilde, Krimhilde, and Sieglinde. Its author is unknown; the best modern editions are those of Simrock (Berlin: 1827) and Lachmann's. " *Zu den Nibelungen, und zur Klage* " (Berlin: 1836). It is perhaps in the form of Richard Wagner's opera, *Der Ring des Nibelungen*, that this epic is best known in this country.

ing images, have obeyed at once the laws of physical antipathy and the suggestions of a *reasonable* piety. As for the poetry of which all cultivation has need, and which they could not find in artistic imagery, they found it in the Holy Writings translated as they are, by their efforts. The Catholics, on the other hand, have not caused their books to be translated, so that they continue to pray in a tongue of which they are ignorant ; but their *art* speaks a language which they understand ; and the art works with which they fill their churches, suffice, in the absence of poetic elocution, to keep their minds in a continual state of enthusiasm and fervour. They appreciate to a higher degree than the Protestants, the *sentiment* of religion, but the *idea* of religion is more highly developed among the Protestants. Protestantism gives birth to more *teachers*, whilst Catholicism produces more *saints*. The former dispenses *justice*, whilst the latter bestows *charity;* the former must be understood by the mind, it speaks to thinkers and to men of action and of intelligence ; the latter must be understood by the heart, it speaks to dreamers and to men of simplicity, and of resignation. Heaven is the domain of the latter ; of the former, the earth.[301]

[301] I am inclined to look upon the foregoing as the most interesting and the most talented chapter in M. d'Arpentigny's work. It appears to me to exhibit marvellous powers of analysis and of thought, and as a comparison of the Catholic and Protestant religions it is perhaps unrivalled.

CONTINUATION.

SPATULATE HANDS [*continued*].

English Hands.

¶ 391.
Scandinavian
colonies.
Effect on the
parent countries.

IT is noteworthy that a grand and profound silence reigns over the kingdoms of Scandinavia and the Cimbrian Chersonese, from the moment that the most robust and most restless portion of the populations of these hardy countries had at last gained a permanent footing in England [and in some of the other quasi-southern tracts]. The present inhabitants of Norway, of Sweden, and of Denmark are the descendants of men who were comparatively weak and of peaceable dispositions, and whom their brethren, the pirates emigrating in search of adventures and conquests, left behind them. These freebooters, who without doubt were all of the hard, spatulate-handed type, mixing their blood with that of the old Breton folk, communicated to these last, their ardour for locomotion and action. Of all the nations of the earth the English are the people who love most to take active walking and riding exercise, to cross the seas and to travel about, and it is among them that the spatulate type is the most freely multiplied.

¶ 392.
Origin of the
Irish race.

It is not without some show of reason that the Irish, loving as they do festivals and quarrels, votaries of the dance and of the bottle, excited or depressed by the merest trifles, a race gifted with an imagination which is at the same time active and highly coloured, and a mind which is not deep, but prompt, keen, and subtle, pride themselves upon a southern origin. I conclude that the conic type is extremely common among them. [302]

The astonishment which the English display at our love of ornament, and of that which appeals to the imagination and to the taste, is not greater than that which we experience at their everlasting search after the comfortable and the useful. They combine art with nothing,—art, I say, which is a means of enhancing one's appreciation of nature ; fashion is to them all-satisfying ; that is to say, the necessarily evanescent authority of a material formality, reigning, as one might say, by itself, bereft of all reasoning acquiescence. Their houses, their furniture, their jewels, their accessories of the table, of the toilet and

¶ 393.
English
materialism.

[302] It is certainly true that the Irish scout the idea of their having sprung from a common origin with the English, and, from what we know of ancient Irish history, it seems practically established that the old Irish stock is of distinctly southern or Oriental origin. Still, as Justin H. McCarthy says in his " *Outline of Irish History from the Earliest Times to the Present Day* " (London : 1884), "As we peer doubtfully into the dim past of Irish history, we seem to stand like Odysseus at the yawning mouth of Hades. The thin shades troop about us, and flit hither and thither fitfully in shadowy confusion. . . . Groping in the forgotten yesterday of Irish legend is like groping in an Egyptian tomb—we are in a great sepulchral chamber ; " and he brings before us in picturesque review the Nemedhians, the Fomorians, the Firbolgs, and the Milesians, with all their weird legends and warlike traditions. " Modern historians, however, prefer to leave the Tuatha de Danaan and the Milesians undisturbed in their shadowy kingdom, and content themselves with suggesting that Ireland was at first inhabited by a Turanian race, and that there were Celtic and Teutonic immigrations " [*op. cit.*, p. 21].

[¶ 393]

of the chase, their mathematical, musical, and astronomical instruments, display in their arid perfection such a pre-occupation concerning the possibly hostile influences of physical nature, such a poverty of artistic invention, and an imagination so prosaic and so dull, that we cannot choose but look upon them as a people distinct in themselves, specially devoted to the enterprises and to the struggles commanded by the taciturn requirements of material existence.

¶ 394.
Town and country life contrasted.

The English are commended for their love of a country life, as if it were with them a taste acquired by education : it is nothing of the sort. They like the country, because in the country, more easily than in town, they can satisfy their craving for the fatiguing exercises which are necessary to their natures. The Spaniards, to whom corporeal action and agitation are highly antipathetic, prefer, to a greater extent even than we [inclined as we are by our climate and our organisation to a moderate form of locomotion], the life of the city to that of the country.

¶ 395.
Gesticulation in speech.

Speech alone, is not, for nations of an artistic temperament, an all-sufficient medium for the expression of their thoughts ; they accompany every word as it were, with a gesture intended to pourtray clearly and rapidly shades of thought which mere words are powerless to convey ; the more artistic the people the more do they gesticulate in speech. Thus the English, who in conversation adorn nothing, and among whom enthusiastically to express a sentiment is looked upon as an affectation, move their entire bodies all in one piece as it were, and hardly ever gesticulate in speaking. They have so little sense of the fitness of things as regards the outward form and inward signification of the embodiment of an idea, that they see nothing ridiculous or false in the spectacle of

[¶ 395]

a clergyman dancing,—a thing which is very common in England.[303]

In the matter of costume, and again as regards deportment [setting aside what is sumptuous and "correct"—qualities of which their natures afford them an infallibly sure instinct], they never fail to confuse singularity with distinction, ostentation with grandeur, coldness and insolence with dignity. They pride themselves upon their love of strange feats and grotesque wagers ; upon their taste for strong meats, strong wines, and foreigners of eccentric behaviour; upon that calm ferocity with which they can find pleasure in the sight of two men fighting for a few shillings. Europe, which is kept awake by the continual hubbub of their clubs, their receptions, and their workshops, gazes from its windows to see them drinking huge bumpers, and becoming purple in the face, feasting to excess, and exhausting themselves and their horses in the everlasting fox hunt and the never-ending race for money ; and they mistake the gloomy and silent astonishment which they provoke for admiration !

¶ 396.
English characteristics

"The English," says Bulwer, "make business an enjoyment, and enjoyment a business ; they are born without a smile ; they rove about in public places like so many easterly winds—cold, sharp, and cutting ; or like a group of fogs on a frosty day, sent out of his hall by Boreas for the express purpose of *looking black at one another*. When they ask you, 'How do you do ?' you would think they were measuring the length of your coffin. They are ever, it is true, labouring to be agreeable ; but they are like Sisyphus. The stone they roll up the hill with so much toil runs

¶ 397.
English recreations.
Bulwer Lytton

---

[303] One is reminded in reading the above passage of Dr. Johnson's remark, when he saw certain reverend gentlemen enjoying themselves, "This merriment of parsons is extremely offensive" [!]

down again and hits you a thump on the legs. They
are sometimes *polite*, but invariably *uncivil;* their
warmth is always artificial, their cold never; they
are stiff without dignity, and cringing without
manners. They offer you an affront, and call it "a
plain truth;" they wound your feelings, and tell you
it is manly "to speak their minds;" at the same
time when they have neglected all the graces and
charities of artifice they have adopted all its falsehood
and deceit. While they profess to abhor servility,
they adulate the peerage; while they tell you they
care not a rush for the minister, they move heaven
and earth for an invitation from the minister's wife.
There is not another court in Europe where such
systematised meanness is carried on, where they will
even believe you when you assert that it exists.
Abroad, you can smile at the vanity of one class and
the flattery of another; the first is too well bred
to affront, the latter too graceful to disgust; but *here*
the pride of a *noblesse* [by the way, the most mush-
room in Europe] knocks you down with a hailstorm,
and the fawning of the bourgeois makes you sick
with hot water."[304]

¶ 398.
The artistic type
in England.
        The conical artistic type is so rare in England that
the higher development of its instincts and of its
reason, shocks the feelings of the masses. Byron, who
belonged to this type, was obliged to seek among the
poetic races of the East the justice, the esteem, and
even the peace, that his compatriots, urged by the

[304] This passage—the concluding paragraph of chapter
lxvi. of Bulwer Lytton's "*Pelham*"—is reproduced above
as Bulwer Lytton wrote it, not as M. d'Arpentigny translated it. It is, I think, one of the most offensive
speeches which the arch-puppy Pelham delivers during
the course of that instructive work, and its use against
us by a foreign author "points a moral," if it does not
"adorn his tale."

hard and prosaic spirit of their latitude, obstinately
refused to accord to him.[305]

Our nation owes to the artistic type, which is extremely widely diffused among us, the caprice and the
brilliancy which characterises it; but as regards
the disdain which this type evinces for all that is
merely useful, we owe to the artistic type the spirit
of frivolity for which we are reproached.

¶ 399.
Artistic type
in France.

The English, continually pre-occupied concerning
their *material advantage*, continually alter and improve
their machines and their industries; to us, who are
blest with a less inclement atmosphere, material
innovations are as repugnant as moral innovations
are to our neighbours. The reason of this is, that
material improvements require a continual physical
labour, whilst the moral ones require a constant intellectual labour. We are progressive in *ideas*, they
in *things*. Our ability exhibits itself in the logic of
theories; theirs, in utilitarianism and the opportunity
for applying their faculties. We sacrifice interests to
principles, *they* sacrifice principles to interests.

¶ 400.
French and
English minds
contrasted.

The expansion of the English mind proceeds like
that of water, outwards rather than upwards; whereas
our intellectual progress proceeds like that of fire,
upwards rather than outwards. The English aim
at well-being by the increase of the domination of the

¶ 401.
The same.
Illustrations.

---

[305] Our author seems to disregard, or to be unaware
of, the fact that Byron first went to Greece in 1807,
in the midst of the storm of praise and adulation which
assailed him on the appearance of his answer to Lord
Brougham's attack in the *Edinburgh Review*, upon his
first volume of verse, "*Hours of Idleness.*" It was
after his return in 1812, after the publication of "*Childe
Harold*," "*The Giaour*," "*The Bride of Abydos*,"
and "*The Corsair*," and after his marriage with Miss
Milbank. and his subsequent separation under the most
disgraceful circumstances, that he left England and
lived in *Italy* till 1823, when he sailed to Greece, to find
death at Missolonghi in April 1824.

[¶ 401]

Bacon.

man over physical forces. They gave birth to Bacon, and they carry on the plan of the Romans; they people and they cultivate the world, *we* civilise it.[306]

Descartes.

We gave birth to Descartes, and carry on the plan of the Greeks; we aim at good fortune by the multiplication and the progress of things which interest the mind.[307] Where our neighbours send traders, we send missionaries; where they carry utensils, we carry books and art-treasures.

¶ 402.
Effect of art in England.

Let the artistic type multiply itself in England, and we shall see the last [up to a certain point] of its eccentricity, and, as a natural consequence, of a great portion of its power. The governing principle being left without universal acquiescence, she would

---

[306] " The *Kingdome* of *Heaven* is compared, not to any great Kernel or Nut, but to a *Graine* of *Mustard seede;* which is one of the least Graines, but hath in it a Propertie and Spirit hastily to get up and spread. So there are States great in Territory, and yet not apt to Enlarge or Command; And some that have but a small Dimmension of Stemme, and yet apt to be the Foundations of Great Monarchies." F. BACON, " *Of Greatnesse of King-domes and Estates* " (London : 1625).

[307] Réne Descartes [or Renatus Cartesius], born in 1596, has been justly called one of the reformers of philosophy. At an early age he became dissatisfied with all the accepted teaching of the schools, and all the methods and dogmata of existing philosophy, and set himself resolutely to discard the teaching of the *sçavants* of his day, with a view to developing a new method of study and analysis. His processes and their results appear in his " *Discours sur la Méthode* " (Amsterdam : 1637) and the bases of his philosophical system may be found in his " *Méditationes de Primâ Philosophia* " (Amsterdam : 1641) and his " *Principia Philosophiæ* " (Amsterdam : 1644). From 1621 to 1649 he lived in Holland, in which latter year, having repaired to the court of Sweden at the invitation of Queen Christina, he died in 1650. In " *An Oration in Defence of the New Philosophy,* " spoken in the Sheldonian Theatre at Oxford, in July 1693, by Joseph Addison [printed, in English, at the end of the third edition of Gardner's translation of Fontenelle's work " *On the Plurality of*

then have, like us, more nationalism than patriotism, that is to say, more inert force, than power of action.

In countries where activity and handiness clearly characterise the spirit of the masses, and shine at the head of all their attributes, people are ashamed of poverty, because it indicates, up to a certain point, the absence of these qualities.   This is what one sees in England, where the avowal of misery is painful, and equivalent to the confession of a crime ; every man thinks that he raises himself in the estimation of others in saying that he is rich.   In Spain, where neither activity nor handiness are inborn characteristics of the people, poverty is for no one a brand of disgrace.   In France, where knowledge is more esteemed than handiness ; meditation than action ; and where intellectual capacity need only demonstrate itself to gather riches, poverty is acknowledged without much scruple.

¶ 403.
National feelings with regard to poverty.

Goodwill [*i.e.*, liberty of action] and liberty, says Bœhme, are the same thing, but goodwill and sociability are two very different things.[308]   One is more free but less sociable where liberty of action is strongly de-

¶ 404.
Liberty of action
Bœhme.

Worlds" (London : 1738), p. 198] we find the following :—" At length rose Cartesius, a happier genius, who has bravely asserted the truth against the united force of all opposers, and has brought on the stage a new method of philosophising. . . . A great man indeed he was, and the only one we envy FRANCE.   He solved the difficulties of the universe almost as well as if he had been its architect," etc., etc., etc.   [A most interesting oration on the reformation of science which occurred in the seventeenth century.]

[308] I do not know exactly *whence* M. d'Arpentigny takes this quotation from the lucubrations of Bœhme. The 1764 edition of his works—the only one I have by me—consists of 200 folio pages of " dissertation " upon Will and Liberty, of which the following is not only an excellent sample, but seems near the point which our author desires to illustrate.   " And in the breaking of the Darkness, the reconceived Will is free, and dwells

[¶ 404]

veloped, as in England, where originality is highly esteemed; one is more sociable but less free where liberty of action is restricted, as in France, where *English drama.* conformity is appreciated. The English drama is passionate and farcical, whereas ours is restrained and humorous; they exhibit more poetry and contrast, whilst we exhibit more art and harmony.

¶ 405.
Application of the above principles. Germany.

You can apply these principles of cheirognomical philosophy to the study of other nationalities; as for instance Germany, a blonde and cold country, which extols the triple intoxication of contemplation, music, and tobacco. There people live seriously, and dream enormously; there they drink out of huge goblets and read out of huge books. The enormous folios of the Encyclopædia open uniform with equally minute dissertations upon the words God, universe, and dandelion. It is the country of inflated poetry, of rigid military minds, of enthusiastic metaphysicians, and of phlegmatic postillions. The ideas which are honourable there, are too positive for us, or are not sufficiently so, for we experience as much repugnance for people who are absorbed in the esoteric essences of things and whose minds attach themselves only to the incomprehensible, as for people whose thoughts cannot soar above the levels of absolute *matter*. Germany would not offer the afflicting spectacle of a noble and learned nation governed by absolutism, if the reasoning portion of its population was more

without the Darkness in itself; and the Flash which there is the Separation and the Sharpness and the Noise (or Sound) is the Dwelling of the Will free from the Darkness. And the Flash elevates the Will, and the Will triumphs in the Sharpness of the Flash, and the Will discovers itself in the Sharpness of the Sound in the Flash of the Light, without the Darkness, in the Breaking, in the Infinity," etc., etc., etc. "*The Works of Jacob Bœhmen*" [or *Behmen*], "The Three Principles of the Divine Essence" (London: 1764), vol. i., p. 136.

capable of action, and if the active portion of its population was more capable of reason. She is full of worthy folk who, gifted with more soul than intelligence, are more fit for good fortune than pleasure. Self-contained in their joy, and lyric in their moments of intemperance, they surpass all other nations in freedom, in innocence, and in good nature.

With them comedy is a matter of *sentiment;* pathetic and expansive, it pourtrays man directed by his heart and by his instinct. With us it is a *matter of judgment;* discreet and restrained, it exhibits man as he is formed by education and by society.. Romantic and synthetic on that side of the Rhine, historic and analytic on this. In France it aims at the *true,* and proposes to redress evils by mockery and laughter; witty rather than tender, it *amuses,* and, appealing to the mind, it *instructs.* In Germany it aims at the *beautiful,* and aims at the redress of evils by tears; tender rather than witty, it *interests,* and, appealing to the heart, it *improves.*

¶ 406.
French and German comedy.

From which data I conclude that comedy is the domain of conic hands in Germany, and of square hands in France. See only, in the vestibule of the Théâtre Français, the busts of Molière, of Regnard, of Dancourt, of Lesage, of Marivaux. They have all of them the aquiline nose, which almost invariably accompanies a square formation of the fingers.

¶ 407.
Comedy writers' hands.

SPATULATE HANDS [*continued*].

The Hands of the North Americans.

¶ **408.**
American
character.

To a higher degree even than the English, from whom they originally sprung, the inhabitants of North America pay the greatest attention to that instruction which teaches them how to act upon matter so as to *utilise* it. Listen to the description of them by Michael Chevalier in his " Dixième Lettre sur les États Unis : "—

Chevalier

¶ **409.**
American
manners and
morals.

"The Yankee is reserved, concentrated, defiant. His humour is pensive and sombre, but unchangeable. His attitude is without grace, but modest and without baseness. His manner of addressing one is cold, often but little prepossessing ; his ideas are narrow but practical ; and he appreciates what accords with the fitness of things, but not what is grand. He is absolutely devoid of chivalrous feeling, but yet he is adventurous. He delights in a wandering life, and has an imagination which gives birth to original ideas,—ideas which are not poetic, but eccentric. The Yankee is the prototype of the hard-working ant ; he is industrious and sober, economical, cunning, subtle, cautious, always calculating, and vain of the tricks with which he takes in the inattentive or confiding purchaser. He is rarely hospitable ; he is a ready speaker,

[¶ 409]

but at the same time he is not a brilliant orator, but a calm logician. He lacks that largeness of mind and of heart, which, enabling him to understand and appreciate the natures of his fellow-men, would make him a statesman; but he is a clever adminis-trator and a man of huge enterprise. If he is but slightly capable of managing men, he has not his equal in the administration of things, in the arts of classifying them, and of weighing them one against the other.

"Though he is the most consummate trader, it is pre-eminently as a colonist that the Yankee excites our admiration; impervious to fatigue, he engages in a hand-to-hand struggle with Nature, at the end of which his tenacity always renders him victorious. He is the first mariner of the world, the ocean is his slave. The most tender passions are slain in him by his religious austerity, and the pre-occupations of his worldly profession. To the spirit of trade, by the aid of which he derives advantage from what he gets out of the earth, he joins the genius of labour which makes it fruitful, and of mechanics which give form to the fruits of his labour."[309] In a nation such as this there cannot exist any but hands which are spatulate, and fingers which are square.

¶ 410.
American colonists.

The good which has been done to the poor by spatulate hands in Russia, in England, and in America has been very small. In Sclavonic Russia, where they have reigned uncontrolled since the invasion by

¶ 411.
The poor among spatulate hands
Russia.

---

[309] The above passage is considerably condensed from pp. 168—72 of Michael Chevalier's "*Lettres sur l'Amérique du Nord*" (Paris: 1836, 2 vols.), of which there exists a translation, entitled "*Manners and Politics in the United States, being a Series of Letters on North America*," translated from the 3rd Paris edition (Boston: 1839), on pp. 116—17 of which the passage quoted may be found.

[¶ 411]

the Scandinavian Rurik,[310] and where the elementary hand, which is that of the masses, is in a state of slavery, the soldier, harshly subjected to the punctilious exigencies of an iron-handed discipline, harassed in turn by the evil genius of barbarism and by the evil genius of civilisation, dares not to allow his glances to wander beyond the limit proscribed by the ever-visible shadow of the knout. In England, where the great majority of the people have no other pole-star than the *commissariat*, the insatiable voracity of the great, leaves but very inadequate relief for that of the humble. In puritanical America the workman's life is a free one, but repose and pleasure are interdicted to him. The life of a Catholic convent is not more gloomy or rigid than that of the manufactories of Lowel for instance. At Pittsburg, work ceases only at necessary intervals for food; and the longest meal during the day does not last more than ten minutes. Man there is reputed only to possess a stomach and two arms. The rest of him does not count.

England.

America

¶ 412.
The dominant class in America. In the same way that the Laplander could not form any idea of a paradise without snow, the Yankee could not understand happiness without labour. Pursuing the same logical method as ours, when we relegate political power to *thinkers*, *i.e.*, to the people of physical leisure, who are always few in number,

---

[310] It was at the call of the Slavs in 861 that Rurik, Sineons, and Trouvor [whose names signify the *Peaceful*, the *Victorious*, and the *Faithful*] crossed the Baltic from Scandinavia, and established in Russia a rule which lasted down to the reign of the son of Ivan the Terrible in the sixteenth century. *Vide* Alfred Rambaud's "*History of Russia*," translated by L. Lang (London : 1879), vol. ix., p. 56, and compare " *La Chronique de Nestor* " [a Russian monk of the eleventh century], traduite en français par L. Paris (Paris : 1834), vol. 1., p. 20.

[because in our opinion pre-eminence belongs to thought], the Yankee has granted the same power to *workers*, who always constitute the larger half of the community, because, in his opinion, pre-eminence belongs to labour.

In his love of labour he has branded as immoral everything that could hinder his work, even to the pastimes which we regard as the most innocent and the most allowable.

¶ 413.
American love of work.

As yet the Americans are simply a nation of super-cargoes, of pioneers, of farmers, of mercers; their laughable pretensions to high tone and elegance of manner are a sufficient proof of this. Cooper has written whole pages· where this punctilious self-sufficiency is spread out in long and heavily-worded periods, in a manner which, symmetrical and pedantic, is strenuously opposed to the freedom and toleration of good society.

¶ 414.
American refinement.

Whilst the French, a nation relatively lazy, deprive themselves of a quantity of material comforts so as to bequeath to their children the means of living an idle life, the English, on the other hand, spend and consume their substance with a commensurate lack of scruple; having no fear of penury themselves, they do not consider it a misfortune for their children to be made heirs of necessity.

¶ 415.
French and English providence.

From which facts I conclude that our good qualities as well as our vices [if such we have], oppose themselves with equal force to the increase of the productive instinct among us. We consume—and, by consequence, produce by manual labour—less than the English, but more than the Spanish, who, in their turn, do less work than us, but more than the Arabs. It has always been thus; and in the same way we must seek in the instincts peculiar to our nation, and *not* in the continually recommended imitation of the materialistic procedure of the English, the counter-

¶ 416.
French, English and Spanish labour contrasted.

[¶ 416]

poise of the remarkable increase of power and prosperity which they have derived from the more
complete development of the genius which is peculiar
to them.

¶ 417.
Necessity of
proper
environment.

The same thing may be said of the types as of
plants—*i.e.*, that they do not shine forth in all their
brilliancy, and do not bring forth all their fruits,
excepting beneath certain latitudes.   Where *moral*
requirements are more regarded than *physical* necessities [and this is what happens in the rich and fertile
countries of the South] nature has given pre-eminence
to the types of men characterised as "southern."
Where, on the other hand, *physical* necessities require
to be more attended to than *moral* [and this is what
occurs in sterile countries and in the North], nature
has given the pre-eminence to "northern" types of
character.

SPATULATE HANDS [*continued*].   CONTINUATION

The Veneration of All People for Pointed Fingers.

ALL nations, however different [physically and morally] they may be from one another, whatever may be the form of their government, the spirit of their culture, or the nature of their ideas upon beauty, worth, truth, and usefulness, agree unanimously in giving pointed or conical fingers to pictorial or sculptured representations of angels or good genii, with which each race, according to its education, considers the heavens to be peopled. Even down to the Chinese and Japanese, unprogressive nations which measure beauty, goodness, and good fortune by the extent of the corporeal development,[311] and for whom the fine arts and liberty, as we understand them, do not exist, all peoples are agreed in this common admiration. The entire human race sees nothing but beauty and elegance in a pointed-fingered hand.

¶ 418. Universal love of beauty as exemplified by fine hands.

[311] "The people of China are divided into four classes or categories of citizens, according to the merits and honours that custom and the law of the land attribute to each. These classes are the literary, the agricultural, the manufacturing, and the commercial. Such is the order of the social hierarchy in China. . . . In point of fact, the two classes esteemed and honoured are the two

¶ 419.
Leisure as the
emblem of
mental
superiority

The fact is that our one great loss [sanctioned by the necessity of labour] is above all things exhibited to us *by* this cruel and humiliating necessity; and it is hence that arises our instinctive respect for leisure, and the sentiment which impels us to presuppose, as the attributes of the beings which are the objects of our love and adoration, hands which to our minds convey the impressions of the ideality, and quasi-divine instincts of contemplation and repose.

¶ 420.
Respect for
idleness.

Among barbarous tribes, as among civilised nations, in the eyes of the masses, the man who does *least* for himself excites a respect which is secondary only to that paid to the man who does *nothing.* Is this a circumstance without significance ?

first; they constitute the aristocracy of mind and of labour. Our nobility could only inscribe upon their blazon a pen—*i.e.*, a paint brush—or a plough : in one, Heaven for field ; in the other, Earth. . . . The Chinese hierarchy is not founded upon seniority, but upon merit. The degree fixes the position, and the higher the position, the higher must be the merit of its occupant." TCHENG-KI-TONG, " *The Chinese Painted by Themselves* " (London : 1885, pp. 61 and 69). The national characteristic of the Chinese hand is its extreme pointedness,—a formation [vide ¶¶ 136-7] entirely in accordance with the contemplative psychological mind of the Chinaman.

SPATULATE HANDS [*continued*].

Roman Hands.

Such hands as I described in the last chapter were not, such could never be, the hands of the Sovereign People. Devoted to war, and to continual movement by the peculiar organisation transmitted to them by the heroes and the warriors who came together at the call of the child of the Brazen Wolf, the Romans have received as their portion, the talent of the arts necessary to men of action ; they excelled in bodily action and in the handling of arms, in the construction of aqueducts, of bridges, of high roads, of camps, of engines of war, and of fortresses. For poetry they had but a reflex and passing fancy, for the fine arts merely a taste born of vanity, despising speculative notions, and respecting nothing but war, political eloquence, history, jurisprudence, and sensual pleasures.

¶ 421. Energy and activity of Roman hands.

As soon as the powerful hands which they had kept for so long gripped upon the enslaved world, turned at last from their specialty by Christian spiritualism, began to be raised towards heaven, the world escaped from their dominion. And it is mere repetition of an opinion which has frequently been expressed, to say that Platonism was not more fatal

¶ 422. Fatal influence of religion and culture upon warlike hands.

18

to the Greeks as a nation devoted to the cultivation
of the beautiful, and governed by ideas proper to
artistic hands,[312] than was Christianity to the Romans
as a people governing the universe by the power of
ideas proper to square and spatulate hands. Speaking
politically, the murders of Socrates and of Christ were
necessary acts of justice. Regard being had to the
good which their systems of morality have since
produced, the human race takes no thought of their
socialist incendiarism ; but Athens and Rome, mortally
wounded in the ideas upon which they existed by the
ultramontane spiritualism of the principles of these
two reformers,—principles which aimed at nothing
less than the substitution of individual intelligence
for that of the masses, aristocracy for democracy,—
were bound to condemn them to death. Among the

[312] This is not a place in which to discuss the effects
of Platonism, or even of Neo-Platonism, upon the Greek
character and constitution. " The ethics of Plato," as
Mr. G. H. Lewes says in his " *History of Philosophy* "
(London: 1880, vol. i., p. 271), " might suit the inhabitants of another world ; they are useless to the inhabitants of this." It was, however, *Neo*-Platonism
which led to the downfall of the Greek Empire, rather
than Platonism : for the latter, following as it did, and
becoming to a certain extent mingled with, the tenets
of the Stoic philosophers, produced a system of philosophy than which little could have been more perfect.
" Stoicism," says Lecky [" *History of European
Morals* " by William Lecky (London : 1877, vol. i.,
p. 325)], " placed beyond cavil the great distinctions
between right and wrong. . . . The early Platonists
corrected the exaggerations of Stoicism, gave free scope
to the amiable qualities, and supplied a theory of right
and wrong, suited not merely for heroic characters and
extreme emergencies, but also for the characters and
circumstances of common life." It was when the
Neo-Platonists, such as Apuleius, Ammonius Saccas,
Plotinus, Porphyrius, and the rest of them, began to
preach their doctrines, that the evil influence of the new
school began to make itself felt,—doctrines which as
Lecky says [p. 328], " made men credulous, because they
suppressed that critical spirit which is the sole barrier

thirty tyrants which Anytus, the enemy of Socrates,[313] assisted Thrasibulus to overthrow, there were *three* disciples of that philosopher ; well, is it not well known that these thirty spilled in eight months more innocent blood than the people had shed during many centuries ?

Xenophon has written upon the Athenian republic some reflections which, to my mind, utterly refute the opinion which he has expressed upon the injustice of the condemnation of his illustrious master :—" There are men," says he, " who are astonished to see that, as a rule, they favour [*in Athens*] artizans, the poor, and the common people more than the honest citizens. It is, however, the surest way to preserve the popular condition of things ; in fact, if the poor, the plebeian, and lower classes are happy, they increase in numbers, and there you have the *strength* of a democracy. If, on the contrary, it is the rich and the well-born who are the most considered, democracy raises against herself an inimical power. They say that we ought

¶ 423.
Xenophon on
Socialism.

to the ever-encroaching imagination ; . . . because it found a nervous, diseased, expectant temperament, ever prone to hallucinations, ever agitated by vague and uncertain feelings that were readily attributed to inspiration. As a moral system, indeed, it carried the purification of the feelings and imagination to a higher perfection than any preceding school, *but it had the deadly fault of separating sentiment from action.* . . . The early Platonists, though they dwelt very strongly on mental discipline, were equally practical " [*i.e.*, as practical as the Stoics], so surely it was *Neo-Platonism* which was to blame.

[313] It was Anytus, who with Melitus and Lycon,—encouraged by the delight with which the Athenians had received Aristophanes' comedy, " *The Clouds* " [which had been written at their instigation in ridicule of the teaching of Socrates],—first stood forth and accused Socrates of making innovations in the accepted religion of Greece before the tribunal of the five hundred,—the accusation which brought forth the glorious "*Apology of Socrates.*"

not to allow all men indiscriminately to harangue
and to be members of the council, but only those
who are distinguished by the most talent and the
most virtue.   There is, however, nothing more
wise than to allow even the lowest plebeian to
speak in public.   If the first citizens only had the
exclusive right to harangue and to sit in council,
it would be a benefit for their class, but not
for the populace; instead of which the humblest
artizan, being allowed to rise and harangue the
assembly, brings before it ideas and suggestions which
conduce to the welfare of himself and of his class.

¶ 424.
Value of popular
oratory.

"But," people reply, "what will a man of this sort
say which can be useful either for himself or for the
people of his class?   In the opinion of the masses,
this man, be he whom he may, with his ignorance,
but with his zeal for the democracy, is worth more
than a well-to-do citizen with grandiose views and keen
penetration, but perfidious intentions."

¶ 425.
Its advantages.

"Perhaps this plan is not the best possible; at
least it tends to the perpetuation of the democracy.
The people requires, not a learned administration
which would enslave it, but liberty and sovereignty
in itself.   Given this, the constitution may be vicious,
but that is the least of its troubles.   What seems to
you to be defective in the political system, is precisely
what makes the people powerful and free."[314]

[314] The above paragraphs are condensed from the
" *Atheniensium Respublica* " of Xenophon, and con-
sist of the passages cited below.[*]

[*] "Ἔπειτα δὲ, δι ἔνιοι θαυμάζουσιν ὅτι πανταχοῦ πλέον νέμουσι,
ἢ τοῖς χρηστοῖς, ἐν αὐτῷ φανοῦνται τὴν δημοκρατίαν διασώζοντες.
Οἱ μὲν γὰρ πένητε καὶ οἱ δημόται καὶ οἱ χείρους, εὖ πράττοντες, καὶ
πολλοὶ οἱ τοιοῦτοι γιγνόμενοι, τὴν δημοκρατίαν αὔξουσιν·" etc., etc.
"'Εἰ μὲν γὰρ οἱ χρηστοὶ ἔλεγον καὶ ἐβουλεύοντο τοῖς ὁμοίοις σφίσιν
αὐτοῖς ἦν ἀγαθά, τοῖς δὲ δημοτικοῖς οὐκ' ἀγαθά," etc., etc., *usque ad*
" ὃ γὰρ σὺ νομίζεις οὐκ εὐνομεῖσθαι, αὐτὸς ἀπὸ τούτου ἰσχύει ὁ δῆμος,
καὶ ἐλεύθερός ἐστιν."—ΑΘΗΝΑΙΩΝ ΠΟΛΙΤΕΙΑ, κεφ. ά.

A nation ought to strive to discover what are its special aptitudes, and this having been done, to act upon this knowledge, and not to embark upon lines which are foreign to its special powers. The North Americans are especially fitted for the industrial arts which assist the lesser sciences; the French excel in the industrial arts which lead to the liberal ones; the English are fitted for industrial arts which assist the higher sciences,— is there not enough in these specialties to satisfy the activity of these three nations? The *moral* graft taking effect as it does upon certain individuals whom it causes to produce fruits of their labour foreign to their natures, could not obtain a hold upon the masses of a nation. The spirit of a nation may be modified, but it cannot be completely transformed. Then let England, rich in spatulate hands, cover continents with her colonies, and seas with her ships; let France, rich as she is in artistic and philosophic hands, scatter broadcast her ideas as the former nation does her men! But the practice of the fine arts by the survivors and perpetuators of the Carthaginian spirit, and the practice of industry by the perpetuators of the Greek genius, will be, for both of these nations, for a long time at all events, nothing more than an indigent source of negative glory, of equivocal success, and of doubtful profit.

¶ 426.
Development of national aptitudes.

There are faculties and qualities which are held in high esteem, which we do not all of us possess, but which we should all like to appear to possess. There are others which we practically do already possess, but of which we dare not to be proud, as if everything was not worthy of consideration when kept in its proper place. This is the sunken rock on which those who, after studying these cheirognomical theories, may undertake to put them into practice are likely to strike. They will often have to rely upon the assurance of third persons to the effect that they have

¶ 427.
Ignorance of self and neglect of specialist talents.

properly pronounced upon the genius which is attached to any particular hand.   Often this denial of one's true character is made in good faith, for how many men are, and always will be, ignorant of themselves ; but often it will result from wounded self-esteem.   No one, for instance, in the polite world expects to be told that he has the talent of manual labour and not of the fine arts, and that the temple of the muses is closed to him.   There must intervene

Prometheus. Prometheus, who, having stolen fire from heaven, *Dædalus.* taught the use of it to man ;[315] Dædalus, who invented the saw, the axe, the sails and masts of ships, and by these means added wings to the shoulders of *Papin and Fulton.* human genius ;[316] and Papin[317] and Fulton,[318] those *Cæsar.* modern Promethei ; and Cæsar, who in his Commentaries dwells more lovingly over his labours of engineering skill than over his strategic powers as a general and his prowess as a soldier ;[319] and

[315] Vide Apollodorus, i. and ii.

[316] M. d'Arpentigny's remark in this place coincides with the opinion which has been expressed by many learned commentators ; *i.e.*, that the flight of Dædalus and Icarus from Crete with wings may be accounted for by the fact that, being the first to use sails, it is not at all improbable that the ancients, seeing them for the first time or from a distance, took these contrivances for wings.

[317] *Vide* note [297], p. 248.

[318] An American engineer, born of Irish parents in Little Britain, Pennsylvania. He may well be compared, on account of the versatility of his genius, to Prometheus, having been in turn a landscape painter, a watchmaker, and a mechanician, and having been equally successful in each vocation. He came to London at the age of twenty-two. He was the inventor of many great mechanical works, and was the first projector of the "Nautilus," or submarine boat, and also of the torpedo system of warfare. In 1807 he started the first steamer on the Hudson. He died in 1815.

[319] This can hardly be said without reservation ; but certainly the minuteness with which Cæsar continually

Charlemagne, who used a doctrine as if it had been ⟶ *Charlemagne.*
an axe;[320] and Peter the Great, who used the axe as ⟶ *Peter the Great.*
if it had been a doctrine;[321] and Macchiavelli, who ⟶ *Macchiavelli.*
teaches us to make use of men as if they were so
much working material, and in whose eyes any-
thing is justified by final success;[322] and Diderot ⟶ *Diderot and Arago.*

enters into particulars of the construction of various
appliances of his campaigns is one of the great charms
of his work. *Vide,* for instance, the description of the
Venetic ships ["*De Bello Gallico,*" iii., 13], of his
bridge over the Rhine [*op. cit.,* iv., 17], his plans
for the special ships to be used against the Pirustæ
[*op. cit.,* v., 1], the description of Trebonius' engines
of war ["*De Bello Civili,*" ii., 8], and a number of
minor descriptions.

[320] We find a good instance of the proselytising operations of Charlemagne in his secretary's account of his
life ["*Vita et Gesta Karoli Magni per Eginhartum
ejus Secretarium descripta*" (Lipsiæ: 1616), p. 11], in
the account of how he bound over the Saxons, after he
had conquered them, to abjure their worship of dæmons
and the superstitions of their ancestors, and to embrace
the Christian religion.*

[321] As, for instance, in the punishment of the inhabitants of Strelitz, who had revolted against him. "Le
cruel, du haut de son trône, assiste dun œil sec à ces
exécutions; il fait plus, il mêle aux joies des festins
l'horreur des supplices. Ivré de vin et de sang, le verre
d'une main et la hache de l'autre, en une seule heure
vingt libations successives marquent la chute de vingt
têtes de Strelitz, qu'il abat à ses pieds, en s'enorgueillis-
sant de son horrible addresse. Quatre-vingts Strelitz ...
sont trainés à Moscou; et leurs têtes, qu'un boyard
tient successivement par les cheveux, tombent encore
sous la hache du Czar."— M. DE SEGUR, "*Histoire de
Russie et de Pierre le Grand*" (Paris: 1829), liv. viii.,
ch. ii., p. 328.

[322] No one ever lived who more than Nicolo Macchiavelli governed his life on the Horatian text "Rem, facias
rem—recte si possis; si non, *quocunquo modo* rem."
It is probable that his unenviable reputation for absolute

* "Eaque conditione a rege proposita, et ab illis suscepta,
tractum per tot annos bellum constat esse finitum, ut abjecto
dæmonum cultu, et relictis patriis cerimoniis, Christianæ fidei
atque religionis sacramenta percipiant."

[¶ 427]

and F. Arago, who resolutely deny, the one, everything that his subtle logic cannot explain,[323] the other,
everything that his winged figures cannot reduce;[324]

Monge and Wast. and Monge[325] and Wast, κ.τ.λ., all those giants of the
racial hand among spatulate subjects; these glorious

unscrupulousness and unrestrained cunning is due
almost entirely to the seventh and eighth chapters of "*Il
Principe*" (Rome: 1532), in which he lays down the
axioms categorically that any means may without
scruple be adopted for the attainment of power and the
government of states,* that good governors are often
established on their thrones by evil means,† and that
honesty and good faith are merely "spectres, raised to
frighten fools."‡

[323] Denis Diderot, the illustrious author of the "*En-
cyclopédie*" [1749-67], born in 1713, died in 1784, was
perhaps one of the sincerest and most pronounced
atheists that ever lived. One of his most celebrated
sayings sufficiently illustrates the above citation of his
name, viz.:—"The first step towards philosophy is
incredulity."

[324] I presume our author means D. [Dominique] Arago,
the celebrated astronomer and natural philosopher
[born 1786, died 1853], who, with Gay Lussac, confirmed
the undulatory theory of light which had been stated
and expounded by Young and Fresnel [*Vide "A Manual
of Cheirosophy*," ¶ 63].

[325] Gaspard Monge, Comte de Peluse, an eminent

* "Dipoi, gli Stati che vengono subito, come tutte le altre
cose della natura che nascono e crescon presto, non possono aver
le radice e corrispondenze loro, in che il primo tempo avverso
non le spenga; se gia quelli tali, come é detto, che si in un
subito son diventati principi, non sono di tanta virtu, che quello
che la fortuna ha messo loro in grembo sappino subito prepararsi
a conservare; e quelli fondamenti che gli altri hanno fatti avanti
che diventino principi, gli facciano poi. Io voglio al' uno e l'altro
di questi modi, circa il diventar principe per virtù o per fortuna,
adurre duoi esempii stati ne' di della memoria nostra; questi sono
Francesco Sforza e Cesare Borgia."—"*Il Principe e Altre Scritti
Politici di Nicolo Macchiavelli*" (Firenze: 1862), ch. vii., "De
principati nuovi, che con forze d'altri e per fortuna s'acquistano,"
p. 41.
† *Op. cit.*, p. 54., cap. viii. "Di quelli che per scellera-
tezze sono pervenuti al principato."
‡ *Op. cit.*, p. 112, cap. xviii. "In che modo i principi
debbiano asservare la fede."

names, I repeat, must be cited to many subjects before they will recognise in themselves the talents which are indicated by these spatulate fingers.

It is the same with nations and with individuals; here you find the same ignorance, and there the same sensitiveness. The Italians of the present day, do they know what they want, what is best suited to them? The Belgians, a nation of shopkeepers, dull, restless fetich-worshippers, perpetually on their knees before the golden calf, do they not look upon themselves as a model nation? And the Russians, because they understand the use of paste jewels and pinchbeck, and because they cheat at cards and disdain women, look upon themselves as civilised!

¶ 428. Misapprehension of national characteristics.

This must be understood to refer to masses, for everywhere, and among all types, there exist individual exceptions—Pindar, Hesiod, Plutarch, and Epaminondas were Bœotians.[326] The great Corneille was born in Rouen,[327] the town of mean and materialistic interests; there are people of a generous, brisk,

¶ 429. Great exceptions to local and national rules.

French mathematician and physicist, founder of the *Ecole Polytechnique*, which gives him a claim to the above citation. Born in 1746, died in 1818; for an account of his life and labours *vide* Dupin's "*Essai Historique sur les Services et les Travaux Scientifiques de Monge*" (Paris: 1819).

[326] Pindar was born at Thebæ, Plutarch at Coronea, Hesiod at Ascra, and Epaminondas at Thebes. Notwithstanding the fact that these men were born, and Helicon, the mountain *par excellence* of the Muses, was situated, in Bœotia, the inhabitants of that province were always looked upon as particularly boorish and illiterate, an anomaly which has been recorded by Horace ["*Epistolarum*" ii., 1., 1. 244] in the lines

> "Quodsi
> Judicium subtile vivendis artibus illud
> Ad libros et ad hæc Musarum dona vocares
> Ad Bœotum in crasso jurares aëre natum."

[327] On 6th June, 1606. The only qualification which this town has for the contempt with which our author alludes to it is the fact that it is perhaps the greatest

[¶ 429]

and subtle disposition to be found in Belgium. Some of
us have heard an Englishman sing in tune! Lesage [328]
was born in dreamy and romantic Brittany—the province which the wits of the seventeenth century nick-
named, on account of its want of practical intelligence,
the French Bœotia. [329]

¶ 430.
Civilisation dates
from the
invention of
windows and
chimneys.

Glory be to spatulate hands! Without them solid
and powerful society could not exist. Without the art
of glass manufacture, to go no farther than that; [330]
without the invention of chimneys as we have them
to-day [an invention which goes no farther back
than the fourteenth century], [331] both of them the
productions of spatulate hands, we should still
be hardly more than semi-barbarians. Keeping it,
as it were, in a forcing-frame, these two inventions
have placed civilisation, the spontaneously generated
flower of warm climates, beyond the reach of the
external influences which in our inclement latitudes

---

manufacturing town in France. *Vide* Guizot's
"*Corneille et son Temps*" (Paris: 1851); English
translation (London: 1852), and E. Desjardin's "*Le
Grand Corneille*" (Paris: 1861).

[328] Alain Réné Lesage, the author of "*Gil Blas de
Santillane*" (Paris: 1715), and "*Le Diable Boiteux*,"
born at Sarzeau 1668, died 1747.

[329] *Vide* note [327].

[330] M. le Capitaine d'Arpentigny takes us, in this paragraph, far enough up the vista of past time, seeing that
glass manufacture is represented pictorially on Egyptian
mummy cases of the twelfth dynasty [B.C. 1800], and
is mentioned in papyri of the *fifth* and *sixth* dynasties.

[331] It has always been a matter of dispute whether any
artificial means of drawing off smoke from their houses
was known to the ancient Romans. Our author is wrong
in attributing the invention of chimneys to the fourteenth
century; in Rochester Castle there are elementary but
true chimneys dating from [*circa*] 1130, and after this
date there are many perfect specimens of the chimney
proper. The beautiful stacks of clustered chimneys,
such as are seen to such perfection on some of the old
Kentish houses, date probably from the fifteenth century.

would have been fatal to its culture and its development.

In a word, it is Heracles whom spatulate-handed
men must set up as a model for themselves. They
will then have nerves in harmony with their temperament, which is sanguine, and with their bones,
which will be big and strong. Let us take a
couple of illustrations. Chopin, the pianist, with his
spatulate hands and *small thumbs* did not fulfil
these conditions. His nerves, of an extreme fineness, were not in proportion to his powerful frame-
work; he was like a violoncello strung up with
violin strings. Thus he failed to produce the
tones which expert physiologists felt that he ought
to have produced. They expected vigour, energy,
and precision, and instead he wrapped himself up,
like velvet-fingered artists, in small and checked
harmonies. A prey to two tendencies, which were
constantly dragging him in opposite directions, he
knew not to which to pay attention. What his blood
urged him, his nerves would not allow him to do;
he aimed at activity, and enervated himself in repose;
he cried after chargers, and rode upon clouds; he
longed to howl like the tempest, but an inward
monitor, the sheet lightning of beautiful faces, and I
know not what yearning after the crested panels of
the heraldic tabard, compelled him to chaunt in a
sotto voce. Courteous and smiling, with a strange
shadow in his eyes, he was one of those creatures
whom the slightest thing startles. He had, with
regard to the world, to the bosom of solitude and to
the loneliness of the centre of the earth, aspirations
replete with restless hopes, with dreamy sadness,
and tender prostrations, which reproduced themselves
with chaste and poetic grace in his compositions;
with a finer organisation he would have been
happier, but he would have had less talent. Its

¶ **431.**
Necessity of
harmony in
organisation

Chopin.

[¶ 431]

charm proceeded from his sufferings; and, as the principle of vitality is in the nerves, Chopin died young.[332]

¶ 432.
The same.
L'Abbé de
Lammenais.

The Abbé de Lammenais was another anomaly of the spatulate and large-thumbed type; but he was wanting in physical construction and not in nerve power. His nerves were as strong and solid as his body was thin, mean, and deformed.[333] He had in his brain, the activity and combativeness which the well-formed and highly-endowed spatulate subject has in his blood. He loved perpetual intellectual strifes and contentions; he fenced with his pen, since he could not do so with his sword. His eyes sparkled at the recital of a battle-scene, and his spatulate hands would keep beating a tattoo during the narration. Mathematician, and highly-versed in all the higher sciences,—litterateur, theologian, and philosopher, he had [like M. de Cobenzl[334]] "horses ready-saddled for

[332] Frédéric François Chopin, the celebrated Polish pianist and composer, was born near Warsaw on the 8th February, 1810. His biographer, in Fetis' "*Dictionnaire des Musiciens*," speaks of "the delicacy and grace of his execution, the result at the same time of his physical constitution and of his sentimental organisation." It was in 1837 that his fatal malady made itself for the first time seriously felt, and from this time until his death it continually increased in virulence. In 1848, after the outbreak of the French revolution, he visited England, and returning to Paris in a dying condition, ended his days there on the 17th October, 1849. "There was in the genius of Chopin," says his biographer, "here and there an energy, but it always seemed to exhaust itself, and his delicate constitution continually recalled his talent within the small compass intended for it."

[333] Larousse ["*Dict. du XIXᵉ Siècle*"] describes him as "né avec un vice de conformation," but neither Castille ["*Portraits Historiques*" (Paris: 1857)] nor A. Blaise ["*Essai Biographique sur M. de Lammenais*" (Paris: 1858)] record that he was actually deformed.

[334] It is not clear to my mind which member of the

every kind of exercise." But,—like "woman and the
art of lying," [335]—knowledge and doubt were born on
the same day; and the end of it was that, turned away
from his faith in his old standard, by science, the good
Abbé one morning found that he no longer knew
from whom,—from the individual man or the community of men, from the pope or the people,—to derive
the principle of authority, having found as many distinct asseverations of the human and conventional
infallibility of the pope, as of the Divine and real infallibility of the people. I should add that at the same
time he was as obstinate as a mule and as sober as a
camel, and that by his figure and style of eloquence
he recalled those haggard and vehement prophets that
in old times Judæa, always greedy of burning phrases,
harboured among her rocky wastes and dusty fig

Cobenzl family is here referred to. Hœlfer, in Didot's
"Biographie Générale" (Paris: 1853-66), mentions
three of the family to whom, on account of their rare
versatility, M. d'Arpentigny might refer with the simile
that "they had horses ready-saddled for every kind
of exercise." Charles de Cobenzl (born 1712, died
1770) was the founder of the "Académie des Sciences"
of Brussels and of the free Art School; his son, Louis
de Cobenzl (born 1753, died 1808) is, however, most
probably the one who is here referred to; he was minister
successively to the courts of Copenhagen [1774], Berlin
[1777], and St. Petersburg [1779-1797]. "He insinuated
himself," says his biographer, "into the good graces of
Catherine II. as much by his diplomatic ability as by
his amiability; ... he composed dramas for the imperial
theatre, and even acted in them himself." The author
of the *"Mémoires de Ségur"* (Paris: 1827, vol. ii.,
pp. 265 and 280), makes him out to be a kind of
Macchiavelli in politics, whilst paying a high tribute to
his social qualities.

[335] *Vide* also ¶ 662, and note [235], p. 202. "No
religious ceremony for women should be accompanied
by *Mantras*—with these words the rule of right is fixed;
for women being weak creatures, and having no share in
the *mantras*, are falsehood itself."—*"The Ordinances
of Manu"* [*vide* note [235], p. 202], lect. ix., ver. 18.

[¶ 432]

trees.   He died old; When he was young the chase
and fencing had been his chief delights.[336]

¶ 433.
Inevitableness of
anomalies in the
types.

Anomalies of this description exist in all the types,
and it even sometimes occurs, rarely no doubt, but
often enough to make this warning necessary, that
the principles laid down in this volume are absolutely contradicted and denied by startling exceptions,
so far is physiology from being even a *comparatively* exact science.[337]   Thus : conical were the hands
of " l'Homme de Brumaire," [338] the enemy of liberty,
whose instincts were shocked by psychological ideas ;
who took the public instruction out of the hands of
the great Fourcroy, to confide it to those of the inept
Fontanes, who substituted discipline and money, the
means of Cæsar's successes, for enthusiasm and
glory, the mainsprings of the republican power, and
who, in a word, left France as poor, as ignorant,
and less important than he found her.   For the rest,
this man's sadly overrated hands were neither fine
nor delicately cut ; they were, on the contrary, strong,

Napoleon.

---

[336] " Pale, mean-looking, and sickly," says Eugène de
Mirecourt ["*Les Contemporains,*" pt. v. (Paris : 1856),
p. 36], " M. de Lammenais was born to be the Bossuet
of our century.   God had crowned his brows with the
aureole of genius.   All the splendours of intellect illumined his soul " [p. 94].   He was born 19th June, 1782,
and died in February 1854.

[337] I have only on three occasions found the science of
cheirosophy to be *absolutely* at fault in every particular,
in reading a hand.   In each of these instances the
"subject" was a prominent member of the dramatic
profession,—in two instances gentlemen, in the third
a lady.   It is not without interest and significance that
I have only found the science powerless to interpret
characters in the cases of members of a profession whose
sole *raison d'être* is the concealment of their own per-
sonalities and the assumption of characteristics not
their own.   It is more significant and interesting when
we bear in mind that "habit is second nature."

[338] *Vide* note [263], p. 217

[¶ 435]

thick, and very short.   He was also a man of detail, and one who, if his objects were grand, aimed at obtaining them by petty means.[339]

[339] Compare John Gibson Lockhart's "*History of Napoleon Buonaparte*" (London : 1871), *passim*, and Thiers' "*History of the French Revolution*" (London : 1877), pp. 587 *ad fin.*

The Conic Type.

and then the lover

Sighing like furnace,

with a woeful ballad

Made to his mistress' eyebrow

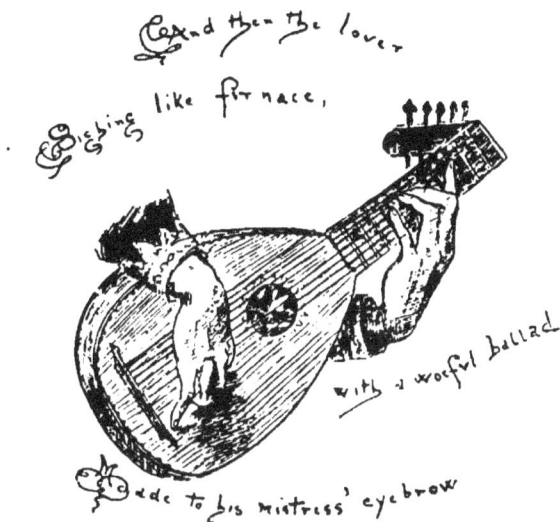

SUB-SECTION XVI.

⸢THIS hand, according as we find slight modifications in   **¶ 434.** Its variations.
its formation, betrays three very different tendencies :
—supple, with a small thumb and a palm fairly, but
not excessively developed, it has as the object of its
endeavours beauty of form ; large, thick, and short,
with a large thumb, it seeks after wealth, greatness,
and good fortune [Napoleon's hand was like this :
Vide ¶ 433]; large and very firm, it has a strong
tendency to fatalism.   All three act by inspiration,
and are relatively unfitted for mechanical arts.   The
first proceeds by enthusiasm, the second by cunning,
and the third by the suggestions of pleasure.

The large, short, thick hand is very common in   **¶ 435.** First variation. Normandy.
Normandy, the country of legal quibbles, where the
judgment is cold and the imagination warm [for
imagination is after all the special characteristic of
every artistic hand, whatever may be its developments].
I refer to the chapter concerning mixed hands the

[¶ 435]

consideration of the Norman hand, and will only deal in this place with the racial artistic hand, which is the most gifted with the true instincts of the type,—of the hand which has as its object beauty of form.

¶ 436.
Its appearance.
Its fingers, thick and large at the first phalanx, become gradually thinner up to the third, which presents the appearance of a more or less drawn out cone. Its thumb is small, as I have just said, and its palm is fairly highly developed.

¶ 437.
Impressionist school.
Whoever has a hand thus formed will attach himself instinctively and without reflection to the picturesque aspects of ideas and of things ; with him the *principle* will be swallowed up by the *form* of a thing. He will prefer, as Montaigne used to say, "that which pleases to that which pays." So long as a thing is beautiful, it does not matter if it be true or not ; greedy of leisure, of novelty, and of liberty, at the same time ardent and timid, humble and vain, he will have more energy and enthusiasm, than force and power. He will pass suddenly from the loftiest exaltation of mind to the profoundest despair. Incapable of command, and still more incapable of obedience, *attraction* will be his guide through life, rather than *duty*. Inclined to enthusiasm, he will live in constant need of excitement, and the activity of his mind will render regular domestic life heavy and uninteresting to him. In a word, he will be a man of sentiments rather than of ideas, appreciating the colours of a thing rather than its features ; he will be light in character, he will have ingenuousness and eagerness, an imagination of fire, and, too often, a heart of ice.

¶ 438.
Second variation.
Sensualist school.
Its characteristics.
A largish palm, smooth fingers, a weak thumb, still more conical finger-tips, *i.e.*, large appetites for sensual pleasures, without sufficient moral control, and a mind lacking the strength to subject the senses to its dominion, the whole built upon a foundation of only

slightly spiritual ideas,—such is, unless I am deceived, the character of artists as a general rule. Beauty is the only thing they can prefer to pleasure ; thus the nymphs only withdrew the muddy reed curtains of their native swamps so as to gaze upon the sun. Subjects of the artistic type do not share the ideas of the other types either upon right or wrong, upon what is good or what is 'useful ; they have *faith,* because it saves them the trouble of *reasoning,* without robbing their senses of any pleasurable feelings, but they will not brook political despotism, because its essential principles are the levelling of ranks, uniformity, and quietude,— conditions, all of them, strongly opposed to their natures ; theirs therefore is generally *relative* liberty, such as that which is found under aristocratic governments, rather than under others ; for such powers have always used as one of the levers of command luxury, pleasure, magnificence, display, art, natural capacities, talents, and high birth.

The artistic instinct is essentially and singularly exclusive and autocratic. Among some nations, as among certain individuals, it manifests itself before any other instinct. Travellers have found sculpture in high esteem in countries where the most elementary principles of agriculture are ignored, notably among the aboriginal negroes of Australia, and among some of the still savage tribes of North America. The artistic instinct is particularly rife among the South Sea Islanders.[40]

¶ 439.
Exclusiveness
of the type.

[40] "We cannot pass over without a word the taste and inclination for sculpture which is shown in the ornamentation of their piroguas, their drums, and even among certain tribes of their habitations, not only by the Tasmanians, the Tahitians, the Hawaians, the natives of the Pelew Islands, the Carolines, and the other islands of Polynesia, but even by some of the aboriginal negroes of Australia, particularly the inhabitants of the Archipelago of New England and of Solomon's, whose

¶ 440.
Defects of the type.

It is among people of artistic organisation that one finds the most subjects possessing only the *defects* of their type; which defects are, sensuality, idleness, egotism, singularity, cynicism, dissipation, intellectual ineptitude, cunning, and a tendency to exaggeration and falsehood.

¶ 441.
Artistic hands in the French army.

Our armies are full of artistic hands of all kinds; they owe to these hands their venturesome, unquestioning, and picturesque activity, and the energetic and prompt enthusiasm which distinguishes them. They are governed by eloquence :—"Pleraque Gallia," says the elder Cato, "duas res industriosissime persequitur, rem militarem et argute loqui." [341]

¶ 442.
The German army.

Inert and gluttonous, the German army is full of elementary hands, and its apathy can only successfully be coped with by brandy and corporal punishment.

Elementary hands.
An Illustration.

One day in a blazing sun, when Holland, delighted at the warmth, had thrown open its windows, I had the good fortune to catch "in the act" a major of the pure local type undergoing the process of digestion after his midday meal. Even at Rotterdam the formidable development of his corporation was a theme for admiration; half asleep he smoked, filling with clouds of tobacco the room where continually seated he passed his life with the crushing immobility

sculptures are sometimes masterpieces of elegance, —a singularity which we have had occasion to remark in speaking of the savage tribes which live on the west coast of North America." The above passage, from Adrien Balbi's "*Elemens de Géographie Générale*" (Paris : 1843, "Océanie," p. 497), is evidently the source of the above paragraph. The learned author goes on to speak in the highest terms of the decorative effect of the tattooing of the natives of Tasmania and Polynesia. Compare Levaillant's remarks, note [168], p. 156.

[341] M. Porcii Catonis "*Fragmenta*," ex libro ii., fragm. 34. "*Charis*," ii., p. 202, k. M. Cato, "*Originum*," ii. *Vide* H. Peter's "*Historicorum Romanorum Fragmenta*" (Leipsic: 1883), p. 49.

ot an Egyptian monolith. This vegeto-military phenomenon was said to absorb six thousand pints of beer per annum, and was said only to become conscious of the existence of his soul when he had drunk largely; at other times this same soul lay curled away somewhere, inert and dead, in the abysses of this huge ganglion, like a ship in the basin of a dock when the sea is at low water.

Governed by the instincts of material advantage and heraldic fetichism, the English army is full of hands which present scarcely anything but the defects of the spatulate type, which are, coarseness, intemperance, moral inertia, temper, and so on. For them war is but a trade, pay is its sole object, and the appetite its motive force; it is by the merit of hecatombs of the slain that it expects to be victorious, certain of defeat if the roaring of the bull Apis be not heard amid the blare of its trumpets. Subjected on account of its brutality to a degrading discipline and corporeal services, it would stand in danger of destruction in the atmosphere of liberty and gentleness which surrounds our armaments.[342]

¶ 443.
The English army.
Spatulate hands.

Beneath our flag the soul sustains the body, beneath the English and German flags the body supports the soul; we obey the *spirit* and act with intelligence, the Germans obey the *letter* and act automatically. We are the first marchers of the world; is it not a recognised axiom [of Maurice de

¶ 444.
English, French, and German armies contrasted.

[342] The whole of the above paragraph is absurd, and its presence is especially to be regretted as, taken as a whole, M. d'Arpentigny's chapter on " English Hands " [xii.] is a clever and interesting piece of analytical writing. What he means by his allusion to the bull Apis I am completely at a loss to comprehend. The passage runs in the original :—" C'est par le mérite des hécatombes qu'elle prétend à la victoire ; certaine d'être vaincue si les mugissements d'Apis ne se mêlent au bruit de ses clairons." (!?!)

[¶ 444]

Saxe] that battles are won not with the hands but with the feet?[343]   As a nation we are warlike rather than military;[344] the Germans are military rather than a nation of warriors.

¶ 445.
Generals of the elementary hand.

Generals who have the ˙ elementary hand pique themselves upon little points of discipline and management; they know how many blades of grass go to a truss of hay; they attach importance to the manner of carrying the arms, and the perfect condition of the uniform; they admire a harsh voice and clownish manners; like the spider in the shady corner of a dusty barn, they are only happy where there is no splendour.   They tend to Cæsarism and their doglike loyalty recognises, as Tacitus says, only the hand which feeds, which fattens them.[345]

[343] I do not find this axiom recorded as coming directly from the Maréchal Saxe in any of the accepted biographies of this great soldier, such, for instance, as those of Scilhac, Néel, Espagnac, or Pérau.   In his own work "*Mes Rêveries*," par Maurice, Comte de Saxe (Amsterdam: 1757, vol. i., book i., cap. vii., p. 144), we find the following:—"M. de Turenne has always gained a superiority with armies infinitely inferior to those of the enemy, *because he could move about with greater facility.*"

[344] Ivan Tourguéneff has placed practically the same sentiment into the mouth of M. Francois in his " *Œuvres Dernières,*" " M. Francois; Souvenir de 1848 " (Paris: 1885, p. 103), when he says:—"We are not a military nation.   That astonishes you.   We are a brave nation, very brave, but not military.   Thank God, we are worth something more than that."   It is interesting to note the parallel which exists between these passages of d'Arpentigny and Tourguéneff and the dissertation of Bacon upon the same subject in his Essay " *Of Greatnesse of Kingdomes and Estates,*" to which the reader is particularly referred [W. Aldis Wright's ed. (*Macmillan*), 1883, pp. 125-127].

[345] M. d'Arpentigny refers, I presume, to the remarks which Tacitus makes concerning mercenaries, and upon the methods of buying the favour of the common soldiers.   *Vide* " *Historiarum* " lib. iii., c. 61, and lib. iv., c. 57.

Tactics, manœuvres, encampments, sieges, estimates, military and naval architecture, the strategy of temporisation and delay, are the especial qualities of hands which are spatulate and square. They have theory, method, and science, and they care more for success than for glory.

¶ 446.
Square and spatulate military hands.

Generals of the conical artistic hand proceed by inspiration, and move by sallies; they are gifted with prowess, promptitude, passionate instincts, boastfulness, and the talent of acting *impromptu;*—they attach equal importance to success and to glory.

¶ 447.
Conical military hands.

Murat, at the battle of Smolensk, commanded a regiment of heavy cavalry, mounted on a grand black stallion, full of strength and grace, calm, caparisoned in gold and covered with the long shining locks of his mane. The king wore a helmet whose golden crest was ornamented with a white plume; immobile, he watched the battle from afar, letting his jewelled sabre trail in the roadway with an air of haughty indifference. Suddenly he becomes excited, his eyes flash, he raises himself in his stirrups, and cries in a loud voice, " Left turn ! Quick march ! " Then the earth trembled, and a noise as of thunder was heard, and those black squadrons, flashing as if with lightning, rushed forward like a torrent, as if they had been dragged forward by this slender white plume. The victory of that day was in great part due to this movement.[346]

¶ 448.
Illustration.
Murat.

[346] Joachim Murat, Maréchal de France and King of Naples, born 25th March, 1771, and shot at Pizzo 13th October, 1815, was renowned throughout Europe for the magnificent manner in which he handled his cavalry. " Il contribua," says Larousse, " à forcer l'Autriche de demander la paix par l'audacieuse manœuvre qu'il fit executer sa cavalerie le 13 Mars, 1797." He distinguished himself similarly at the taking of Alexandria and at the battles of the Pyramids and of Ostrowno, Smolensk, and Moskow in the Russian campaign.

¶ 449.
Murat and
Junot.

Murat, the most lyrical of the warriors of the imperial epoch, had, like Junot,[347] the other hero of audacity, hard conical hands.

¶ 449a.
Greek battles.

In their calm audacity the Greeks of the brightest days of antiquity, before attacking the enemy, made solemn sacrifice to the Muses,—*i.e.*, to the gentle deities who open to men the holy gates of persuasion, of concord, and of harmony.[348]

¶ 450.
The Etruscans
and their music.

The Etruscans chastised their slaves to the sound of the hautboy, so as to soften their anger, and prevent them inflicting a severer punishment than justice required.[349]

¶ 451.
Alexander
Dumas.

Alexander Dumas, one of our greatest word-painters in the matter of descriptions of battle-scenes, has also (as I have said elsewhere) an artistic hand, only, being a poet of infinite variety, his hands are very supple.

[347] Andoche Junot [b. 1771, d. 1813] was a native of the Côte-d'Or, who, raising himself in the French army —which he entered as a volunteer in 1792—by his grand audacity, became successively Marshal of France [1807] and Duke of Abrantes. It was in the Holy Land in 1798 that he particularly distinguished himself, routing a force of 10,000 Turks with a body of 300 cavalry, after a conflict of fourteen hours' duration. He was defeated in the Peninsular war [at Vimeira] by the Duke of Wellington. Having been disgraced in the army, he went mad and killed himself in July 1813. He was the husband of Laure Perron, Duchesse d'Abrantes.

[348] This custom of the Greeks, especially observed by the Spartans, is recorded by Pausanias in his ΕΛΛΑΔΟΣ ΠΕΡΙΗΓΗΣΙΣ.*

[349] *Vide* the "*Fragments of Aristotle*," ii., 606. Julius Pollux also tells us that the Etruscans not only fought, but inflicted punishments under the same influences. *Vide* Julii Pollucis "*Onomasticum*" (Amsterdam: 1706). †

---

* ΛΑΚΩΝΙΚΑ, I. Ζ'.—"Ἐν ἀριστερᾷ δὲ τῆς χαλκιοίκου Μουσῶν ἱδρύσανμο ἱερόν, ὅτι οἱ Λακεδαιμόνιοι τὰς ἐξόδους ἐπὶ τὰς μάχας οὐ μετὰ σαλπίγγων ἐποιοῦντο, ἀλλὰ πρός τε αὐλῶν μέλη καὶ ὑπὸ λύρας καὶ κιθάρας κρούσμασιν."

† Τυρρηνοὶ δὲ τῷ Ἀριστοτέλους λόγῳ, οὐ πυκτεύουσιν ὑπ' αὐλῷ μόνον, ἀλλὰ καὶ μαστιγοῦσι, etc.—Segm. 56, lib. iv., cap. vii.

In 1823, before Pampeluna, the Spanish army passed their nights playing the guitar, smoking cigaritos, chanting litanies, and telling their beads ; by day lying in the dusty grass they chatted, chewing the young shoots which grew around, or slept in the sun. At the screech of the mitrailleuse you might have seen them like a flock of frightened geese flying in all directions, crying out and reproaching their saints. In vain did their officers try to rally them ; old threadbare red capes, old three-cornered hats of bloodstained leather, white shakos with red tufts, leaders shirtless and perspiring beneath the weight of their ornaments, olive-hued and sun-dried almoners, haggard *vivandières*, stunted scribes and soldiers in tatters,— all disappeared in an instant, swallowed up by the clouds of dust. The Comte d'Espagne, who commanded them, avowed that this rabble, at the same time eager and feeble, ferocious and cowardly, could only be terrible to the enemy on account of their robberies, and their piratical and bohemian instincts.

¶ 452.
The Spanish army in the Peninsular War

The Comte d'Espagne was himself a little stunted man, broadshouldered, and wiry, dressed more or less like a butcher-boy, very active, very cruel, very courteous, who saluted with an air every time he heard the cannon of the town, and whom we saw everywhere half bully and half captain, always discoursing, lecturing, and violently agitating his little tuft of feathers. Artistic hands.

¶ 453.
Le Compte d'Espagne.

The reign of hard spatulate hands was a reign of materialism and of hatred; that of the artistic, directed by the psychic, hands was an epoch of relative spiritualism and of love. Inaugurated by Abeilard and St. Bernard, it commenced with the twelfth century, and lasted down to the close of the thirteenth.[350]

¶ 454.
The reign of hard spatulate and artistic and psychic hands. Twelfth to fourteenth century.

[350] *Vide* note [350], p. 241.

Psychic hands spread throughout the nation, and
scattered everywhere the torrents of enthusiastic
mysticism which they had amassed in the contemplative leisure hours of the cloister, under the reign
of the turbulent and spatulate hands.   Everything is
taken up and worshipped with ardour,—God, woman,
and war; but war for a great and pious idea, and
not for a material interest.   An enthusiastic poetic
spirit coursed through the veins of the nation, and
as she had the temperament of heroism and of art,
art and heroism became part of their manners and
customs.   It was the era of splendid costumes, of the
courts of love, of the crusaders, and of chivalrous
epics ; like real life, history became tinged with the
colours of romance.   Dante,. Petrarch, and Gerson
were pre-conceived,[351] the ideal woman, veiled until
then, revealed herself, and the worship of the Holy
Virgin was established.[352]   By the establishment of the
orders of begging friars, the Church is opened to the
lower orders, whose instincts, better understood than

[351] Dante Alighieri was born in Florence in 1265 ;
Petrarch and Gerson, however, were neither of them
born until the fourteenth century, the former in 1304, and
the latter not until 1363.

[352] M. d'Arpentigny gives way to a regrettable loose-
ness of diction, when he says that the worship of the
Blessed Virgin Mary was established in the twelfth
century.   What he means, of course, is, that it was in
this century that the worship of the Virgin, already
long established, received an enormous impetus from
the establishment of the orders of Mendicant Friars,
of whom especially, the Cistercians at the beginning
of the century, introduced their custom of dedicating
every church to the Blessed Virgin Mary, and this
introduction of the worship of the Virgin was enormously assisted and supported at the end of the same
century by the introduction and propagation of the
rosary by the Dominicans.   For the rest, the worship
of the Virgin Mary, which had been seriously weakened
by the invasions of the Goths and Vandals, flourished
again when the Franks got once more the upper hand

[¶ 454]

formerly, were powerfully rehabilitated.[253] Artistic hands, happy and triumphant, although constrained within the region of psychological ideas, excite themselves in all directions; the statues of angels and virgins with which the *imaginations* of the people were gorged, spread their wings of stone, and, palm in hand, flew in clouds to the churches built to receive them; all men realised the grace and the beauty of the variegated garments, the dazzling stained glass, the triangles of light which at the feet of the sunbeams passed across the shades of the cathedral, symbols of a worship of pastoral origin.   Ecclesiastical architecture, which flourishes only when a whole nation feels, thinks, and believes as one man, suddenly obtained a sublime development; the grandeur and the spirituality of the sentiment of the times found its reflection in the grandeur and spirituality of the magnificent cathedrals of the period.

[A.D. 420]. Clovis built the first church to the Virgin at the eastern extremity of Paris, as also the Church of Our Lady of Argenteuil.   Excepting Chilperic, all the Merovingian kings encouraged Her worship, especially Clothaire, whose wife Waltrude and daughter Engeltrude founded the Church of Notre Dame de l'Escrignol in the year 600.   Full particulars of these periods of Catholic History may be found in Dr. F. C. Husenbeth's *"History of the Blessed Virgin Mary"* (London : *n.d.*).

[253] M. l'Abbé C. Fleury gives the following account of the institution of Mendicant Friars in his *"Discours sur l'Histoire Ecclesiastique"* (Paris : 1763, Discours viii., ch. 8).   They arose principally after the Council of the Lateran of 1215, and were the outcome of, in the form of a protest against, the luxury of many of the monasteries of the time.   "They thought that they must seek the remedy in the opposite extreme, and renounce the possession of all temporal advantages, not only by following the rule laid down by St. Benedict, so severe on this point, but altogether, so that the monastery should have no fixed income whatever."   The monks

¶ 455.
The construction of colossal monuments. National characteristics.

It has been said it was by means of the violence exercised by kings upon the people, that the gigantic monuments of Memphis, of Thebes, and of Æthiopia, of Babylon, and of India, were raised; but in the first place they took centuries to build, and violence, as I understand it, is not compatible with the idea of long duration.   I would explain these marvels as they have explained the existence of the mediæval cathedrals, viz., by the general consent of all to an unique idea.   These peoples pursued architecture, because it was the talent of their day,˙ just as to-day industry is the talent of ours.   Strange people that we are ! we can conceive a nation of haberdashers, but not a nation of masons.

¶ 456. .
The artistic and philosophic types contrasted.

I have said that the artistic mind is essentially exclusive, and when it is common to an entire nation, and nothing stands in the way of its full exercise, it develops such extraordinary manners, and such eccentric customs, that they become incomprehensible to the other types. Voltaire, who alone in himself represents exclusively in its highest development the philosophic type, as presented to us by the eighteenth century, denies Babylon and her customs as they have been described to us by Herodotus ; he denies the existence of the ancient Egyptians, as we

had, therefore, two alternatives, viz., to work or to beg.
" The minor brethren, and the other new religious orders of the thirteenth century, chose to beg : they were not monks, but were doomed to wander about the earth, working at the conversion of sinners from whom they might expect to receive alms; and besides, their wandering life, and the necessity of preparing what they proposed to say to the people, did not seem to them to be compatible with manual labour. . . . The Venerable Guigues, in the constitutional code of the monks of Chartreux, and in the Council of Paris in 1212, desires men to give errant friars the means of existence, so that they need not go begging, to the disgrace of their order."
We have seen [note [169], p. 156] what was Erasmus' opinion of mendicant friars.

know of it circumscribed within a winged circle by the hands of her priests, hovering on unprogressive pinions within a circle of bull-headed gods ; he denies the Indian nations, who without doubt would in their turn have denied *him* by reason of *their* not being able to understand him.[354]

To every century its own generation of men, to every generation of men its own physical organisation, to every organisation its own peculiar talents. Each

¶ 457.
Predominating characteristics of generations.

[354] Speaking from a fairly intimate knowledge of the works of Voltaire, I think that M. d'Arpentigny has entirely misconceived the opinion of Voltaire upon the architectural phænomena of which he speaks, as any one who will take the trouble to glance through the 1837 edition of the " *Œuvres Complètes de Voltaire*" will readily agree. It is true that he ridicules the accounts which are given us by Herodotus, especially in " *Le Pyrrhonisme de l'Histoire*" (ed. 1837, vol. v., p. 73), where he says of that author, " Nearly all he has told us on the authority of others is fabulous ; . . . when Herodotus retails the stories which he has heard, his book becomes nothing but a romance resembling the Milesian fables." Again, in the " *Fragments sur l'Histoire*" he says, " Amateurs of the marvellous say, ' These facts must necessarily be true, because so many monuments support them.' I say, ' They must necessarily be false, because the vulgar have believed in them.' A fable is told a few times in one generation, it establishes itself in the second, it becomes respected in the third ; the fourth raises monuments to it. There was not, in the whole of profane antiquity, a single temple that was not founded on a folly ; " and again he cites the history of Herodotus as an example. Voltaire had the greatest respect for the monuments mentioned by d'Arpentigny. In the " *Dictionnaire Philosophique*" [ed. 1837, vol. vii., p. 681], he says, " To know with a certain amount of certitude something of ancient history, there is only one method, *i.e.*, to see whether there remain any incontestable monuments ; " and he then quotes as authentic instances Babylon and the monuments of ancient Egypt, and in his " *Essai sur les Mœurs*" [ed. 1837, vol. iii., pp. 13, 15, 22, and 27], he discourses interestingly and at length concerning Babylon, Egypt, and the monuments of ancient India.

[¶ 457]

age, therefore, as it extinguishes itself, carries neces
sarily away with it the secrets of a notable portion
of the ideas which animated it.

¶ 458.
Herculaneum
and Pompeii

Herculaneum and Pompeii, rediscovered after
seventeen centuries with their obscene signs and
audacious frescoes, beneath the ashes under which
they both found a living tomb upon the same day,
have furnished us with more details concerning the
inner and familiar life of the ancients, their tastes,
their current ideas, than all the books which they
have left to us.[355]

¶ 459.
Details of
domestic life.

The most important details of the habits and
customs of an epoch are the points which are the
least remarked by the people whom they signalise;
they do not record them, and posterity can only obtain
information concerning them, as it were by accident.

¶ 460.
Distinctive
systems of
government.

From an artistic point of view, what could be
finer than the organisations of Sardanapalus, of Nero,
of Heliogabalus, of the Borgias [father and son], and
of Catherine II.? As they remained to the last day of
their lives faithful to the logic of their type, I do not
think they can ever have known remorse.

¶ 461.
The Gnostics.

Carpocras of Alexandria, and Basilides, the founders
of the sect of the Christian gnostics[356] [a species of

---

[355] Those of my readers who have visited Herculaneum
and Pompeii, and who are *familiar* with the museum at
Naples, will recognise the exactitude of all that our
author says in this paragraph.

[356] I should hardly have cited either Carpocras [or
Carpocrates] or Basilides as founders of the enormously
divergent creeds known as Gnosticism. Gnosticism
was the immediate outcome of the universal and highly-
organised systems of dissent from the old narrow
Judaism, which obtained in the first and second centuries after Christ. The old pagan creeds and the old
philosophies made—to use the words of the writer of the
article in " *Chambers' Encyclopædia* "—" a last stand,
and produced in their and the ancient world's dying
hour *gnosticism.* The wildly-opposite ideas of polytheism, pantheism, monotheism, the most recondite

illuminati], so far from proscribing sensual pleasures, looked upon them as a direct means of communication with God, and ranked them among the acts recommended as wholesome and meritorious.[357]

Some of these ideas exist to this day in Abyssinia,[358]

philosophical systems of Aristotle, Plato, Pythagoras, Heraclitus, Empedocles, etc., together with the awe-striking mysticism and demonology which after the Babylonian captivity had created in the very heart of Judaism that stupendous and eminently anti-Jewish science of Kabbala—all, it would appear, had waited to add something of their own to the new faith which could not hold its own under all these strange influences." Simon Magus was probably the earliest recorded gnostic, and after him Menander, Cerinthus, Nicolaus [the founder of the Nicolaitans], Saturninus of Antioch, Bardensanes of Edessa, Tatian and Valentinus of Rome, and a many others founded more or less celebrated schools of gnosticism. Basilides of Alexandria [125-140 A.D.] founded a school which inculcated principles strangely in accordance with those of the modern cultus of Theosophy, as laid down in A. P. Sinnett's remarkable work "*Esoteric Buddhism*" (London, 5th edition, 1885), and founded mainly on the Kabbala. Carpocras was the chief of a non-localised school of gnosticism, so I presume that the affix "of Alexandria" in the above paragraph is given him accidentally instead of to Basilides. *Vide* Lewald's "*De Doctrinâ Gnosticâ*" (Heidelberg: 1818); Möhler's "*Ursprung des Gnosticismus;*" Malter's "*Histoire du Gnosticisme*" (Paris: 1843).

[357] M. d'Arpentigny has illustrated his work in this place with an account of some of the more immoral rites and beliefs of certain of the schools of gnosticism, of which the most immoral was perhaps that of the Nicolaitans. The passage, which I have omitted as unnecessary, comes from L'Abbé Fleury's "*Histoire de l'Eglise.*" "*Mœurs des Chrétiens,*" chap. xiii., "Calomnies contre les Chrétiens" (Paris 1739).*

[358] *Vide* James Bruce's "*Travels to discover the Source of the Nile*" (Edinburgh: 1790, vol. iii., p. 292), or, for

---

* "Ces soupçons étaient appuyés par les abominations que les gnostiques, les Carpocratiens et d'autres hérétiques commettaient dans leurs assemblées et que l'on a peine à croire même sur le récit des pères."

[¶ 462]

and they have been re-discovered among the nations who inhabit the South Sea Islands.

¶ 463.
Platonism.

Platonic affections, filial, paternal, and parental, are much less accentuated among artistic hands than among the square and spatulate types.

a minute if somewhat unpresentable compendium of all that has been said upon the manners and customs of Abyssinia in this connection, *vide* a rather rare work, entitled "*Les Abyssiniennes, et les Femmes du Soudan Oriental, d'après les Relations de Bruce, etc., etc.*" (Turin: 1876).

ARTISTIC HANDS [*continued*].

The Artistic Hands of the Sixteenth Century.

WHEN one says that an æra is essentially religious, it is synonymous with a statement that it is only slightly sensual, and consequently more poetic from the esoteric than 'from the exoteric point of view. Thus the cathedrals of the thirteenth century are more remarkable in the ideas revealed by their plans, than in their execution. Like many of the barbaric lyrics, whilst they strike the soul with wonder, they offend the taste; grand as a whole, they err in points of detail, and are more pleasing to synthetical than to analytical minds; for the hands of the masses make themselves apparent in their construction, rather than the hands of the individual. They glorify the whole body of their constructors, and not individual workmen; at the same time it is the psychological hands which have made themselves apparent rather than the other types; as for the artistic hands, they have evidently only worked under the direction of the others.

¶ 464. Monumental architecture of the thirteenth century.

But at the period known as the Renaissance these last hands took their revenge. The art which had been attracted by Greek sensualism, suddenly seized

¶ 465. The Renaissance.

[¶ 465]

with an intense desire of liberty, made a rapid transition from the symbolical immobility in which it was kept enchained by mysticism, into the world of palpable realities and of purely human fantasies. Like the turbulent barons of the time, art professed an absolute independence; it was no longer beauty which touched the soul, and the glory which comes from God, that the sculptor and the warrior sought for in the solid stone and on the field of battle; it was beauty from its *picturesque* aspect, and glory such as the world can give and as it is comprehended by sensualism. Directed by psychological hands, art produced nothing but temples; under the free artistic hands it produced only, or excelled in producing only, palaces. Without folding its wings, it moderated its flight, and forsook the gods to pay tribute to men; it became less grand, but more eloquent, more graceful, more brilliant, and more exquisite than in the middle ages. It passed from public to private decorations, from the service of the masses to that of individuals.

¶ 466.
Sculptures of the
Renaissance
period.

Finally, if, as in the times which preceded this epoch, art did not glorify any particular individual, it was because, as formerly, it entered more or less into the organisation of all. A considerable number of excellent specimens of the sculpture of this period remain among us to this day, of whose sculptors the names are unknown; if by the incontestable superiority of their talents they had *astonished* their contemporaries, it is to be presumed that their names, consecrated by gratitude and by admiration, would have been transmitted to us by historians or by tradition.

¶ 467.
Art in the
sixteenth
century.

Art, therefore, in the sixteenth century, was the gift of the majority; but it was not practised for the benefit of a single idea. It had the enthusiasm, the movement, and the individuality of this epoch of duels and of civil wars, of wild love-adventures and of glittering cavalcades, of carousals and of deeds of daring, performed by little bands of adventurers. Formerly, art had had more solid basis than visible exterior, now the conditions were reversed ; the people of whom it was almost the unique industry, loved it for its own sake, and also for that of the material good of which it was the source. Women, whom it adulated, gave it their love in return ; and it afforded the main delights of those kings clad in velvets and satin so spruce and so sensual, by whom the France of that period was so well represented.

Art, by reason of its variety, and of its immensity, not being a thing that one can teach, that one can inculcate into the masses like a common industry, it is necessary, in order that a whole generation should highly appreciate art and practise it with success, that it should be *born artistic;* and I say that, having minds organised for the development of a single end, viz., *art*, its hands will also be constituted for the same end. **¶ 468.** Artistic instinct of a community

The hand of Francis I. was artistic in this sense, that its palm was large, the thumb small, and the fingers smooth [*vide* the statue of this prince at St. Denis], but the fingers were quite appreciably spatulated. Such is the hand of men who are active and fond of horses,[359] who are governed by their own fantasies, and whose changeable humours have no other motors than the suggestions of their temperaments. These inconstant minds submit more than others to the influences of their surroundings. Well, the sixteenth century being essentially the age of art and literature [by reason of the immense number of artistic hands that existed then in the south of Europe], Francis I. encouraged the individuals who practised these callings, not on account of the **¶ 469.** Francis I.

---

[359] *Vide* ¶ 164.

[¶ 469]

possibly resultant intellectual progress,—that was the least in his thoughts,—but on account of the pleasure which he expected from them, and which in effect they gave him.[360]

¶ 470.
Tastes of the sixteenth century.

The sixteenth century was the epoch of splendid oligarchies, of baronies and of grand seigneurs, of aristocratic republics and monarchies, of holy wars,— *i.e.*, of strife concerning *forms* of worship,—of political tricks and ambuscades, of bold voyages and discoveries, of sorcerers and of astrologers, of enormous vices and debaucheries, and of knights without fear rather than without reproach ;—in a word, of horrible slaughters, where the manner of killing was of greater importance than the mere death itself; an epoch full of contradictions, at the same time serious and bantering, clad in embroideries and in rags, running after finely-executed missals and chalices, reading Rabelais and Gerson, surrounded by outlaws and artists, ivory crucifixes and mythological nudities, whilst miserable and deformed dwarfs, and beautiful girls played about upon gorgeous carpets with tame tiger-cubs.

¶ 470a.
Failings of the artistic type.

Passionate love of order, prudence, and usefulness is the gift neither of nations nor of individuals who are governed by the artistic instinct.

¶ 471.
Architecture of the Renaissance.

In its capricious grace and its florid opulence a palace of the Renaissance is a sort of temple raised to the glory of some deity, incarnate indeed, but inaccessible to the wants of human nature, the influences of cold and of heat, of shadows, and of dampness. It suggests grandeur, power, and riches rather than contains them, windows, staircases, galleries, colonnades, terraces, and porticoes,—all are arranged for display and nothing for comfort, as we understand it to-day, when the humblest

---

[360] *Vide* notes [291], p. 242, and [366], p. 311.

peasant, more refined than the nobles of the time of Charles VIII., eats with a fork rather than with his fingers. So the independent classes have long since abandoned to the lower orders those abodes so richly sculptured, carved, starred, and emblazoned, in which, among delicate statuettes and beneath aerial turrets, the aristocracy of brilliant appanage, but still of the rude and coarse hands, of the sixteenth century used to struggle and fight as was their · wont.

Under the last of the Valois it was the same of costume as of architecture; it was more important to be elegantly than to be thoroughly dressed. They paid more attention to their ornaments than to their clothes, and they preferred the bodily discomfort of an awkward fashion of garments to the offence which would be caused to their taste by an inelegant vesture.

¶ 472.
Costumes of the period.

Still, costumes are not invented, they are born spontaneously, and, like legal institutions, are the necessary results of the nature of things.[361] Francis I. and Voltaire—typical representatives of their respective centuries—were in turn so completely expressed by their costumes that one could not imagine Francis in the costume of Arouét, or Arouét in the costume of Francis.[362]

¶ 473.
Francis I. and Voltaire Costume.

[361] Compare the explanation given to Colonel Cheng Ki Tong by the lady to whom he complained of the plainness and uniformity of modern dress, to the effect that "a plain coat is much more convenient to *turn*. She observed that formerly the costume designated a political party, and if the fashion remained to-day, men would ruin themselves with dress!"—"*The Chinese painted by Themselves*," pp. 161-2.

[362] There are but few works on costume that do not cite François I. and François Marie Arouét [better known as Voltaire] as exponents of the costumiers' art in their respective centuries; but both overdid the matter, and dressed with an extravagance, in advance even of the extravagance which reigned supreme around them.

¶ 474.
Changes of
costume.

Constant to type among peoples who are unchangeable and in institutions which pride themselves upon non-progress, like the Romish Church, costume is continually changing among changeable peoples—not by reason of a concerted will, but by the necessary effect of a contemporary moral state ; for man alone is capable of thought, man alone is gifted with the sense of decency, and man alone dresses himself.

¶ 475.
Fashions.

There is this to be said in favour of the fashions : they encourage uniformity.

¶ 476.
The negroes.

The nakedness of niggers, regard being had to their colour, which clothes them as it were in a shadow, and serves them as it were as a garment, is less immodest than that of white people.   The negro does not come into the world wholly clad like the animals, but quasi-clad,—a circumstance which places him between the ape and the man in the scale of creation.[263]   The Hindoos, almost as dark in colour as the negroes, are similarly nearly as stationary as the latter.   Immobility is the supreme attribute of the animal world.[264]

¶ 477.
National
costume.

Uniformity, so dear to the Russians, is disdained by free nations, because it classifies and restrains. *Our* costume, because it represents the most en-

[263] Dr. Benjamin Moseley ["*A Treatise on Tropical Diseases*" (London : 1803), p. 492] has observed that "Negroes are void of" [bodily] "sensibility to a surprising degree. They sleep soundly in every disease, nor does any mental disturbance ever keep them awake. They bear chirurgical operations much better than white people ; and what would be the cause of insupportable pain to a white man, a Negro would almost disregard,"— a curious illustration of the theory that high development of one sense is generally accompanied by absence or deterioration of another.

[264] The colour of the negro is not without its constitutional advantages. Attention has been drawn by Dr. William Ogle to the fact that pigment occurs also in the

lightened social state in the universe (‼), is precisely the one which is the least suited to the blacks, and to people who have remained primitive.

¶ 478.
Good taste in dress.

Spatulate-handed subjects viewing things as they *are*, and conic-handed subjects viewing things as they imagine them to be, the former dress themselves in the manner which actually becomes them best, the latter in the manner which they think suited to their imaginary attributes ; whence we get the extravagant costumes of the flowing-haired artist, the provincial poet, and—the restaurant ham! all *three* of a race which is eccentric and conic-handed.

¶ 479.
Common sense.

People of real action and of clear good sense have, as we have said, as a rule, spatulate hands.

¶ 480.
Greek costume.

Under the later Valois, costume suggesting always the nude, the art and literature of the time recalled those of the Greeks, who lived semi-nude [as a matter of taste, not on account of coarseness of mind], and were always ignorant of the prudery and false modesty inherent to the peoples of our colder climes.

¶ 481.
Poetry of the epoch.

Flaunting, alert, sensual, and scantily clothed, the French Muse of that century, richer in words and idioms, more highly coloured, less bashful and less pedantic than she of the seventeenth century, teems with the strong and lusty youth of a nation predestined to every class of success.

¶ 482.
French ideas in art.

In those days the ideas of the South being in a majority, France naturally sought her models of art in Spain and Italy. Now, Northern ideas being in

olfactory regions, and he traces to this fact an increase in the acuteness of smell. Dr. Ogle attributes the acuteness of the smell of the negroes to their greater abundance of pigment [*vide* " Anosmia," by Dr. William Ogle in "*Medico-Chirurgical Transactions*," vol. 53]. Albinos and white animals neither see nor smell so delicately as creatures that are dark-coloured. —ALEXANDER BAIN, "*Mind and Body, the Theories of their Relation*" (London : 7th edn., 1883).

[¶ 482]

the majority, we seek them in England and in Germany; French genius is like the Janus of the
ancients, it has a dual intelligence and two faces.

¶ 483.
Paris.

The Paris of to-day, regard being had to the intelligence proper to the nations of the North, is well
off as she is. This intelligence gives to the nation
the kind of moral strength which suits the manners
of the present day. In the sixteenth century she
would have been better off had her situation been
more southerly.

¶ 484.
The Girondins
in 1793.

In 1793 the Revolution, turned by the brilliant
oratory of the Girondins from the object prescribed
to it by the needs of the epoch, would infallibly have
perished, had it not been for the Mountain, whose
members, born nearly all of them in the North, saved
the situation by dissipating the clouds of romance in
which it was becoming enveloped, and by restoring
in its midst the sentiment of reality.[365]

¶ 485.
Origin and
impulse of art.

Art, among highly-civilised nations, emanates from
the individual or from universal reason. Among
nations governed by instinct it emanates from God or

[365] The "Mountain" [*Les Montagnards*] were the
"ultra" party in the Convention during the French
Revolution. "The partisans occupied the right side of
the assembly, the national guard, and the club of the
Feuillans; the Girondins possessed the majority in the
assembly, but not in the clubs, where plebeian violence
carried the day; and, finally, the most extravagant
demagogues of this new epoch, seated on the highest
benches of the assembly, and thence named "The
Mountain," were all powerful in the clubs and with the
mob. "The moderate party" were called the *Plain* in
opposition to the left side, which was styled the *Mountain*, where all the Jacobins were heaped up, as one
may say, one above the other. On the graduated
benches of this mountain were to be seen all the deputies
of Paris, and those of the departments who owed their
nomination to the influence of the clubs, and of those
whom the Jacobins had gained over since their arrival,
by persuading them of the necessity of giving no quarter
to the enemies of the revolution. In this party were also

by inspiration. Immature in the eighteenth, which was the most humanly-intellectual century, it was highly developed in the sixteenth, which was the most divinely-intellectual century. Art flourishes particularly at the periods when the nations on the march of progress have one foot upon the territory of barbarism, and the other on that of civilisation, when they believe as much in miracles and in occult sciences as in daily occurrences and exact sciences. Art is then sufficiently human and sufficiently Divine, and develops itself by reliance equally upon science and upon inspiration.

Barbarous nations affect festivals and splendour, civilised nations substitute luxury and good taste. Such was the court of Francis I. at a time still rude, but already civilised ; and such were the accompanying appliances of civilisation, that it took five days to go from Paris to Fontainebleau.[366]

¶ 486. *Barbaric splendour.*

In the centre of an oval lawn, surrounded by trees of graceful growth and massy foliage, there rises in the gardens of the Tuileries a pedestal on which is

¶ 487. *The Dioscuri in the Tuileries gardens.*

included some men of distinguished abilities, but of precise, rigorous, and positive characters, who disapproved of the philanthropic theories of the Girondists, as mystical abstractions."—Thiers, "*History of the French Revolution*" (London : 1877). "The Legislative Assembly," ch. i., and "The National Convention," ch. i.

[366] *Exempli gratiâ*, the meeting of Francis I. and Henry VIII. of England, when, as Adolphus says in his "*History of France*" (London : *n.d.*) "the magnificence which was displayed by two princes equally splendid, profuse, and vain, made the spot on which they met retain the name of the "Field of the Cloth of Gold" [vol. ii., p. 75]. "The magnificence which accompanied him through life deserted him not at his death ; his funeral obsequies were performed with unusual pomp ; and the proclamation which announced his death displayed his character; 'a prince mild in peace and victorious in war; the father and restorer of learning and the liberal arts.'"

an excellent replica in marble of the beautiful group
of the Dioscuri. These immortals are undraped,—the
elements had no power to injure the gods,—their
movements have the appearance of being slow and
graceful, time, and the causes which drive men to
hurry and to physical exertion, being non-existent for
them. They are equal in age, as in beauty, but one
of them, more self-contained than the other, wears a
more imposing air ; it is the one who reverses his
torch at the moment of descent into the kingdom of
the dead. Farewell, for a time, to the fleet coursers
and the native stream, beloved of swans and of
rose-laurels. Exempt from our cares, freed from our
solicitudes, their life, very different to ours, prolongs
itself in the absolute calm which is given by their
æthereal natures, and in the cultivation of the attributes of unending youth; carried away by an unlimited
and mutual love, they enjoy it without reflection in an
indolence full of security, with a gentleness which is
as innocent as it is profound. They are naked, as I
have said, but their heads are covered with flowers,
as if in symbolism of eternal happiness, and the
eternal fruitfulness of the race of immortals. From
whatever position one contemplates this group, we
find nothing but harmonious lines accentuating their
calm, their eloquence, and their suppleness ; but
strength makes itself startlingly apparent, underlying
this attitude of repose, and one realises that these
are indeed the tutelary deities of manly exercises,
in whom the Greeks honoured the celestial pro-
tectors of her athletes, of her horsemen, and of her
mariners.[367]

[367] This group, which will doubtless be remembered by
many of my readers, stood near the Terrasse du Bord de
l'Eau, close to the Orangeries ; it was destroyed by the
Communards in May 1871. *Vide* note [140]; on Castor
and Pollux, p. 130. Those who know the Eternal City

The inspired, reflective, logical, enthusiastic, exact artist to whom we owe this masterpiece had without any doubt fingers with developed joints, a large thumb, and conical finger-tips. Nor was it an ordinary man, who, going back to the rustic cradle of the gods of the heathen mythology, and inspiring himself with memories of antiquity, placed this group upon a spot recalling the umbrageous arenæ of Olympus and the holy' pastures of " the verdant Elis, abounding in horses."

¶ 488.
The group and its arrangement.

In France the action of the Southern conical type upon those of the North, is naturally less than that of the Northern types upon the Southern. The result follows that this artistic type, too much modified, has not among us the value of specialism which distinguishes it in countries where, instead of being merely tolerated, it is encouraged; as for instance in Italy.

¶ 489.
Reciprocal effects of art.

. Nations among whom—as among us—*all* the types are largely represented have more shades and gradations of character than fixed and determinate characteristics. Those among whom two types alone form an immense majority have more fixed peculiarities, and more originality of procedure than shades of character. ·We are more easy-going and tolerant than these last, because it has been given to us to identify ourselves without effort with all characters — a thing impossible to the masses of a population, who are carried away by the ascendency of a too exclusive genius. The *usefulness* of what strikes one at first sight as being merely *beautiful* entirely escapes the observation of the spatulate nations of the North ; and conversely, what is *beautiful* in things which appear

¶ 490.
Nations of mingled types.

will remember the original group, which, taken originally from the Villa of Pompey, stands now at the top of the steps of the Church of Sta. Maria Aracœli, in the Piazza del Campidoglio, on the Capitoline.

at first to be merely *useful*, escapes the comprehension
of the nations of the South.[268]

¶ 491.
Incompatibility
of types.

If you hate interminable wars of words, those
loquacious and sterile battles, you will avoid throwing
together, not only people of different types, but even
people representing two distinct shades of an identical
type. Each of them being permeated with sentiments
of which the other knows not how to form an exact
idea, the misunderstanding will be unresolvable.
Thus among persons of an identical type small hands
generalise too much, whilst large hands do not do so
sufficiently. In the eyes of Victor Jacquemont, the
naturalist and geologist,—for whom art and poetry are
as nothing, and who, surrounded by the luxury of the
nabob, regrets the little chamber of his father, where
he partook of the humble family dinner,—Asia, the
vaunted of poets and mystics, is the most miserable
and unfortunate continent of the globe. In like manner œconomists judge of the prosperity of a country
by the number of its machines, and of its artists by
the *number* of its monuments, and so on with the
rest of it.

¶ 492.
French and
English
character.

There are more elements of contradiction, of discussion, and consequently of *moral agitation* in France
than in England, where the quasi-similitude of tendencies is proven by the quasi-uniformity of types ; and
as a too great conformity of ideas is not a slight cause
of boredom, it follows that the English, who at home
are verbose only upon matters of interest, are, when
not surrounded by the turmoil of voyages and
of business, the most bored people in the world.

¶ 493.
Causes of French
perfection.

If we were not the most civilised and the most

---

[268] This reminds me of a celebrated axiom of one of
the most celebrated leaders of modern æsthetic taste,
to the effect that " a mind cannot be said to be really
artistically appreciative, until it can see beauty in the
perfect construction of the common wheelbarrow."

[¶ 493]

cultivated nation of Europe, *i.e.*, the most voluntarily subjected to the rules agreed upon by reason, we should be the most turbulent and the most divided. And it is this high state of civilisation, this lofty abnegation of our individual instincts in favour of reasonable measures, which causes less advanced nations than us to look upon us as over-refined almost to the point of being factitious.

If it is true that we enter more readily than any other people into the characteristics of other nations; if it is true that there exists no nation that does not prefer us to any nation other than itself, it is evidently because beneath our medium sky there is no type either of the North or of the South which is entirely foreign to us. A point of moral conformity reveals our relationship with all peoples—a relationship which the Romans, having mingled with all races, have transmitted to us with their blood. Why, such is the case even to the savages, whose fantastic humour we have understood, together with their bizarre instincts, and this to such a point that our fortunate colonists, for the purpose of establishing and extending their influence in the New World, have not been obliged, like the Anglo-Americans, to come to the terrible expedient of a war of extermination. English approval is sufficient for an Englishman; as for us, our consciences are uneasy if we do not obtain universal approbation, whence our nation derives the generous duty, which she imposes upon herself, of referring more to the inspirations of universal chivalry than to those which are suggested by national individualism; and whilst, after having inoculated Italy with the sacred fever of liberty, and having broken the fetters of America and of Greece, we conquer Algeria only to regenerate her, the English have never interfered in the affairs of foreign nations save to render them tributaries to their commerce and

¶ 494.
France the most
perfect nation
in the world !
Causes of this
(continued).

[¶ 494]

their industry. It has been said that Germany is
the *heart* of Europe; so be it, but *we* are the *head*.
Our sun illumines the march of the civilisation of the
whole world, and this continent, of which we are at
once the hope, the light, and the joy, acknowledges
that it has made a new conquest, when art, science,
or liberty have advanced a step among *us*, for alone
of all nations we know how to impress upon our
conceptions the seal of universality (!!).

The Square Type.

Then a soldier
Full of strange oaths & bearded
like the pard,
Jealous in honour sudden & quick in quarrel.

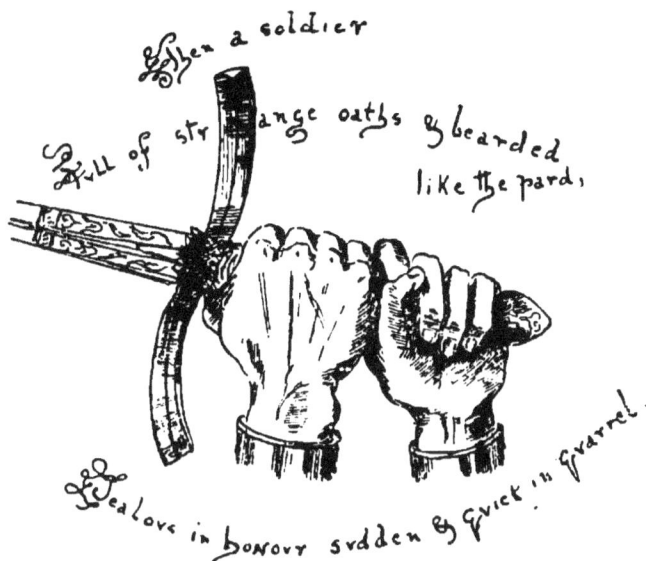

SUB-SECTION XVIII.

USEFUL HANDS.   [Plate VII.]

THE USEFUL TYPE.

¶ 495.
Its appearance

IT is of medium size, but large rather than small, knotty fingers, the outer phalanx square, that is to say, its four sides extend parallel to the tip [you must not take any notice of the curve which nearly always finishes off the points of the fingers], a large thumb, with the ball thereof well developed, the palm of medium dimension, hollow and rather firm.   I do not propose to consider these hands with a *small* thumb, for the reason I have given at the head of Sub-Section IX.

¶ 496.
Square and conic hands contrasted.

If I have made myself clearly understood, the reader will have gathered that a type characterises itself no less by its repugnances than by its inclinations ; by its defects than by its qualities.   Well then, perseverance, foresight, and the spirit of order and conventionality, which I have pointed out as being nearly absolute strangers to artistic hands—

hands which the beautiful and the pleasant occupy
far more than the serviceable and useful,—*abound*,
on the contrary, in the intelligences signalised by
square fingers.

¶ 497.
Characteristics
of the square
type.

To organise, to classify, to arrange, and to render
symmetrical are the mission, the duties of the useful
hand.   It has no conception either of beauty or truth,
apart from theory and conventionality.   They have
for similitude and homogeneity, the same love which
conic hands have for contrasts.   They know wherein
things which differ are similar, and the points in
which things that seem similar in reality differ ; which
faculty, as Montesquieu says, constitutes the spirit
in which the various degrees of hierarchy range
themselves in clearly-defined lines, and in which,
according to them, lie the principles of political power
and wisdom.   They intentionally confound discipline
with civilisation,[369] *i.e.*, prescribed with agreed order.
They feel things harshly, or at least severely, ranging
all things as duties, subjecting *thoughts* to *thought*,
men to *man*, and only tolerating such impulses of the
mind, the soul, and the heart as reason [considered
from its narrowest aspect] accepts and permits.

¶ 498.
Conservatism.

One law of all others is dear to them, that of continuity, and it is above all things according to that
rule, that is to say, by tradition and transmitted
law, that their extension takes place.

¶ 499.
Limited
intelligences of
the type.

Such intelligences, otherwise vigorous, have no
wings ; they can expand, but they cannot rise.   They
are shod with seven-leagued boots, but the fiery chariot
of Elijah is foreign to their natures.   The *earth* is

---

[369] "If the people is kept in order by fear of punishment, it will be circumspect in its conduct, without
feeling ashamed of its evil actions ; but if it is kept in
order by principles of virtue and the laws of social
politeness, it will feel ashamed of a culpable action, and
will advance in the path of virtue."—(*Confucius*).—
AUTHOR'S NOTE.

pre-eminently their abode, they can see nothing beyond the *social* life of man; they know no more of the world of ideas than what the naked eye can know of heaven. Beyond this they are always ready to deny all that they cannot feel or understand, and to look upon the limits of their understanding as the limits of nature.[370]

In France it was not until the seventeenth century, the period devoted especially to method and etiquette, which were at this time reduced to a science, that the ideas of which square hands are the almost exclusive active instruments began to manifest themselves in the usages of society.[371] Architecture under their sway no longer represented themes of poetry or of imagination, as it had in the fifteenth and sixteenth centuries,[372] but displayed a tendency to symmetry and material usefulness; like the man of the world who marries and settles down, architecture cut her connection with fantasy, and settled down into the cold lines of stern reality.

¶ 500.
Square hands of the seventeenth century.

The monuments of Louis XIV., stripped of all architectural idealism, half palaces, half convents, and part barracks, grand in surface but not in character,[373] suggested by their uniformity and by their aridity the spirit of the inexorable and vain despot, whom the care of his own person and of his false splendour kept all his life far from the battlefield, far from the ways of heroism and popularity,[374] and to whom toleration and clemency, those virtues of great souls and of

¶ 501.
Architecture under Louis XIV.
Louis XIV. and Philippe II

[370] How often, whilst putting into practice the science of cheirosophy, will the cheirosophist recall this paragraph!

[371] *Vide* ¶¶ 112 and 502 and the notes to those paragraphs.

[372] *Vide* ¶ 471.

[373] *Vide* note [272], p. 243.

[374] M. d'Arpentigny is not quite accurate here, for Louis XIV., as is well known, led his army in person in

[¶ 501]

great minds, were as unknown as they were to Philippe II.[375]

¶ 502.
St Simon on
Louis XIV.

"The mind of Louis XIV.," says St. Simon, "was below mediocrity, glory was for him throughout his life more a foible than a taste. Born moderate, secretive, and complete master of his tongue, his love of order and regularity were immense, and he was always on his guard against high merit and superiority of mind, talent, or sentiment. He judged men by their love and aptitude for detail, sunk as he was himself in trifles, and losing his time, as he did, in the examination of minutiæ. Loving symmetry, he fancied himself gifted with an appreciation of the beautiful. He settled every morning the work of the day, and gave his orders with precision, keeping punctually to the hours which he had prescribed. Whatever happened, he took physic once a month, heard mass every day, and took the Sacrament five times in the year. He liked walking and riding; he sat his horse well, shot magnificently, danced well, and played racquets and billiards to perfection. His smile, his language, and even his looks were always under control; his politeness, always full of gradations, was unvarying; to ladies he removed his hat entirely, but

Holland and in the reduction of the Franche Comté. [*Vide* (Adolphus') "*History of France*" (London: *n.d.*), vol. ii., p. 443]. Still, his biographer [*op. cit.*, p. 502] says :—" Though he frequently took the field, and reduced in person Franche Comté, and several of the strongest towns of the Netherlands, yet his personal courage has not escaped imputation; and in repeated campaigns he never exposed his life or reputation to the hazard of a battle."

[375] It is true that the discipline and inflexibility of Philip II. were such as to have become proverbial, but at the same time France had not seen, until the time of Louis XIV., a ruler whose statesmanship, or whose patriotism and splendour, were greater than those of this monarch.

removed it more or less far according to their rank.
To the lower orders of nobility he held it in the air
or close to his head for a few moments, carefully
pre-considered. To knights or persons of lower rank
he contented himself with touching his hat with his
hand; to princes of blood royal he uncovered as if
to a lady; at meals he half rose for each lady-in-
waiting who arrived. He *required* that his mistresses
and the ladies of the court should eat heartily—for
no one was ever less *romantic* than he—when it
was his pleasure to see them *feed*. When he was
travelling he did not like them to feel either heat or
cold; they ingratiated themselves with him by an
ever-equable temperament, and he required that they
should be always gay, and ready to walk, dance, or
follow him wherever he pleased to go. He was
always dressed in more or less quiet colours, with
very little ornamentation ; he never wore rings or
jewels of any kind save upon the buckles of his shoes,
₰ garters, and hats; he wore his orders *underneath* his
coat, excepting on special or festive occasions, when
he wore them outside, jewelled to a value of from
eight to ten millions of francs." [376]

If the large-thumbed square type of hand [which
is the only one to which these characteristics can
belong] had not been in an immense majority in
France during this reign, the name of Louis XIV.
would not have come down to us surrounded by such
a mass of eulogy. The men of this generation, similar
in organisation and in temperament, were also similar,
not only in mind, but also to a great extent physically ;
if we read their works, and look at their portraits,

¶ 503.
Historians and
courtiers of the
period.

---

[376] The above is extracted from a quantity of different
places [pp. 32, 35, 118, 119, vol. i., etc., etc.] in the
" *Mémoires de M. le Duc de St. Simon, l'Observateur
Veridique sur le Règne de Louis XIV.*" (Londres: 2ᵉᵐᵉ
edition, 1789).

[¶ 503]

they might all be members of a single family. They have all the same large aquiline noses and stern mouths, the same positive, methodical, reasonable, and restrained minds.

¶ 504.
Physiognomy of the eighteenth century.

Round faces, tip-tilted noses, free glances, lively behaviour, both physically and morally, belong essentially to the philosophic race of the eighteenth century.

¶ 504a.
Ditto of the nineteenth century.

And noses like eagle's beak, faces like the lion's muzzle, round eyes like wild beasts, and glowering eyebrows, belong to the warlike and turbulent times of the Empire.

¶ 505.
Literature of the type.

Even in literature what of ideality these useful hands can comprehend stops short at a very narrow limit. They keep away from idealism just as they do from boldness of thought or novelty of form. Their timorous and conceited muse, whose knowledge is far in advance of her seeming innocence, never adventures herself save upon well-worn roads, preferring to proceed by memory, rather than by sentiment, being spirited rather than imaginative. Such subjects prefer words which describe objects with reference to their *use*, rather than those which merely describe their *form* [thus, for instance, they will use the generic terms " bark," or " ship," rather than junk, sloop, brig, etc.]; whence comes, of course, the want of local colouring that we find in their writings.   In the *style* of what they call poetry, they like above everything else clearness and correctness, rhythm, the balance which results from careful arrangement and combination ; in social relations they require security and exactitude ; in life, moderation.

¶ 506.
Mental attributes of the type.

Circumspect and far-seeing, they like what is clearly known, and suspect the unknown. Born for the cultivation of medium ideas, they pay less attention to the actually real than to the apparently real ; they recommend themselves by their good sense, rather than by their genius; by their spirit and cultivated talents, rather than by the faculties of imagination. In their eyes—eminently fitted to give an opinion on the point—the most sociable man is not the one who appreciates most the good qualities of his fellow-men, but the one who cares least about their defects. They do not seek after *beauty*, which is a requirement of the soul, but after *good*, which is a requirement of the mind.

This kind of teasing despotism, which has its source in the love of order and regularity ; hypocrisy and conceit, which result from an exaggerated love of reserve and appreciation of good behaviour ; this sort of pedantry, which is born of respect of persons ; coldness, which resembles moderation ; flattery and adulation, vices peculiar to spirits endowed with the hierarchical instinct; stiffness of manner and bearing; harshness of punctuality; and abject submission in view of the objects of one's ambition—these are the principal defects of persons belonging to the useful type.

¶ 507. Defects of the type.

Such subjects will accept none but the man who is cultivated, well taught, disciplined, moulded, and trimmed upon a certain pattern. Where the man of learning shows himself in all his glory, they go to seek their models and their examples. When the nation, like an open-handed parvenu, desired at length to speak a language worthy of her fortunes and the height of her glory, it was not to the hidden sources of the chivalrous epics of the middle ages, that the square hands of the seventeenth century appealed ; they turned their eyes towards Athens and Rome, towards the sanctified names of Euripides, of Virgil, of Demosthenes, and of Cicero. From that time forth power, reinforced by the talent of the universities, had its own literature, just as it had its own architecture, both of them characterised by imitation

¶ 508. Language of the type.

Seventeenth century.

[¶ 508

and reflection of other styles.[377]   But in the country, the natural stronghold of elementary liberty and truth, men held fast to the ingenuous poetry of the land, and despising the strange gods of the capital remained Gaulish, Christian, and romantic.[378]

¶ 509.
Nationalism.

"Every country," says Philip de Commines, speaking in one of his works of the long sojourn of the English in France, and of the Germans in Burgundy, " whatever men may do, ends invariably by remaining the property of the *peasantry, i.e.,* to the nationalists." [379] It is the same of the schools of literature.

[377] Voltaire says of this century :—" In eloquence, poetry, literature, and in books of morality, the French were the legislators of Europe. . . . Preachers quoted Virgil and Ovid ; barristers quoted St. Augustin and St. Gerome. The mind that should give to the French language the turn, the rhythm, and the clearness of style and grandeur had not yet appeared. A few verses of Malherbes showed that it was capable of grandeur and of force, but that was all. The same genii who had written in excellent Latin, as for instance the President de Thou and the Chancellor, were no longer the same when they dealt with their own language;" and he goes on to point out that it was in this century that the works of Voiture, of Vaugelas, and of La Rochefoucauld, really commenced the regeneration of the French language. —" *Siècle de Louis XIV.*," chap. xxxii.

[378] An interesting and curiously parallel passage to this commences the opening chapter of Walter Pater's recent work, " *Marius, the Epicurean* " (London : 1885, ch. i., p. 1), which reads as follows :—" As in the triumph of Christianity the old religion lingered latest in the country, and died out at last as but paganism—the religion of the villagers—before the advance of the Christian Church ; so in an earlier century it was in places remote from town life that the older and purer forms of paganism survived the longest. While in Rome new religions had arisen with bewildering complexity around the dying old one, the earlier and simpler patriarchal religion, 'the religion of Numa,' as people loved to fancy, lingered on with little change amid the pastoral life, out of the habits and sentiment of which so much of it had grown."

[379] Compare the last note. Philippe de Commines

Townsmen rather than citizens, men of the square-handed type, prefer certain *privileges* to absolute *liberty*. Authority is the base of all their instincts, the authority of rank, of birth, of law, and of custom; they like to feel, and to impose the yoke. " Anything which hinders a man," say they with Joseph de Maistre, " fortifies him; he cannot obey without improving himself, and by the mere fact of his thus conquering himself, he is the better man."

¶ 510.
Socialism of the type.

See, for instance, what a deplorable influence the troubles of the Fronde exercised upon the square-handed type of men under Louis XIV. In the fear of seeing their troubles recommence, they hastened in their fanatic love of order to invest the king with autocratic powers; they intoxicated him with splendour and voluptuousness; they lost themselves in the dim necessitudes of servilism; they raised the throne to the level of the altar, and proudly proclaimed themselves to be the apostles of monarchical fetichism.[80] From the gentle and convenient God of Montaigne and of Rabelais they passed to the iron-handed, harsh-voiced, and intolerant God of Pascal [381]

¶ 511.
The type under Louis XIV.

Preachers.

[born 1445, died 1509] entered the service of Charles, Duke of Burgundy, in 1464, and left him to enter that of Louis XI. in 1472. The earliest known MS. of his work entitled " *Cronicque et Histoire faicte et composée par feu Messire Philippes de Commynes, Chevalier, Seigneur d'Argenton*," dates from the commencement of the sixteenth century. The first printed edition appeared at Paris in 1523, and complete editions appeared in 1528 and 1546. The best modern edition is that entitled " *Choix de Chroniques et Mémoires sur l'Histoire de France* " (Paris: 1536), edited by J. A. C. Buchon. The remark quoted by M. d'Arpentigny occurs in book iv.

For an account of the " Guerre de la Fronde " *vide* Voltaire's " *Histoire du Parlement de Paris*," ch. lvi. At the end of the disturbances the parliament ceased practically to exist, and Louis XIV. held autocratically the reins of government [*op. cit.*, ch. lvii.].

[381] Pascal, the philosopher and Jansenist, the author

and of Bossuet.[382]   They lacked the gift of persuasion
and their preachers, highly appreciated at court on
account of their magnificent language, could only convert souls in the Cevennes with the aid of the

Literature.   arquebuse and halberd.[383]   We find no impartiality,
liberality, or appreciation of antique times in the works
of their historians ; no lyricism among their poets ;
beneath the learned tissue of their style, beneath the
parsimonious sprinkling of metaphors and ornamental
Latin quotations which decorate it, one catches a
glimpse of souls trembling beneath the vigilant scru-
tiny of literary, religious, and political pedantism, to
which, by virtue of the organisation which governs
them, the sublime flights of enthusiasm and liberty
must remain for ever strangers.

¶ 512.
Art in the same
period

Thus, on the one hand, plastic art no longer existed,
for plastic art is as nothing where the consent of the
masses is not regarded as important ; and, on the
other, real poetry was not yet existent, for the grand
Angel, whose forehead is diademed with stars, and
who sits at the right hand of God,—the angel of lyric

of the celebrated "*Pensées sur la Réligion et sur
quelques autres Sujets*" (Paris : 1669) [complete edition
(Paris : 1844)], was the author of the "*Lettres Provin-
ciales*," preached under the pseudonym of Louis de
Montalt, in 1654, which Voltaire calls the first book of
real genius which the century had seen.

[382] M. d'Arpentigny is wrong in attributing these
characteristics to the dogma of Bossuet, whose style of
preaching was always sublime and poetic to the point of
real pathos.   Our author would have done better in
citing his successor Bourdaloue, of whom Voltaire tells
us:—"He was [1665] one of the first that introduced
reasonable eloquence into the pulpit. . . . In his style,
which was more nervous than florid, without any
imagination in its expressions, he seemed to desire to
convince rather than to touch, and he never aimed at
pleasing his congregation."—"*Siècle de Louis XV.*,"
ch. xxxii.

[383] *Vide* note [296], p. 247.

poetry and sublime thoughts, the angel whom Racine
would have invoked had his age permitted it, had not
yet hovered on his fiery wings over France, pre-
occupied as she was by the cares of government,
and the religious scruples of her king.   The artistic
hand had departed, and the psychic had not yet
arrived ; it was the æra of letters and of wits, of
shades, of sentiment, and of literary subtlety ; of
great *talents* no doubt, but not of great *lights*.   Coustou,   Sculptors
Coysevox, and Puget—the latter especially—still gave
life to the dull cold marble, but they did not give
to it beauty.[384]   Poets, too great not to be misunderstood, Le Poussin exiled himself in Italy, Lesueur   Painters
shut himself up in a cloister, and Claude Lorrain in
the contemplation of nature.   No official of this
period possessed the true sentiment of the beauties
of nature; in the laying out of gardens, geometry
was substituted for design, and symmetry for grace.
At the theatre excessive restraint, conventionality,   Drama
and artificiality, proved fatal to the drama and chilled
it.   The tragic muse—like the nation—seriously
shackled, refused to move forward, and seemed to
realise the fact that beneath the crushing load of
its innumerable rules, she must act without *grace;*
at the same time comedy, fable, and moral romance,
compositions of a medium nature, and left more
free, probably for this reason attained during this
epoch their highest limits of perfection.   France,
which together with the spirit of the artistic hands
had lost her hatred of conventionality and regularity,

[384] This is of course a matter of opinion ; there are no
doubt those among my readers who have looked with
genuine admiration upon the statues of Coustou in the
Luxembourg, and his group " *The Rhone and Saone* "
at the fountain in the Tuileries Gardens ; as also
Coysevox's groups of " *Mercury and Fame* " on the
pillars of the gate which leads into the same gardens from
the Place de la Concorde.

[¶ 512]

her brilliant energy, her light refrains, and the caustic wit so dear to the contemporaries of "le sire de Brantôme,"[585] advanced only with heavy and measured tread; she had disciplined even gallantry, and had theoretically turned the course of the stream of *Le Fleuve de Tendre;*[586] she had assumed a grave and magisterial air, and a voluminous peruke; she was busy, and must not be disturbed. Upon science, whose bold and logical deductions are regarded askant by the authorities; upon history, whose inquiries cause them annoyance; and upon warlike courage, the blind ardour and adventurous enthusiasm of which have so often compromised them,—square hands impose as an untransgressible limit, as an inevitable starting-point, and as a curb, official confidence, tradition, and tactics. If the instincts of recalcitrant and innovating minds urge them towards the sublime horizon of the world of ideas—so much the worse for them; this terrible square type count among their arguments spoliation, exile, fetters, and the scaffold.

¶ 513.
Illustrations.
Baville.

Such was Baville, who only saw one bad point in the public tortures and slaughter of the Protestants

[585] Pierre de Bourdeilles, Abbé and Seigneur of Brantôme, chronicler and writer, born in 1527, died in 1614. He passed the greater part of his life in the professions of soldier and courtier. Gentleman of the bedchamber to Charles IX. and Henry III., he seemed destined, by his character as much as by the adventures which befell him, to become the chronicler of his epoch. He has been called "Le Valet de chambre de l'Histoire," a title well earned by his works, of which perhaps the best known and most notorious are, "*Les Vies des Hommes Illustres et Grands Capitaines de son Temps*" (Leyden: 1665); "*Mémoirs contenans les Anecdotes de la Cour de France sous Henri II., Francois II., and Henri III. et IV.*" (Leyden: 1722); and "*Memoirs contenans les Vies des Dames Galantes de son Temps*" (Leyden: 1693).

[586] The "*Pays de Tendre*" was an allegorical country, a good deal referred to by Romancists of the seventeenth

[¶ 513]
Dominic.

at Montpellier,[387] and that was—the compassion with
which they inspired the mob. Such was Dominic,
whose fanaticism suggested the idea of the extermina- D'Albe.
tion of an entire people ;[388] such was the Duc d'Albe,
who vaunted himself as having caused 18,800 men
to perish upon the scaffold.[389] Such was the calm Robespierre.
Robespierre, petrified by logic, legality, and incor-

century,—a country entirely devoted to the pleasures of
love, a kind of elysium of lovers and their lasses. This
is the stream referred to by Boileau, in his tenth satire
" Des Femmes," in the lines :—

> " Puis bientôt en grande eau sur le *fleuve de Tendre*
> Naviger à souhait, tout dire et tout extendre," etc.

[387] Montpellier was one of the most redoubtable
strongholds of the Protestants, by whom it was fortified
in the seventeenth century. It was, however, besieged
by Louis XIII. in August 1622, and fell in the following
October. The massacres of Protestants which followed,
and the atrocities committed by the conquerors of
Montpellier, form one of the blackest pages of the
religious history of the times.

[388] Dominic, the founder of the monastic order that
bears his name, having despatched 1,207 monks into
Languedoc to collect proofs of the heresy of the Albigenses, and having found contradictions in the evidence,
besought the authorities *to exterminate them* [EYRE
CROWE, " *History of France* " (London : 1858), vol. v.,
p. 175]. Martin says of him [" *Hist. de Fr.*" (Paris :
1878), vol. iv., p. 25], " Un immense anathème poise sur
la tête de ce moine, qui passe pour le génie de l'inquisi-
tion incarné ; . . . il s'imagine servir le genre humain en
poursuivant sans pitié les 'suppôts de l'enfer qui
perdaient tant de milliers d'âmes,' et crut obéir à la
voix de Dieu en étouffant les murmures de sa conscience
et le cri de ses entrailles."

[389] This is what we find laid down in so many words
in Joannis Meursi " *Ferdinandus Albanus, sive de
Rebus ejus in Belgio per sexennium gestis Libri IV.*"
(Amsterdam : 1638).*

---

* "Gloriatus alibi dicitur ; uno illo sexennio suo, lictoris
manu, xviii. ciɔ ac ciɔ mori jussisse, præter eos quos bella aut
præsidia hausissent."—Lib. iv., p. 86.

[¶ 513]

Louis XIV.   ruptibility.[390]   Such was Louis XIV., whose cramped spirit could never rise to doubt.[391]

> . . . "non men che saver, dubbiar mi aggrata." [391]

¶ 513a.   Richelieu. An excellent portrait of Cardinal Richelieu by Philippe de Champagne, belonging to the Museum of Caen, represents that prelate with pointed fingers. It is a gratuitous piece of flattery, if indeed it was so meant, for the hand being represented in profile the fingers could not appear otherwise than pointed.[393] Cardinal Richelieu, who recommends in his will that all men of a too delicate sense of honour should be banished from the conduct of public affairs, had, as we all know, a comprehension of social and politic ethics as broad as his moral sense was narrow.[394] More attentive to the interests of heaven than to

[390] "Robespierre, who distinguished himself in the constituent assembly by the severity of his principles, was excluded from the legislative by the decree of non-election. He entrenched himself now among the Jacobins, where he domineered with absolute sway, by the dogmatism of his opinions, and a reputation for integrity which gained for him the epithet of the *incorruptible*" [Thiers, "*History of the French Revolution*" (London: 1877), "The Legislative Assembly," ch. i.]. Perhaps the minutest and most unbiassed account of the life and character of Maximilien Robespierre is to be found in Lamartine's "*Histoire des Girondins*" (Paris: 1847), of which an English translation was made by H. T. Ryde: "*History of the Girondists, or Personal Memoirs of the Patriots, etc.*" (London: 1849), from which to a great extent Mr. Lewis' "*Life of Maximilien Robespierre*" (London: 1849) is taken.
[391] *Vide* ch. xxxvi. of Voltaire's "*Siècle de Louis XIV.*"
[392] Dante, "*Inferno*," canto 21, *fin.*
[393] *Vide* ¶ 495.
[394] "*Testament Politique d'Armand du Plessis, Cardinal Duc de Richelieu*" (Amsterdam: 1788), chap vi. [p. 211]:—"C'est ce qui fait qu'au lieu de lui représenter les avantages que les princes réligieux ont pardessus les autres je me contente de mettre en avant

those of the earth, he only took up arms against the Protestants to deprive them of their material strength.[305] Though he had the mania of verse, *i.e.*, of rhythm and measure in speech, he had not the sentiment of poetry developed to the slightest degree. He was a deeply considered, and bold enemy of independent and audacious instincts, and he was a leveller in whose eyes two things alone were sacred—unity and authority.

Certainly, like Aristotle, that paragon of the square type; like Boileau, the prototype of rhythmical poets; like Turenne and Vauban, generals of the scientific school, Richelieu had square finger-tips, and not pointed ones.

¶ 514.
Parallels.
The hands of Richelieu.

Versailles, where everything is arranged in straight lines, and effaces itself in a tyrannical and wearisome symmetry; where the houses like stately dowagers stand in rows, cold and stiff, uniformly ornamented with massive facings of red brick in imitation of porphyry;—Versailles, where one feels that one ought only to walk in one's best clothes, and at the pace of a procession; where, to analyse the feelings to which the splendid poverty of its bastard architecture gives birth, the mind turns to arithmetic rather than to poetry;—Versailles, I say, with its gardens and its palace, will always be for large-thumbed useful hands

¶ 515.
Versailles of the square type.

que la devotion qui est nécessaire aux Rois doit être exempte de scrupule : Je le dis, sire, parce que la delicatesse de la conscience de V. M. lui fait souvent craindre offenser Dieu en faisant certaines choses dont assurement elle ne sçauroît s'abstenir sans péche."—Chap. viii., sect. iii. [p. 246.]:—"La probité d'un Ministre Public ne suppose pas une conscience craintive et scrupuleuse : au contraire il n'y a rien de plus dangereux au Gouvernement de l'Etat," etc ; and the same spirit continually pervades the maxims contained in this extraordinary little volume.

[305] *Vide* ¶ 384

the most perfect exemplification of monumental
beauty, as it is understood by them.

¶ 516.
The bureaucrat
and civil servant. The man who is a publican by nature, the bureaucrat moulded to the true type of his race, has
necessarily square fingers. Satellite of the science
of arithmetic, he gravitates around its arid sphere
and draws from it its faint refulgence. His harsh
and sullen pen deals only with matter of law and
rule, for he is permeated by the fiscal instinct, which
with him takes the place of all natural feeling. Living
outside the pale of thought and of events, far from
the clashing of opinions, of interests, and of swords,
all his sensibilities are concentrated upon himself, as
is also everything in the way of combinations that
his mind can grasp. He is so constituted that he
cannot feel a passion for anything ; the crowd surges
and murmurs in the street,—it is the king who is
passing ; immediately he wreathes his lips into a smile
of compulsory cheerfulness, which will disappear like
the flame of an extinguished candle when his master
has gone by. In his eyes the best government is the
one under which he can live his careful life ; he
knows beforehand at what age he will marry, and
how many children he will have. Having no hope
of achieving glory or distinction, innovation of any
kind being absolutely forbidden him, he never loses
an opportunity of decrying it when it is being vaunted
by popular enthusiasm. For all professions which
are not literary, he has the same contempt that the
peasant has for the trades which do not entail hard
work. In his estimation man has only been the
superior animal since the invention of paper, and he
considers as problematical and hypothetical the renown
of the pretended great men who have never learnt to
read or write. He considers the ranks of the priesthood and the classifications of the social scale to be
far more worthy of consideration than the poet, the

artist, or the philosopher, patricians by Divine right; he expects that imitation and assimilation will give him a factitious aristocracy, a rank, an importance that he can never expect to derive from the eminently plebeian nature of his labours.

In France the uniformed writer whose duty it is to regulate the renovation of the soldiers' shoes, and to count the survivors after the battle, has a right to the same decorations, distinctions, and emoluments as the general who has headed the soldiers into victory; in this case ink has the same value as blood. We have gaugers and caterers, who hold the rank of colonel, who wear decorations, stars, and ribbons.

¶ 517.
Military scribes.

In China there is an insignium of honour conferred upon men of letters, and upon the highest state officials, which France should adopt for the distinction of her bureaucrats—the peacock feather.[396]

¶ 518.
Chinese insignia of rank.

The soldier appreciated by the subject of the spatulate type is tall, with broad shoulders, a ruddy complexion, an equable temperament; he is gay, frank, martial, and deliberate in manner. In the country, the chickens come of their own accord to forage for themselves in his haversack; like the veterans of the Empire, he will not believe in the traditional goose with the golden eggs, or the cricket of good omen, but he has implicit confidence in the powers of his sword and his brandy flask. So long as he is strong and valiant, one may forgive his intemperance.

¶ 519.
Soldiers of the spatulate type.

The general whose large hands have square fingertips requires that he shall be exact and orderly, always clean, carefully attired, and scrupulously neat. He will not allow him the pleasure of an occasional dissipation; he must not be hungry or thirsty excepting at the regimental and prescribed hours. No doubt he will be brave and robust, but before all he

¶ 520.
Soldiers of the square type

---

[396] *Vide* ¶ 418, and notes to the paragraph.

[¶ 520]

will be obedient and submissive to discipline; he will have clear judgment, but his wit will be neither brilliant nor refined.

¶ 521.
Constitution of an army.

According as a man belongs to these two types, as they at present stand in France, an army ought only to be an instrument whose perfection consists in being strong and supple, that is to say, of soldiers more vigorous than intelligent, and of inferior officers of greater docility than capacity. A highly-developed capacity in subalterns is to them more a drawback than an advantage; it is pretended that it brings them infallibly to a state of scorning the details of things, to presumption and disregard of discipline; and while in this connection one may cite Sallust, who has written that the soldier who affects the fine arts has necessarily a feeble understanding.[397] A high compliment, indeed, to the independence of artistic natures.

¶ 522.
The army under Charles X.

In the barracks, towards the end of the reign of Charles X., the guards with their elementary hands lived by themselves, calm, empty-headed, and characterised by a vulgar and inoffensive appearance. Corpulent and inert, they sat their horses boldly, but without grace; whilst some regiments made an ornament of their uniforms, for them it was but a garment. They rose at daybreak, and went to bed

---

[397] The passage referred to is the one in which Sallust, describing the army of L. Sulla in Asia, states that they were completely demoralised by luxuries, and by a taste for the fine arts.*

* SALLUST, "*Bellum Catilinarium*," cap. xi :—"Huc accedebat quod L. Sulla exercitum quem in Asia ductaverat, quo sibi fidum faceret, contra morem majorum, luxuriose nimisque liberaliter habuerat loca amœna voluptaria, facile in otio feroces militum animos molliverant. Ibi primum insuevit exercitus populi romani amare potare; signa, tabulas pictas, vasa cœlata mirari. . . . Igitur hi milites, postquam victoriam adepti sunt, nihil aliqui victis fecere."

early; their windows were garnished with flowers; they used to go for long silent walks two and two, from which they would bring back a bouquet of violets, or a branch of May-flowers. When they came in they passed a wet comb through their hair, and set to work to brush their clothes. A wicker cage shaded by a bunch of chickweed, their portraits painted in oils by a house-painter, a net for quail-catching, and a green china parrot, with a curtainless bed, a deal table, and the chest of drawers, ·and three chairs provided by government, completed their modest furniture. They married women of their kind, *i.e.*, women with large feet, of masculine proportions, of equable temperament, drinking everything and knowing like them how to march in line, and who shared with them their admiration for the great porringer of the Invalides.

These useful-handed guards went in for the free educational courses open to them, in the murky domains of the Latin tongue; they frequented laboratories, amphitheatres, and libraries, and always absorbed sciences with avidity; they numbered herbals and collections of insects among their possessions. One might consult them as to the time, as if they had been dials; as to the date, as if they had been calendars; the day of the week, as if they had been almanacks;—it flattered them. They prided themselves on being trussed at all points conformably to the prescriptions of the regimental code. They could stammer dead languages, and on great occasions they would fire great names at you like bonbons out of a pea-shooter.

¶ 523.
Their mental characteristics.

Enthusiasm, turmoil, and the outdoor forms of life, on the contrary, were the domains of the younger guards with smooth spatulate fingers; lovers of the bottle, hardy and elegant horsemen, they were pre-eminent for the graces of their persons and appoint-

¶ 524.
The younger soldiers.

¶ 524]

ments; they dressed beautifully, and alone of all men carried their whips " with an air." One met them everywhere : like wasps to the ripened fruit there flocked to them the idlers of the *cafés*, the eternal sippers of absinthe and of white wine ; and the English, grown tired of ale and red-haired beauties ; and the baronesses of the Holy Empire with their insatiable requirements ; and portionless damsels in search of a beau, wearing their hearts on their sleeves ; and the frisky and impudent members of the ballet, and all the rest of them ! In their rooms, rather less bare than those of which I have just spoken, one might see a guitar, some flowers, an assortment of pipes, a volume of Pigault-Lebrun, and silhouettes of women framed in gilt paper.

¶ 525.
All things contribute to national advancements.

Whenever a nation advances in a particular direction, every influence combines to forward it in that direction, even those to the detriment of which, the movement is accomplished. That is what one sees under the reign of Louis XIV., when the spirit of every class tended towards material order above all things, so as to merit the favour of the governing type, which made use of them against themselves just as navigators make use of a head-wind, and make it carry them forward.

¶ 526.
The " Lettres Edifiantes " and their authors.

One is struck by two things in reading the " Lettres Edifiantes ;" firstly, by the strength of will and the spirit of self-abnegation and of patience, by the courage, and the knowledge of the missionaries who wrote them ; and secondly, by the childish faith which they reposed in the efficacy of the last ceremony of the Catholic religion. They seemed convinced that whoever was not of their Church was not only not a Christian, but was a positive Atheist.[398] So narrow-minded, so impious an exclusiveness, and the disdain

Vide note [12], p. 117.

which they always testified for human reason unil-
lumined by revelation, have rendered vain all their
efforts upon the infidels, among whom they have
driven more people to fanaticism and to martyrdom
than to civilisation. To listen to them one would
believe that it is the Church, and not God, who dis-
poses of human souls. One of them steals a baby
from its mother and baptizes it in secret; at the close
of the ceremony the baby dies, and the missionary
weeps for joy at having saved a soul from hell,—
as if by this formality God was prevented from a
possible misapprehension as to the baby's innocence. It is during the seventeenth century that
the greater part of the " Lettres Edifiantes " were
written; they reflect admirably the hard and intolerant, hierarchic, disciplined and obstinate spirit of
the useful hand. But these missionaries were nearly
all gifted with simple and charitable hearts; their
genuine modesty was in no respect " the pride that
apes humility," nor their benevolence towards the
humble and meek, of the kind which has been defined
as "the hatred of the great and powerful." [399] These
qualities have modified the defects of their type ; but,
had their hearts been as exclusive as their minds,
their small successes would have been smaller still.

There is this difference between the love of constituted authority as it is understood by spatulate
hands and by useful hands (free in either case from
the modifying influences of education), that the former
attach themselves to the person of the despot, and the

¶ 527.
Spatulate and
useful hands
contrasted.

[399] " Lettres Edifiantes et Curieuses concernant
l'Asie, l'Afrique, et l'Amérique, etc." (Paris : 1838),
publiées sous la direction de M. l'Aimé-Martin. These
letters, dated principally from the seventeenth century,
constitute a large collection of records of missions
undertaken by French and other priests in all parts of
the known world.

[¶ 527]

latter to the institution of despotism ; for the former the tyrant must be powerful, for the latter he need only be properly constituted.

¶ 528.
Artistic and materialistic order.

Artistic hands observe material order only in so far as it helps and contributes to beauty. Useful hands love it for itself, admitting freely to their lives everything resultant therefrom. Order as it is understood by the English and Americans chills our artistic taste, and is antipathetic to us. Excessive order reduces all principles to the level of methods, a proceeding which almost materialises them, and up to a certain point strikes them with sterility. It is thus, as Madame de Staël said, that analysis kills the spirit of a thing, that chemistry kills its life, and that reasoning kills its sentiment.[400]

[400] " *Œuvres Complètes de Madame de Staël-Holstein* " (Paris : 1856), *passim*. " La précision métaphysique, appliquée aux affections morales de l'homme, est tout à fait incompatible avec sa nature. Le Bonheur est dans le vague, et vouloir y porter un examen dont il n'est pas susceptible, c'est l'anéantir comme ces images brillantes formées par les vapeurs légères qu'on fait disparaître en les traversant."—" Essai sur les Fictions," *op. cit.*, vol. i., p. 62. " L'imagination a peur du reveil de la raison comme d'un ennemi étranger qui pourrait venir troubler le bon accord de ses chimères et de ses faiblesses. . . . Le courage et la sensibilité, deux jouissances, deux sensations morales, dont vous détruirez l'empire en les analysant par l'intérêt personnel, comme vous flétririez le charme de la beauté en la décrivant comme un anatomiste."—" De la Littérature," vol. cit., pp. 317-9.

USEFUL HANDS [*continued*].

Chinese Hands.

SQUARE-TIPPED hands must be in an immense majority in China, and for this reason : the masses defer willingly to the requirements of hierarchy and to the sovereign authority of a single man.[401] They do not weigh reason against logic, but against usage ; they esteem good sense more than genius, things ordinary more than things extraordinary, the real more than the ideal, and the middle course rather than the extremes.

¶ 529.
Square hands in China.

They prefer social and practical to speculative philosophy, history and the other moral and political sciences to metaphysics and abstract sciences.[402]

¶ 530.
Science

The man who governs his family well, who has been a respectful and dutiful son, who has the proper deference for his elders, is judged worthy and capable of governing a province, a kingdom, or an empire.[403]

¶ 531.
Filial affection.

[401] *Vide* note [311], p. 267.

[402] " We have a religion of the literary class which corresponds to the degree of culture of the most enlightened body in the empire. This is the religion of Confucius, or rather his philosophy ; for his doctrine is that of the founder of a school who has enunciated moral maxims, but has not meddled with speculative theories upon the destinies of man and the nature of the Divinity."—Tcheng-Ki-Tong, *op. cit.*, p. 19.

[403] " The family," says Colonel Tcheng-Ki-Tong, " is

¶ 532.
Etiquette.

They have placed placid politeness, tact, and the observance of conventionalities and of rites, at the head of the social virtues.   In China these things regulate the various ways in which each man, according to his age, rank, or profession, must walk about, sit down, enter, leave, listen, look, salute, dress, and move.

¶ 533.
Social qualities
under
Louis XIV.

It was the same thing in France in the seventeenth century, the epoch when passive obedience constituted the first merit of a son or of a subject, and when a knowledge of heraldry, of etiquette, of ceremonies, of formulæ, and of the ways of the world sufficed to endow a man with the epithet of an accomplished gentleman.   They could not write or spell properly in the court of Louis XIV., but they saluted with more grace than in any other place in the world.

the institution upon which is based the whole social and political edifice of China. . . . From the most remote period the influence of the family has predominated every order of idea ; and we say, quoting Confucius, that to govern a country one must first have learned to govern a family.   The family is essentially a government in miniature ; it is the school in which governors are formed, and the sovereign is himself a disciple of it" [p. 7].   " The chief authority is vested in the most aged member of the family, and in all the important con· junctions of life reference is made to his decision.   He has the attributes of the head of a government ; all documents are signed by him in the name of the family " [p. 9].   " *The Chinese painted by Themselves* " (London : 1885).   It is thus that we find in Chinese literature a continual recurrence of such passages as the following, taken from " *The Shoo King, or the Historic Classic* " [translated by W. H. Medhurst (Shanghai : 1846), p. 792], in the sixth book, called the Book of Chow, sect. 2 : " The King said, as it were, ' Oh, Kūn Chin, you are possessed of excellent virtue, being both filial and respectful.   And as you are dutiful and fraternal towards your elder and younger brethren, you can display the same qualities in government, I therefore command you to regulate our northern border," etc., etc.

The moral portrait of Confucius, as it has been transmitted to us by his disciples, affords the perfect model of the superior man as he was understood by the Chinese. Here are some of its salient features : Kung Fu'Tsu, when still resident in his native village, was extremely sincere and upright, but he had so much modesty that he seemed to be deprived of the faculty of speech. When, however, he found himself in the temple of his ancestors, and in the court of his sovereign, he spoke clearly and distinctly, and all that he said bore the evidence of reflection, and of maturity. At court he spoke to the inferior officials with firmness and rectitude, to the superior officers with a polished frankness; in the presence of the prince his attitude was respectful and dignified. When the prince ordered him to court and bade him receive the guests [the more important vassals], his attitude took a new aspect, and his gait was slow and measured, as if he had shackles attached to his feet.

¶ 534.
Manners of
Confucius.

In saluting persons who stood near him, either to the right or to the left, his robes, both in front and behind, always fell straight and in well-arranged folds. His pace quickened when he introduced guests, and he held his arms extended like the wings of a bird. When he entered the gates of the palace he bent his body as if the gates had not been high enough to allow him to pass through erect ; he never paused as he entered, and never stood about upon the threshold. In passing before the throne his countenance changed suddenly : his tread was slow and measured, as if he had been fettered, and his speech seemed as shackled as his feet. Holding his robe in his hands, he advanced thus up the hall of the palace, with his body bent, and holding his breath as if he dared not breathe. In going out, after having taken a few steps, he relaxed his grave and respectful countenance, and assumed a

¶ 535.
The same.

[¶ 535]

smiling expression, and when he reached the bottom of the stairs, letting his robes fall, he extended once more his arms in the wing-like attitude.

¶ 536.
His salutations.

In receiving the distinctive marks of homage [as the envoy of his prince] he inclined his body profoundly, as if he could not support it; then he raised it up with both hands, as if to present himself to some one, lowered it once more to the ground, as if to replace it : wearing on his countenance and in his attitudes the appearance of fear, and displaying in his gait, now slow, now rapid, all the different emotions of his soul.

¶ 537.
His costume.

His costume for sleeping or repose was half as long again as his body. In the house he wore thick garments of fox skins. On the first of every month he put on his court dress, and went to pay his respects to the prince.

¶ 538.
His customs at meals.

He never ate meat that was not cut straight and neatly, nor of dishes which were not dressed with the sauce that suited him, nor if the colour was not right, or the smell did not please him, or if the meats were not of the right season. In drinking alone he restricted himself by no rule, but he never took enough to disturb his mind in the least degree : he always took a moderate quantity of ginger with his food, and whilst eating he never conversed. If the mat on which he was to sit was not carefully laid, he would not sit upon it ; when invited to a feast by the inhabitants of his village, he would never leave the table before the feeble and older men who were present.

¶ 539.
His imperturbability.

His stables having caught fire, he said, when he returned home, "Has the fire killed any one ? I do not mind about the horses."

¶ 540.
His loyalty.

If he was ill, and the prince came to see him, he always had himself placed with his head to the east, and dressed in his court dress, and girded himself with his richest belt. When the prince summoned him to his presence, he went immediately on foot, without waiting for his equipage, which followed him.

¶ 541.
Funeral rites.

If any of his friends died, having no one to superintend his funeral rites, he used to say, "The cares of the funeral will be mine."

¶ 542.
His rules of conduct.

When he went to sleep he never assumed the position of a dead man, and at home he used to lay aside his habitual gravity. If any one called on him whilst he was in mourning, even when it was for an intimate friend, he never failed to change his countenance and to assume a conventional expression. If ever he met any one in robes of state, or who was blind, though he himself wore nothing but his ordinary garments, he never failed to treat him with deference and respect. When he met any one dressed in mourning, he saluted, alighting from his carriage; and he acted in the same manner when he met people carrying memorial tablets bearing the names of citizens.[404]

¶ 543.
The same.

When a thunder-storm broke out upon the town, or a storm wind rose suddenly, he always changed countenance, and looked with an air of terror towards heaven. When he rode in his chariot, he stood upright, holding the reins in his own hands: he never looked behind him, spoke only with a grave accent, and never pointed with his finger.[405]

[404] Tablets upon which are inscribed all the titles which the family has possessed for several generations, are always in China carried in procession whether of marriages or of funerals. It is the custom for passersby to salute such processions when they meet them in the streets.

[405] The above account of Kung-fu-tzu is taken from J. P. G. Panthier's accounts of the Chinese law-giver, contained in his works "*Confucius et Mencius*" (Paris: 1846), and "*Le Ta Hio, ou la Grande Etude*" (Paris: 1837).

¶ 544.
The moral
inculcated.

Here follow some of the maxims of Confucius; they prove that genius, that lofty reason and wisdom are to be found in all countries and among all races; to love one's fellow-creatures is a virtue, to know them is a science,[400] and to despise them is the ruin of virtue.

¶ 545.
Necessities of
government.

Men who can make studied orations are not the right ones to decide in matters of crime; they require only men who are gentle, sincere, and just; who can always preserve the happy medium. If a state is not governed by reason, riches and honours are then only subjects of shame. If, on the other hand, it *is* governed by reason, poverty and misery then become subjects of shame.[407]

¶ 546.
The social
instinct.

The man of superior mind lives at peace with all men, without any fixed rule of invariable conduct. The man of inferior intellect, on the other hand, acts always according to a given plan, without being able to suit himself to his fellow-men. . The former is easily served, but is satisfied with difficulty; the latter, on the contrary, is served with difficulty, but is easily satisfied.

¶ 547
Vox populi vox
Dei.

What heaven sees and hears is merely what the people see and hear: what the people consider worthy of recompense or of punishment is what Heaven will reward and punish; let those, therefore, who govern peoples be attentive and reserved.

¶ 548.
Know thyself.

If in the morning you have heard the voice of Divine reason, in the evening you may be content to die.[408] Study yourself, perfect yourself, be simple in heart, and love your neighbour as yourself. If you do not know the value of men's words, you know not the men themselves.

[400] *Vide* note [99], p. 90.
[407] Compare Giles' " *Gems of Chinese Literature* (London and Shanghai: 1884), p. 3.
[408] Giles, *loc. cit.*

And then the justice

full of wise saws & modern instances

SUB-SECTION XX

PHILOSOPHIC HANDS. [Plate VIII.]

THE army of philosophers is, as we know, divided into two principal camps,—that of the sensualists, and that of the idealists. Among the former we derive our ideas from external influences, whereas among the latter we devolve them from our inner consciousness.

¶ 549.
Divisions of the type.

Locke and Condillac have been the most eminent exponents of the doctrines of the sensualist school, and Descartes, Malebranche, and Leibnitz, the most vigorous champions of the idealist school.

¶ 550.
Illustrations.

Kant, since then, adopting a medium course, has admitted both innate ideas, _i.e.,_ those pre-existent and inherent in the soul, and transmitted ideas, _i.e.,_ those which form themselves in our minds by the prompting of the senses.[400]

¶ 551.
Kant.

[400] Immanuel Kant, who has been justly described as " one of the greatest and most influential metaphysicians of all time," was born at Königsberg in April 1724, and

¶ 552.
Fichte, the idealist.

But quite recently Fichte has raised once more the standard of the exclusive idealists. He is a subtle and abstract metaphysician, very difficult to understand, without warmth, passion, or love; who, isolating thought, or rather idea, from all kinds of covering, demonstrates and expounds it in terms as cold and severe as figures.[410]

¶ 553.
Ideal philosophy in Germany.

The ideal philosophy reigns almost undisturbed in Germany, a country without large towns, without social animation, flat, gloomy, silent, monotonous, where science is held in more esteem than art, contemplation than action, theory than practice; where life manifests itself only by the intelligences of the people, and where for these reasons imaginative men are led and authorised to believe that the real has no

occupied the position of professor in the university of Königsberg, which city he never left for thirty years. The opening sentences of his biography in " *Chambers' Encyclopædia* " are a sufficient commentary on the above paragraph :—" The investigations by which he achieved the reputation of a reformer in philosophy refer not so much to particular sections or problems of that science, as to its principles and limits. The central point of his system is found in the proposition, that before anything can be determined concerning the *objects* of cognition, the *faculty* of cognition itself, and the *sources of knowledge* lying therein, must be subjected to a critical examination. Locke's psychology indeed, at an earlier period in European speculation, had shown a similar tendency; but before Kant, no thinker had definitely grasped the conception of a critical philosophy, and Kant himself was led to it not so much by Locke as by Hume's acute scepticism in regard to the objective validity of our ideas, especially of the very important idea of causality. The Kantian criticism had a twofold aim : first, to separate the necessary and universal in cognition from the merely empirical (*i.e.*, from the knowledge we derive *through the senses*); and, second, to determine the limits of cognition." Kant died in February 1804.

[410] Johann Gottlieb Fichte, born in 1762, was, as our author says, one of the purest idealists that ever lived.

existence save in the ideal, and that all our sensations come to us from our souls.

But this is not the case in France, the land of innovation, of outward movement of active and passionate sociability, where the senses being more excited than the soul [which never raises its voice save when the senses are completely silent] seem to us to be the true source of all our ideas, and it is for this reason that I shall deal only with the hand which is the born instrument of intelligences turned towards the sensualist and rationalist philosophy.

¶ 554.
Social instinct in France.

The palm rather large and well developed; the joints well marked in the fingers. The exterior phalanx half square and half conic, a combination producing with the upper joint a kind of egg-shaped spatule; the thumb large and indicating the presence of as much logic as will, *i.e.*, composed of two phalanges of equal, or practically equal, length.

¶ 555.
Appearance of the philosophic hand.

We have seen that the inclination of the spatulate finger-tips draws them irresistibly towards that which is materially useful; that the inclination of the conic phalanx has as its aim, beauty of form, *i.e.*, art; and that that of the square finger-tips tends towards social utility, medium and practicable ideas,

¶ 556.
Instincts of the philosophic type.

A pupil of Kant, he published under his influence his first work with the startling title, "*Kritik aller Offenbarung*" (Königsberg: 1792). His whole life was devoted to the exposition of what has been described as "a system of transcendental idealism,"—a system which may be gathered from his works "*Die Wissenschaftslehre*" (Königsberg: 1795), "*System der Sittenlehre*" (Jena: 1798), and particularly from his "*Anweisung zum Seligen Leben, oder die Religionslehre*" (Berlin: 1806). He became rector of the University of Berlin in 1810 and died in 1814. "He combines," says his biographer, "the penetration of a philosopher, with the fire of a prophet, and the thunder of an orator; and over all his life lies the beauty of a spotless purity."—*Vide* J. H. Fichte's "*Fichte's Leben und Literarischer Briefwechsel*" (Sulzberg: 1830-31).

[¶ 556]

and realisable combinations. As for the genius which
accompanies phalanges which are quasi-square and
quasi-conic,—it is characterised by the love of and
constant desire for absolute truth.

¶ 557.
Effect of
formation.

By their joints philosophic hands have calculation,
more or less rigorous deductions, and method; by
their quasi-conic tips they have the intuition of a relative form of poetry ; and by the whole combination of
formations, including, of course, the thumb, they have
the instinct of metaphysics. They plunge into the
outer as well as into the inner world, but they seek
less after the form than after the essence of things,
less after beauty than truth ; more than any of the
other types they show themselves greedy of the
severe enthusiasm, which is diffused by the inexhaustible reservoir of the higher moral, experimental,
and philosophical [sensually speaking], and æsthetic
sciences.

¶ 558.
The philosophic
character.

You have a philosophic hand ?   I conclude then that
the philosophic spirit centres in you with greater or less
intensity.   You feel a desire to analyse and account
for your sensations ; the secret of your own existence
occupies your thoughts, as also does that of the origin
of all things.   Your beliefs, your ideas, and your
opinions, are not adopted on the faith of other people,
but only after having examined them from every point
of view.   Reason seems to you to be a more reliable
guide than instinct, than faith, even than love ; it is to
the reasoning faculty, and not to custom, or education,
or law, that according to you everything must be
consecrated ; you think like Socrates, that that which
wounds reason wounds humanity in all that it holds
most holy and best. Above the priest, the interested
propagandist of the dreams of imagination, you place
the philosopher, the apostle of the morality which
draws men together, and dictates to them the law of
loving one another, when all religions separate them

and make them hate one another. You know that doubt is as inevitable to us as death, and neither doubt nor the idea of death can alter your serenity. You proceed by analysis, but you aim at synthesis thereby; you occupy your thoughts at the same time with details and with the mass, with the individual and with mankind, with the atom and with the universe,—in a word, with the exception and with the rule the order, which in the material world others have seen in symmetry, you find in *affinities;* you claim a religious liberty because you feel that God has given you the intelligence of the just and of the unjust. You ignore vain scruples and superstitious terrors, and make use of all the pleasures with moderation. If you do not recognise *all* these characteristics as applying to yourself, you will at least own to possessing most of them.

Subjects of the square type reproach Louis XV. for having allowed himself to be despoiled of the absolute power with which they had armed him, as if the spirit of the times had not always a greater influence upon an isolated individual, than [whatever may be the temper of his disposition] that which that individual can exercise upon the times; as if princes, were not like other men, subject to the irresistible empire of the circumstances among which they live.

¶ 559.
Effect of the type under Louis XV.

At the time of Louis XV.'s accession to the throne, a type of hands, sprung from the masses of the people during the time of the regency,[411] had just risen to the surface of society, with all knowledge of its strength, and the ardent egoism which drives every instinct to

¶ 560.
Birth of the philosophic spirit.

[411] Louis XV. was an infant at the date of the death of Louis XIV., and Philippe, Duc d'Orleans, was elected regent by the Parliament in 1715, a post which he held until 1723, when upon his death the regency was entrusted to le Duc de Bourbon, and after him, in 1726, to the Cardinal de Fleury.

prefer itself openly and ostensibly to every other. They were hands of the philosophic type.

¶ 561.
Their influence upon France.

Contrary to the useful hands, which for upwards of fifty years had appealed to subordination, to authority, to usage, to custom, to conventionality, to faith, and to predestination, the philosophic hands appealed to reason, to examination, to proof, to liberty, and to free thought.   At these words, France, which was being crushed beneath the heavy pall of formalities, raised her head and breathed again.   One saw her, like a becalmed vessel warned at last of a coming wind by an unexpected breeze, spread her sails in haste,—her sails which had so long lain idle,—hoisting her joyous ensigns, and saluting with magnificent flourishes of trumpets. the blessed hands who restored her to wider spaces, to innovation, and to movement.[412]

¶ 562.
Religion of the type.

Attacking first of all the despotism of religion, the philosophers said :—" What distinguishes us essentially from the animals is *reason;* it is, therefore, from reason that the idea of God comes to us, because the animals, who are without it, have no idea of the sort.   If, therefore, our reason is our sole guarantee of the existence of God, it follows that it is reason alone that should direct for us the studies which have Him for their object.   He would not blame us for not holding a faith which is condemned by our reason, the faculty whereby He has been revealed to us, and without which He would be unknown to us."

¶ 563.
Their influence in politics.

The intolerant Catholicism of this epoch, having been threatened by these arguments, and caused to totter at last upon its base, these philosophers then

[412] A condition of things immediately resulting from the weak and sensual character of Louis XV., a monarch who practically took no part whatever in the conduct of the national affairs, and who was in every way supremely unfitted to carry on the system of government which had for its basis, Louis XIV.'s celebrated phrase, " L'Etat, c'est moi ! "

turned the efforts of their aggressive dialectics against the despotism of politics.

"Kings were made for the people, and not people for kings." This maxim, hitherto regarded as impious, seemed just and sacred to a generation which, having arrived at the point of reasoning out its education, and making it conform to its intelligence, considered that it had so much the more the right to reason out its government. Liberty established herself victoriously in men's minds, but it was in the sphere of ideas alone that for a long time she dared to soar with complete freedom. It was not until 1789, the epoch when she took her place in the laws of the land, that she made her way into the sphere of action. Since then art has opened to her her sanctuary; and to-day philosophers are trying to find her a way into morals; they claim for all men electoral rights, divorce, and the freedom of women, and preach the doctrine of individual Protestantism.

¶ 564.
Socialism of the type.

Literature in the seventeenth century, by reason of the imposed and inflexible ideas which governed this epoch, had, and indeed could only have, itself as an object; and all literature which is thus circumstanced must necessarily occupy itself more with style than with substance. For philosophic hands, devoted by their instinct to the search after the absolutely true, literature was but an instrument, with the aid of which they explored the unlimited domains of thought.[413] Their writings are brilliant from their

¶ 565.
Literature of the type.

[413] Madame de Staël, in her treatise "*De la Littéra-ture*" (Paris : 1800), ch. xx., has said in this connection : "In the time of Louis XIV., the perfection of literary art itself was the principal object of men of letters, but already in the eighteenth century literature assumes a different character—*it is no longer merely an art, it is an instrument*, it becomes a weapon of the human mind which until then she had been content simply to instruct and amuse."

[¶ 565]

variety, utility, and extent, and from the profundity of their ideas, whilst those of the useful type shine by form and literary style.

¶ 566.
Drama.

The theatre, of which the eighteenth century [bolder in its words and thoughts than the seventeenth, but not freer in its actions] made a kind of rostrum for popular harangues, did not lose less by this innovation than literature had lost from the artistic point of view, when it was reduced to the secondary condition of an instrument merely. But history and philosophy, set free by the philosophic hands from the shackles of faith, of tradition, and of revelation, infinitely enlarged the radius of their investigations. In the impetus which they gave to the minds of men, new sciences were discovered, and forgotten arts were resuscitated. A monument more worthy of admiration, more gigantic even than the cathedrals of the thirteenth century, and which like them concentrates in itself all the genius and science of an epoch—the "Encyclopædia" was, amid the plaudits of the most sagacious, the most sceptic, the most learned, and the most witty that the world has yet seen, commenced and finished in less than thirty years.

The
Encyclopædia.

¶ 567.
Importance of
the masses.

Until now they had relied upon individuals, now they believed in communities; the State was no longer contained within the royal mantle. A power, until lately unknown, and high-handedly over-ridden, —" public opinion "—inspires at last a wholesome terror; like the chorus in a Greek tragedy, the democracy intermeddles with the actions of kings: in defiance of history and tradition, political innovations are guided by pure reason; liberty, though not yet legally constituted, enlarges the minds of men, just as toleration enlarges their hearts; man becomes for his fellow-man the object of a grand love, of an enthusiastic admiration. Our fleets, directed by philosophers and men of science, plough through the

Discoveries.

tempests of two. hemispheres in search of nations and of islands and of continents hitherto unknown, among which to pour out the excess of our happiness and enlightenment; even upon savages kindly glances are bestowed.

The God of to-day is not the sullen, punctilious, jealous God of former days; it is no longer necessary to construct formulas and perform penances; to be happy is to adore Him; and the nation, less pre-occupied than formerly concerning the ill-defined benefits of the *other* life, scattering everywhere flowers of grace, of talent, and of wit, bearing manfully its load of alternated certitude and doubt, throws itself enthusiastically upon the pleasures of *this* one, but at the same time, neither are the higher sciences nor philosophy losers thereby. The manners of society, which the end of the seventeenth century had left harsh and cruel, softened down; fanaticism disappeared, etiquette relaxed itself; the barriers of the hierarchy were cleared away; the lower orders increased in power, and very soon it will no longer be a question of levelling, but of equality. Our villagers, hitherto so despised, come forward into the light, where the muses love to see them, with their virtues still somewhat tinselled over by the golden age. Virtue, genius, and talent assume the forms of simplicity and good fellowship; such, for instance, as we find them in Malesherbes, Franklin, Turgot, Lafayette, J. J. Rousseau, Diderot, and others: men's minds take a higher flight, lyricism comes to light, and the nation, freed at last, mirrors itself in its own intelligence and in its own beauty; she rejoices in herself, in her moral strength, and in the universality of her genius; the other nations, struck with awe, come to it for legislators, and North America, attentive to the voices of our philosophers, prepares for herself an era of solid happiness and

¶ 568.
Religion.

Manners of society.

[¶ 568]

unheard-of prosperity, realising in its midst the presence of their fruitful theories.

¶ 569.
Appreciation of the type.

The philosophic type alone, because to a certain point both worlds are known to it, can understand and appreciate the other types.

¶ 570.
Necessity of suitable environment for development.

As in the case with nations, individuals do not attain, excepting in ages of greater or less advancement, the perfect intelligence of their philosophical faculties, which, to come into being and develop themselves properly, require at least the light of experience. Philosophic hands, like those belonging to the other types, exist in all classes of society; only the genius which they represent becomes abortive, or manifests itself but very imperfectly among persons who are chained by their ill-fortune to gross labours.

¶ 571.
Philosophy of the type.

The philosophy of spatulate and useful hands studies the problems of facts, of practical ideas, of realities, of politics, and so on.

¶ 571a.
That of other types.

That of conic and pointed hands tends towards strange beliefs, speculative ideas, and so on.

¶ 572.
Analysis and synthesis.

Hands of the quasi-square and quasi-conic formation generally reveal eclecticism, and it is for this reason that I have given them, above all others, the name of philosophic hands. When they are large, they incline to analysis; when they are small, they incline to synthesis; with a small thumb they are guided by heart, with a big thumb they are guided by head.

¶ 573.
Clergymen.

The same may be said of Churchmen, as of philosophers and of artists. The knowledge and direction of *men* is the task of the priests belonging to the types known as northern; the knowledge and the direction of *souls* is that of the priests of the types known as southern. To the first belong sciences and doctrines, to the last, faith; the former have most authority, the latter have most love. Spatulate hands are pre-occupied concerning this world and about their Church; the conic, concerning heaven and their God. For the former the priesthood is merely a trade; for the latter it is nearly always a ministry. The confessional [which in very truth exhibits none but the worst aspects of humanity] increases the severity of the former and the leniency of the latter; I have before me a rough black silhouette of M. le Cardinal de Clermont-Tonnerre, late Archbishop of Toulouse; his disdain of the generation in which he lived, for its ideas and for its works, had surrounded him with an atmosphere of sanctity in the court of Charles X.[414] A priest, said he, debases himself by asking; he must insist. He was a very small man, with a proud, even an arrogant, walk; the gaze of the mob, whatever its expression, did not embarrass him in the least. Haughty, pacing along in scarlet upon his horse, the ornaments of his caparisons glittering with ostentation, he seemed perfectly ready to recommence the old struggle between the spiritual and temporal authorities. He had a large nose, large eyes, and large lips, on a small face; his iron-grey hair, flattened down and dressed like the bristles of the wild boar, and cut straight across his forehead

[414] Anne Antoine Jules de Clermont-Tonnerre, born in Paris 1749, died at Toulouse 1830. In 1782 he became Bishop of Chalons, and took a leading part in the French revolution. He was made Archbishop of Toulouse in 1820, and became Cardinal in 1822. He was renowned for his ultramontane views and for his rigidity in matters of etiquette and social observance. On one occasion, having been engaged in a dispute with Feurtier, the Minister of Public Instruction, he made the following characteristic reply :—"Monseigneur, la devise de ma famille, qui lui a été donnee ent 1120 par Calixte II. est celle ci : '*Etiamsi omnes, ego non;*' c'est aussi celle de ma conscience." After this he was forbidden the court until he should apologise, which he subsequently did by order of Leo XII.

just above his eyebrows, gave him an air of harshness, which the habitual expression of his features,
his brown complexion, his great age, and the depth
of his wrinkles accentuated rather than relieved.   In
the barracks we used to call him *the mitred ourang-
outang;* his great spatulate hands gave freely it is
true, but without grace, without true charity.

¶ 574.
Illustrations.

Hildebrandt [415] and La Rovère [416] undoubtedly belonged to the square or to the spatulate type.

[415] Hildebrandt, a German priest, who, having
achieved great distinction for his activity and diplomatic ability under all the popes from Gregory VI.
to Alexander II., was crowned Pope under the name
of Gregory VII. in 1073.   His continual strife with
Henry IV. of Germany only terminated with his death in
1085.   He was remarkable all his life for his·austerity
and inflexible determination.   A detailed account of his
life and character may be found in Bowden's "*Life
of Gregory VII.*" (London: 1840).

[416] Julien de la Rovère reigned as Pope Julius II.
from 1503—1513, having been born in 1441.   He shines
forth among the potentates of his day by reason of his
vast political capacity and for the marvellous energy
with which he was continually engaged in struggling to
attain the objects of his ambition and to advance the
interests of the Church.   The records of his sovereignty
contain chronicles of wars which he waged successfully
against Cæsar Borgia, Louis XII., and the Emperor
Maximilian of Germany.   He founded the Holy League
of 1511, and his death, in 1513, left a state of things
existent in the affairs of the Holy See which was a fit
basis on which to commence the reign of Leo X.   All
the accounts that we have of his pontiff tell us of his
inordinate pride and of his continual political struggles;
he also was a liberal patron of literature and of the
arts, but this was entirely subordinated to the strife after
his own political ambitions.

The Pointed Type.

The sixth age shifts itself
into the lean & slipper'd pantaloon

spectacles on nose & pouch on side.

SUB SECTION XXI.

PSYCHIC HANDS. [Plate IX.]

THE psychic hand is of all others the most beautiful, and consequently the most scarce, for rarity is one of the conditions of beauty. It is small and fine by relation to the rest of the body. A medium palm, smooth fingers [or fingers with the joints only just perceptible], the outer phalanx long, and drawn out to a point, the thumb small and elegant. Large and with joints it has force and combination, but it lacks ingenuousness.

¶ 575.
Its appearance.

Let common sense be the guide of the useful hands, —hands of which order, arrangement, and unity are the aims and objects; let reason be the solitary beacon of the philosophic hands, carried, as they are, ever towards liberty and truth; these are axioms which I have just been stating. As to psychic hands they bear to those two types the same relation the artistic bears to the spatulate type: they attach, they add to the works of the *thinker* in the same way that the artist adds to the works of the artizan, beauty and ideality; they gild them with a sun-ray, they

¶ 576.
Psychic view
of matters.

raise them upon a pedestal, and open men's hearts to them ; the soul, forgotten and left behind by philosophic hands, is their guide, truth in matters of love and sensibility is their end, and expansiveness of heart is their means.

¶ 577.
Effect of exclusive influence of the types.

You have seen the world under the sway of spatulate hands, and you have found movement, activity, industry, war, tumult, and cultivation of power and material good. You have seen it ruled by artistic hands, and you have found romantic enterprises [that is to say, attainment of an ordinary object by extraordinary means], want of foresight, brilliant folly, splendid misery, and the fanaticism of form. You have seen it ruled by square fingers, and have found fanaticism of method, and narrow-minded and universal despotism. You have seen it governed by philosophic hands, and you have found fanaticism of science, doubt, mobility, and liberty, without a base on which to steady itself.

¶ 578.
Effect of psychic hands on national fortunes.

In Europe, up to the present time, psychic hands have never been able to attain to domination,—perhaps because they have never desired it, disdainful as they are of material interests in the lofty sphere in which they are kept imprisoned by the genius which animates them. Nevertheless their intervention has never failed when the dramas of life, brought to their highest possible state of complication, have required a quasi-divine agency to unravel them. What insults would have been reserved for the intelligence and dignity of manhood, if, in electrifying the cities of Greece, these hands had not obtained for her the victories of Salamis and of Marathon. Spain, the religious and poetic, has never been violently convulsed save under their impulsion ; without them she would have perished in 1812,[417] just as Germany,

[417] M. d'Arpentigny refers, I presume, to the liberal constitution of the Cortes of Cadiz which followed the

which, already vanquished as regards her princes, her crowned fetiches, was only saved by a few young enthusiasts, long-haired idealists, whose hearts were resolute, though their faces were serene, who hymning their God, their country, and liberty, threw themselves into the field of battle, to the strains of the heavenly lyre.

Taken as a whole, these hands love grand struggles and despise little ones. When Greek sensualism was at its height, they were represented by Plato ; when Roman sensualism was at its height, they found their archetype in Christ. They do not struggle save with the grandest athletes ; to Bossuet, the biblical champion of terrorism and of form, they oppose Fénélon, the evangelical champion of the inner spirits of things and of love. Against Voltaire and Diderot, who appeal to the senses and to reason, they excite the psychological opposition of Vauvenargues and of Rousseau. Finally, in our own day, you have seen them holding imperial materialism in check by the aid of Chateaubriand, of Benjamin Constant, and of Madame de Staël.

¶ 579.. Represented in all times and countries.

The psychic hand is not, as writers of romances have pretended, the exclusive privilege of old families. Always scarce, it exists nevertheless, everywhere, even among the most abject classes, where it vegetates ignorant of itself, misunderstood and disdained, on account of its comparative inaptitude for manual labour.[418] Apollo, alas ! at one time was a cowherd.[419]

¶ 580. University of the type.

battle of Salamanca [22nd July, 1812], at which the power of Napoleon was finally broken,—a constitution which was, however, abrogated by Ferdinand VII. some years later.

[418] *Vide* ¶ 284, and compare " *A Manual of Cheirosophy,*" ¶ 76.

[419] When Apollo had been banished from heaven by Jupiter for slaying the Cyclops, he came to Admetus, King of Thessaly, and was employed by him as a shepherd for nine years.

¶ 581.
Motives of the
psychic type.

Artistic hands wish to see imagination and art everywhere ; square hands look for rule and arrangement; philosophic hands for human reason. It is Divine reason which, by virtue of the same law,—the law which is derived from natural and constitutional dispositions,—psychic hands desire to see everywhere. The work from which ideality is absent, in which love and religion have no part, and in which the soul cannot be interested, is for them a dead letter. They pay no attention to form, save in the domain of art, and besides they would not be able to pay attention to it, convinced, as they are, that civilisation is not the absolute consequence of any particular form of religion [as for instance of Christianity, which has checked the races of the Aztecs from their birth, which has remained impotent in Abyssinia against the Barbary States, and in Russia, and Poland, and

Slavery.

America, against slavery] ;[420] that liberty is not the absolute consequence of a democratic form of government; and that slavery is not the absolute consequence of an autocratic form of rule. In their eyes religious faith is a fact as real as rational certitude; so they excuse, even when they do not accept, the peculiarities of all religions,[421] thinking in

[420] Our author seems to have lost sight of the abolition of the slave trade by the legislation which extended from 1811 to 1837. Serfdom was of course not suppressed until after the publication of this work [vide note [283], p. 237].

[421] After all, man is so small as seen from on high, that it matters little in what manner he honours God. God understands all languages, and especially that which is expressed in silence by the movements of the spirit; and, as Whinfield has said in his edition of the " Quatrains of Omar Khayyam " (London: 1883),—

" Pagodas, just as mosques, are homes of prayer ;
'Tis prayer that church bells chime into the air.
Yea, Church and Kaa'ba, Rosary and Cross,
Are all but divers tongues of world-wide prayer." *

زنّار و کلیسیا و تسبیح وصلیب ـ حقّا که همه نشانه، بندگیست *

this matter even like the hatted angels which Swedenborg, rapt in the spirit [according to his own account],
heard lauding the purity of the doctrines of the
Tartars ;[422] and, like the oracle of Delphi, which when
consulted as to the best manner in which to honour
the gods, *i.e.*, as to the religious form best suited
for the morality of nations, answered, " Obey the laws
of the country "[423]—in monarchy they see beauty, in
republics they see good, and the East, dreamy, immobile, and silent as a desert, pre-occupied about
its future state, and governed by a sole individual,
seems to them to be as wise and as happy as the
stormy West, regular and resounding like the ocean,
pre-occupied about the things of the earth, and
governed by its communities.

The two ideas to which the human race owes the
most noble part of its good fortune and dignity,—that of
the beautiful in art, and that of right in politics,—had
their birth, and died in the antique world with the
anthropomorphic polytheism, as it was understood

---

[422] I do not quite follow M. d'Arpentigny's statement
concerning the angels in hats [anges *coiffés de
chapeaux*]; he evidently alludes to Swedenborg's
account of how certain spirits came to him in a
trance, and told him about the worship and manners of
the Tartars, an account of which may be found minutely
given in " *Eman. Swedenborgii Diarii Spiritualis* "
partis tertiæ volumen secundum (London : 1844), p. 186,
no. 6077.*

[423] " You need not pry into the future ; but assure
yourselves it will be disastrous unless you attend to your
duty, and are willing to act as becomes you." †

---

* " De Incolis Tartariæ, prope Chinam, Tartaria Minor " :—
" Erant aliqui inde apud me, venerunt cum dormiebam, et dormiebam tranquille, [cum] evigilatus animadvertunt quod non domi
essent, sed alibi, mirati ubinam, quia non agnoscebant talia quæ
prorsus mundo spiritualium similia essent," etc., etc.

† DEMOSTHENES, ΚΑΤΑ ΦΙΛΙΠΠΟΥ, Α'. " Οὐ γὰρ ἄττα
ποτ' ἔσται δεῖ σκοπεῖν, ἀλλ' ὅτι φαῦλ' ἂν μὴ προσέχητε, τοῖς
πράγμασι, τὸν νοῦν καὶ τὰ προσήκοντα ποιεῖν ἐθέλητ', εὖ εἰδέναι."

by the Greeks; and these ideas have revived in the modern world only at the time of the Renaissance, which was a sort of resurrection of this same polytheism.[424]

¶ 583.
Illustration.

We read in Joinville[425] that during the siege of Damascus a woman having been met between the camp and the city by a priest of the army of the crusaders, the latter asked her what she was supposed

---

[424] The author here appends a reference to LECONTE DELISLE. He alludes, I presume, to the preface to Leconte Delisle's "*Poemes Antiques*" (Paris: 1852), in which the poet advances a theory which is quite on a par with that contained in the above paragraph. "Since Homer, Æschylus, and Sophocles," says he, "who represent poetry in all its vitality, plentitude, and harmonious unity, decadence and barbarism have invaded the human mind; . . . the Christian era is wholly barbarous. Dante, Shakespeare, and Milton have given proofs of their individual genius, but their language and ideas are barbarous. What have we remaining of the centuries which have elapsed since the decline of the Greek constitution?—a few powerful individualities, a few grand works without continuity or unity. And now science and art are reverting to their common origin."

[425] This narration may be found in M. Natalis de Wailly's "*Histoire de St. Louis par Jean Sire de Joinville, etc.*" (Paris: 1868), chap. xxxiv., in the original French of de Joinville,* and an excellent translation into English occurs in the Bohn library "*Chronicles of the Crusades*" (London: 1848), "Joinville's Memoirs of Louis IX.," pt. ii., p. 469.

---

* "Avec les messaiges qui la àlèrent, ala frères Yves li Bretons de l'order des frères Breescheours qui savoit le sarazinois. Tandis que il aloient le leur hostel a l'ostel dou soudanc frères Yves vit une femme vielle qui traversoit parmi la rue et portoit en sa main dextre une éscuellée pleine de feu et en la senestre une phiole pleinne d'yaue. Frères Yves demanda, 'Que veus tu de çe faire?' Elle li respondi qu'elle vouloit dou feu ardoir paradis que jamais n'en fust point et de l'yaue esteindre enfer que jamais n'en fust point. Et il li demanda, 'Pourquoy veus tu çe faire?' Pour çe que çe je ne veuil que nulz face jamais bien pour le gueredon de paradis avoir, ne pour la poour d'enfer, mais proprement pour l'amour de Dieu avoir qui tant vaut et qui tout le bien nous peut faire" [p. 158].

to be doing with the water which she carried in a pitcher, and the fire that she carried in a brazier : " It is," replied she, " to burn up paradise and extinguish hell-fire, in order that in the future men may worship God for love of Him as their sole motive." This reply is said to have enchanted St. Louis, who enthusiastically applauded the sublime piety which had dictated it.

" The soul of God is truth, and His body is light." [420]   ¶ 584.
" Three things came into the world at the same in-   The Druids and Pythagoras.
stant : man, liberty, and light." [427]—Sublime formulæ; ample and serene verification of the strength and of the beauty of human intelligence.

Psychic hands are in an immense majority in   ¶ 585.
Southern Asia, whence comes the essentially religious,   The type in Southern Asia.
contemplative, and poetic spirit of the nations which

[426] This is the basis of a good deal of the Pythagorean philosophy, and sentiments practically expressing the axiom may be found by the scholar in Thomas Gale's " *Opuscula Mythologica*" (Amsterdam : 1688), " Sexti Pythagorii sententiæ a Ruffino versæ," p. 646,* in Conrad Orelli's " *Opuscula Græcorum Veterum Sententiosa et Moralia*" (Lipsiæ : 1819), p. 49, " Aliæ Pythagoreorum Sententiæ."† We find a similar sentiment expressed in the Ruba'yát of Omar-i-Khayyam, in the 328th " Ruba' " of Nicolas' text [" *Les Quatrains de Kheyam*" (Paris: 1867), p. 164], which may be translated " God is the soul of the universe ; the universe is a body, the angels are its senses, heaven, the elements, and all creatures are its members " :—

،حق جان وجهانیست وجهان جمله بدن

و اصناف ملائکه حواس این تن

&c. &c.

[427] The first maxim of the Druidical dogma.

* " Deus sapiens lux est, non capax contrarii."
† " Deus quidem unus est ; non ille autem, ut quidam suspican-
.ur, extra hunc mundi ornatum, sed ipse totus in toto circulo omnes generationes inspicit ; . . . . in cœlo lumen, et pater omnium, mens et animatio omnium, circulorum omnium motus sive motor."

[¶ 585]

inhabit it; whence comes also their respect for maxims [synthesis], and their disdain for methods [analysis]; whence comes the preference which they feel for virtue [source of repose] over science [source of activity]; whence comes the languishing condition of arts, trades, and agriculture, and their theocratic and despotic governments, .which are necessary for nations to whom reason and action are a torment.

¶ 586.
Origin of
religion and
philosophy.

It is from gloomy and somnolent Asia, the continent of huge empires, of opium, and of wild orgies of drunkenness, that all the grand religions have sprung; and it is from mobile and laborious Europe, the land of little republics, of wine, and of moderate intoxications, that all the philosophies which have stood face to face with religions, and have contested them, have sprung. In Asia society is formed by its worship, in Europe the worship is formed by society.

¶ 587.
East and West.

The East, says Aristotle, has always taken a delight in metaphysics [which, applying themselves to logic, teach men to think with subtlety], and the West in morals [which, applying themselves to reason, teach men to live honestly].[428]

¶ 588.
Asiatic genii.

Asia is the land of "genii," as Europe is the land of "fairies." Well, the genii being individualities [or, perhaps, substantivities], gifted with an activity upon which no particular direction is imposed, they can do whatever they please, like the princes of their country, where they have always rejected. any division of power.[429] The powers of our fairies, on the

[428] I am not acquainted with this passage, but I presume it occurs in the "*Problems*" of Aristotle, if anywhere in his works.
[429] On this word "genie" Sir Richard Burton gives a most interesting note in his "*Arabian Nights*" [*op. cit.* (*vide* note [17a], p. 112), vol. i., p. 10], wherein he says :— "Jinni, the Arab singular (whence the French 'génie'), fem. Jinniyah; the Div and Rakshah of old Guebreland and the 'Rakshasa,' or 'Yaksha' of Hinduism.

other hand [which are, after all, purely adjective], are restrained to some solitary attribute : one can give beauty or courage, another strength or wealth ; whenever they dispose of one of these gifts the gift is unlimited. Enclosed between two parallels, they have only one direction, but that *direction* is unbounded. Gifted with every power, the genii are, as it were, hemmed in by a circle in which they can act in any manner whatsoever, but though their power is unlimited as to kind, it is limited as to quantity, their power being subject to that of a genie more powerful than them.[430]

Thus nations are revealed by their fables just as they are by their laws and their religions. With us power has a particular direction and certain defined

¶ 589.
Oriental and Occidental authority.

It would be interesting to trace the evident connection —by no means accidental—of ' Jinn ' with the ' genius ' who came to the Romans through the Asiatic Etruscans, and whose name I cannot derive from ' γιγνόμαι,' or ' genitus.'   He was unknown to the Greeks, who had the dæmon (δαίμων), a family which separated like the Jinn and the genius into two categories, the good (Agatho-dæmons) and the bad (Kako-dæmons).   We know nothing concerning the status of the Jinn amongst the pre-Moslemitic or pagan Arabs ; the Moslems made him a supernatural anthropoid being, created of subtile fire [*vide* Koran, chap. vi., xv., and lv.], not of earth like man, . . . the last being Jan bin Jan, missionarised by prophets, and subject to death and judgment.   From the same root are ' Junun ' = madness [*i.e.*, possession or obsession by the Jinn], and ' Majnún ' = a madman."   *Vide* also vol. iii., p. 225.

[430] M. d'Arpentigny makes in this place a reference to "DE FLOTTE," but I cannot identify either the author or the work referred to.   He cannot refer to Ely de Flotte's work "*Les Sectes Protestantes*" (Paris : 1850) ; I conclude he refers either to *De la* Flotte's "*Voyages en Orient*" (Paris : 1772), which was translated from the London edition of "*Pocock's Voyages*," or else to the same author's "*Essais Historiques sur l'Inde, etc.*" (Paris : 1769), but I cannot find any passage bearing upon the above paragraph.

[¶ 589]

limits, it follows a given line of action; among Oriental nations it has no particular direction, and its only limit is that of a more powerful arbitrament.

¶ 590.
Affinity between German and Sanskrit.

It has been said that there exists an affinity between the German and the Sanskrit languages; there exists also an affinity between the dreamy spirit of the German nation and the contemplative minds of the children of Brahma.[431] Indeed, Germany is in Europe, as India is in Asia, the country where one sees the largest number of psychic hands.

¶ 591.
Every character has its own language.

Spiritualism being the special attribute of this noble type, it has come among us, where well-being, good laws, and liberty have helped it to multiply itself[432] and become understood—it has come, I say, having to express sentiments and ideas of a peculiar temperament, to follow the examples of the artistic, the useful, and the philosophic types, all of which for the same cause, have used in turn a different language, created by each one of them for its particular use. The language of Rabelais and of Montaigne is not the language of Pascal, nor is the language of Pascal that of Rousseau. Whence, then, come the grammatical innovations of Chateaubriand and of Lamartine, those eagles of our psychological literature? have they not caused as much astonishment as scandalisation to our literature? Let new ideas

[431] It is in a great measure since the above was written that the labours of Max Müller and other learned philologists have placed before the world the principal data concerning the Eastern origin of the Aryan or Indo-Germanic [? Indo-European languages]. The reader who feels interested in the subject will find all he wants in Prof. Max Müller's " *Comparative Mythology* " (Oxford: 1856), and " *The Science of Language* " (London: 1681-3).

[432] " In 1722 France contained only 18,000,000 inhabitants. In the reign of Louis XIV. the death-rate among the better classes was 1 in 26, to-day [1843] it is 1 in 52."—AUTHOR'S NOTE.

[¶ 591]

be expressed by new forms; to extend the meaning of a few words is not to alter a language, it is to enrich it, and to condemn this theory is to rebuke the means, which, for their glory as well as for our instruction and amusement, have been used by the greatest writers of all ages.

You will not fully appreciate either the ideas or the language of psychological writers, O ye who have spatulate or square hands! you will find among them neither the precision nor the method which are so dear to you; their perpetual invocations of glorious intelligences,—splendid rivals of the stars,—will bore you. You look towards war and its resulting interests, they take their pleasure in the esoteric dreams of their souls, in the contemplation of intangible realities; your Muse, occupied by the real world, sings of flowery pleasaunces, of the shock of armies, and of armoured fleets; she relates the escapades of young students and the fury of ancient pundits, the graces of an ideal Lisette, and the vulgar tribulations of a contemptible dinner; she appreciates Beaumarchais, the mechanician, the duellist, the pamphleteer, the man of spirit, of wit, of action, of movement and of heart.[433] Theirs, on the other hand, plucks flowers only to scatter them before the saints in the church porch, she takes hold of women only to cast them in ecstasy and palpitation

¶ 592. Language and literature of the types.

[433] Pierre Augustin Caron de Beaumarchais [born January 1732, died May 1799] was a watchmaker, and the son of a watchmaker. He became an accomplished musician, and, having taken to literature, achieved great distinction as a wit, a poet, and a satirist. His four most celebrated works are "*Eugénie*" [1767], "*Les Deux Amis*" [1770], and the libretti of *Il Barbiere de Seviglia* [*Le Barbier de Seville*, 1776], and *Le Nozze de Figaro* [*Le Mariage de Figaro*]. *Vide* L. de Loménie's "*Beaumarchais and his Times*" (London · 1856), translated by H. S. Edwards.

at the feet of God ; lyricism is as natural to her as the song is to the bird, as perfume and attraction is to amber ; hers is the harp of David, and the holy enthusiasms which bear our souls aloft upon the wings of the morning ; the murmurs of the ocean, of the waterfall, of the forest, and of the mountain are for her but the echoes of her own sublime voice ; lover of the ideal and of the infinite, she soars with the angels, following the blazing track of the impetuous comet, and, of all the sounds of earth, listens only to the sighs of a simple and loving heart, which upraises itself to God. You of these practical types live pre-eminently by your heads and by your senses ; they live pre-eminently by the soul, by the heart ; you think, they feel ; you speak, they sing ; you are composed of flesh and blood, they are of flame and light. A great gulph is fixed between you, and two different languages are not too many for two natures so diametrically opposed.

¶ 593.
Influence of the types and their media of instruction.

Such is the intelligence attached to psychic hands ; through the works of Milton, of Klopstock, of Schiller, of Goethe, of Swedenborg, Chateaubriand, Lamartine and Victor Hugo, of Georges Sand, C. Didier, and De Vigny, they hold sovereign sway, if not over the greatest minds, at least over the noblest hearts ; they have given us the highest lyrics, psychological romance, intense poetry, and the inspired odes of the illumined wing and ardent flight. Their influence upon the masses has been enormous, restoring to them the enthusiasm which the analytical philosophy had killed, and rehabilitating for them the God who had been killed by the turpitudes of the sacerdotal sanctuary. Before they had been preached to about the *necessity* of self-abnegation, the psychic type spoke to them of its charms ; to draw them into the paths of their cultus they have garnished those paths with the flowers of a poetry which is almost Divine; like the murmuring pines of the Alpine mountain tops, they have shed in abundance tender shadows and universal harmonies. To be appreciated by the intelligent, they have taken up the lyre; to be understood by simple minds they have plied the abundant jet of their eloquence couched in the simplest forms of words.

Civilisation in Spain and in Italy is born of poetry and liberty, *i.e.*, of artistic and psychic hands, which propose for themselves things that are grand, magnificent, and sublime. In France it is born of science and of authority, *i.e.*, of useful and philosophic hands, which propose for themselves things which are useful and true. Our history is more instructive than interesting; the history of Spain is more interesting than instructive.

¶ 594
Civilisation in France and Spain.

Like the Greeks, who relegated manual labour to the infernal gods, the Spaniards think that it degrades the people and individuals in a direct ratio to the amount of love of it which they evince. The Italians have placed physical repose under the protection of a third of the saints of the calendar, not for the purpose of rendering it more respected by the people, but to arm it against the avaricious and worrying influences of political and fiscal regulations,—a state of things incomprehensible to the English or to the Americans, who, beneath their lustreless sun, can only escape from spleen by sheer hard work.

¶ 595.
Appreciation of physical laziness

The Spaniards of to-day do not understand in the least degree the artistic expression and literary formulation of the poetry which is with them inborn,—a fact which makes them appear, intellectually, so small, though they are really greater than they seem.[434]

¶ 596.
Poetry of the Spaniards.

[434] Compare the opening words of Bacon's essay, " *Of Seeming Wise*," which read, " It hath been an opinion

¶ 597.
The social
instinct in the
South and the
North.

In the South, where the climate itself is nourishing, and one can live on almost nothing,—where, indeed, one can live forty-five days without food,[435] man is not sufficiently necessary to man to prevent the social tie being a weak one; but in the North, where the climate creates hunger, man, at war with a hostile nature [especially since civilisation has tended towards the unhardening of his physique], feels too strongly the need of help and support for society not to be strong.

¶ 598.
In India and
Norway.

In the towns of Norway, all the houses communicate by means of inner doors or subterranean galleries. In the East Indies, on the other hand, families isolate themselves as much as possible from one another.

¶ 599.
The psychic type
in India.

Though poorly-gifted as regards the arts of war, of the chase, of navigation, of locomotion, and the cares of real life as *we* understand these terms, the psychic type has none the less reigned sovereign in India up to the thirteenth century, the epoch when it was dispossessed of the temporal power by the Mahometan Tartars, who drove it into the temples alone. Until that time no one arrived at power and high consideration save by piety, contemplation, and holiness,—qualities and virtues which in our latitudes open for us the gates of heaven, but open no other gates than those. Thus, among us are spatulate and hard useful hands in majority, whereas in India it is the pointed and soft hands which numerically are in the ascendant.

¶ 600.
Character of the
Bedouins.

And the same case probably exists in the heart of

that the French are wiser than they seeme; and the Spaniards seeme wiser than they are."

[435] M. d'Arpentigny quotes the "*Lettres Édifiantes*" [*vide* note [99], p. 339] as his authority for this statement, but does not give the exact reference. I have been unable to identify it.

[¶ 600]

the old tribes of the Bedouin Arabs of the Red Sea, people who,—occupying a country almost absolutely sterile, and where for this reason large gatherings of men are impossible,—cannot be naturally apt at the arts, the trades, and the sciences, which alone the life of cities is able to induce. Of what use to them would be the implements of rural cultivation and mechanical industry, in a country where rural cultivation is impracticable, and where a tent, a camel, and a courser suffice for the physical needs of man,— needs which mother nature has beneficently proportioned to the resources of the soil. But the fœcundity which she has refused to their soil, and to a certain extent to their judgment, she has prodigally bestowed [as if to shield them from the worries adherent to a life without occupation] upon their imaginations ; and as a nation they are poetic, romantic, religious, chivalrous, hospitable, contemplative, and of dignified and sober manners. Their country, which is that of physical mirages, is also that of moral ones : from all these facts I conclude that their hands are psychic, but very hard.

"Like Arabs in general," says le Duc de Raguse, "the Bedouins of the Red Sea have a high idea of the nobility of their blood ; they intermarry only among themselves, and they would consider it beneath them to ally themselves with a stranger. They sometimes buy slaves, but these never have children by them."[436] Thus their instinct tells them that a country which can neither be ameliorated nor modified by human intelligence, cannot be inhabited with any chance of happiness, save by a special race ; and that any other race, however slightly it might differ from theirs, would run a grave danger for want of an organisation

¶ 601.
Exclusiveness of the Arabs.

[436] " *Voyage du Maréchal Viesse de Marmont, Duc de Raguse, en Hongrie, ... en Palestine, et en Egypte*" (Paris : 1834-5, 4 vols.), vol. iv., p. 174

in perfect sympathy with the climate and the country. I have, in another place,[437] condemned this mistrust of foreign blood, but it is evident that the Bedouins of the desert, if all that we are told of their country is true, are quite beyond the pale of the motives which suggested this condemnation.

¶ 601a.
Immobility of Muhammedan-ism.

Muhammad, according to Arabian historians, had also very hard hands. His book teems with action, and has the burning, monotonous, and sterile grandeur of the desert. Empty of ideas and full of repetitions, presumptuous ignorance and a kind of gloomy and solemn poetry, flow through it in a continuous stream. The water contained in this fatal cup is by no means a fountain of life; elaborated as it was among a people necessarily immobile, it has stricken with immobility all the nations among whom fanaticism and war have introduced it.[438] But, if immobility is salutary and logical in the desert, elsewhere it engenders naught save corruption and death.

¶ 602.
Muhammad.

Muhammad, throughout his life, took his pleasure in war, love, and activity; his paradise, peopled by women, is spread out beneath the shadow of the sword;[439] he sought relief from his intellectual labours in sweeping out his tent, in repairing his shoes, and in tending his horses and herds.

¶ 603.
Elementary mechanical talent of the Arabs.

There is no Mussulman town, says the intrepid and judicious Badia-y-Leblic [surnamed Ali Bey], where mechanical arts are as little known as in Mecca. There is not to be found there a workman capable of forging a lock or a key; all the doors are closed by means of wooden pegs, trunks and cases by means of padlocks brought from Europe. The armourers can make nothing but inferior matchlock guns, curved knives, and the lances which are used in the country.

[437] *Vide* ¶ 365, and notes to the paragraph.
[438] *Vide* ¶¶ 317—324.
[439] *Vide* Al Qur'án, chaps. xlvii. and lv.

[¶ 603]

As for the exact sciences, adds the traveller, they are in the same condition as the mechanical arts.[440]

¶ 604.
Palmyra.
Its solidity.

Palmyra was built by Europeans, and if its ruins, like those of Baal-Bec, are still standing, it is only because the Bedouins, who cannot themselves build anything which is solid, cannot even destroy anything which has been solidly built.[441]

¶ 605.
Impracticability
of the type.

The racial psychic hand, liberally endowed as it is, has nevertheless only a mediocre comprehension of the things of the outer world and of real life; she looks at them from too high a point to be able to see them well. Spiritualists have lyricism, mysticism, prophetic ecstasies, luminous, synthetic comprehension of all human knowledge; but the talent of applied sciences, including that of the government of men united in a common society, is wanting among them, unless, as in India, they only have to deal with people belonging to their own type. Again, it would be a great mistake to suppose that the psychic type is more guarded against the errors which are incidental

[440] Condensed from " *Voyages d'Ali Bey el Abassi en Afrique et en Asie*" (Paris : 1814, 3 vols.), vol. iii., chap. xvii., pp. 389-90. An English edition has been published, entitled " *Travels of Ali Bey in Morocco, etc., between the years* 1803-1807, *written by Himself*" (London : 1816), vol. ii., chap. vii., p. 99.

[441] Tadmor or Palmyra, the City of Palms, was, we are told, built by Solomon in the tenth century B.C. [1 Kings ix. 18, and 2 Chron. viii. 4]; but it probably existed before his time. It was enormously enlarged and rendered of paramount importance as a city by the Emperor Trajan, and in the third century was the capital of Syria and Mesopotamia, under Odenathus and Zenobia. It was re-conquered in 275 by Aurelian, and re-fortified by Justinian. It was destroyed in 744 by the Saracens. I presume it is in allusion to its position and structure as a Roman city that our author asserts it to have been built by Europeans. As good a description of the city as any that exists is the one contained in Murray's " *Handbook for Syria and Palestine*" (London : 1858).

Fanaticism.

to the imperfections of our natures, than others ; the world of ideas is not less perilous and deceitful than that of things real.   In the Indies, blinded by religious fanaticism, the worshippers of Siva garland themselves with flowers, clothe themselves in brilliant garments, and cast themselves as prey to the sacred sharks of the Island of Sangor; and mothers even more exalted than they, cast their infants to these same beasts.[442]   But if in their enthusiasm spiritualists are always ready to devote *themselves,* they require also for the triumph of their ideas unlimited devotion *thereto.*   With their synthetic manner of thought, no isolated sentiment, no idea of detail could either touch their hearts, alter their convictions, or turn them from their object ; it is in their eyes above all that the end justifies the means ; if occasion should arise they will shed blood—their own or other people's,—their own without regret, that of others without remorse.

¶ 606.
The horse, and its apparent influence.

The horse being, be it said as concerning the Arab, of all animals the one which impresses us the most with its brute organisation, it follows that we should hold in contempt the understandings of peoples and individuals who love it exclusively.   Nations of horsemen have never freed themselves from the rude and showy shackles of comparative barbarism.   Apt at raids and invasions rather than at permanent conquests, and convinced that the cultivation of the mind destroys the energy of the heart, they have destroyed

[442] Siva (= Auspicious) is the name of the third god of the Hindu Trimûrti or triad, and is looked upon as one of the most awful and venerable of the gods ; and certainly the description given of him in works of Sanskrit theology is of the most terrible description. His worshippers are called Saïvas, and are divided into several classes, of which the Aghorins are the Saïvas most addicted to the sacrifices mentioned above.   *Vide* H. H. Wilson's "*A Sketch of the Religious Sects of the Hindus*" (London : 1862), pp. 188, etc. [vol. i. of his complete works, edited by Dr. R. Rost].

more empires than they have founded ; such for instance were the Parthians, the Tartars, and the Arabs.

The horse was the soul of feudalism,—that system of violence and ostentation which gave to the qualities of physical force, and to the suggestions of family pride that preference which nations who live as it were on foot, have always accorded to solidity of judgment and the enlightenment of the soul. In the ancient world the life of the Centaurs was passed in intemperance, in the midst of quarrels, amid the screams of the women which they captured ; in the same way in the present day brawlers, drunkards, and bullies are more numerous among our cavalry than among our infantry.[443]

¶ 607.
The same.

The special manœuvres in which the horse is a necessary adjunct produce but few generals with broad views ; great warriors have always come from the ranks of the infantry, *the queen of battles*, the intelligent and redoubtable foundress of empires and of durable glory.[444]

¶ 608.
Cavalry and infantry.

In course of time the ox renders the oxherd heavy and slow as he is himself ; the hunter, on the other hand, becomes restless, active, and ferreting, like his dog ; man can only perfect himself by frequenting, by knowing, and by loving his fellow-man ; and it is because in Greece, anthropomorphism was the basis of worship, that this country advanced so quickly and so far ahead of Egypt, a country brutalised by the adoration of animals.

¶ 609.
Effect of association with animals.

[443] This seems rather hard upon the memories of Bellerophon's Pegasus, Alexander's Bucephalus, Orlando Furioso's Brilladore, Reynaud de Montauban's Bayarte, Roderigo's Orelia, or any of the other celebrated horses of history ; but *vide* the next note.

[444] The reader will bear in mind that, as M. Gourdon de Genouillac has told us in his preface, M. le Capitaine d'Arpentigny was always attached to regiments of the line.

The Mixed Type.

Last scene of all that ends this strange eventful history

Is second childishness & mere oblivion, sans teeth, sans eyes, sans taste, sans everything

SUB-SECTION XXII.

MIXED HANDS.

THE MIXED TYPE.

I GIVE this name to the hand whose undecided out-lines appear to belong to two different types.[445]

¶ 610.
Its definition.

Thus, for instance, your hand is of the "mixed type" if, being spatulate, the formation is so little accentuated that it might be mistaken for a hand of the square type. Again, a conic elementary hand might be taken for an artistic hand, or an artistic hand may be taken for a psychic, and so on and *vice versâ.* Or, in like manner, a philosophic hand may be mis-taken for a useful hand, and *vice versâ.*

¶ 611.
Examples.

The intelligence which is revealed by a mixed hand is one which partakes of the nature of the intelligences

¶ 612.
Its indications and value.

[445] It was without considering this sub-section and the next that Adrien Desbarrolles made the remarks which we find on p. 275 of the 15th edition of "*Les Mystères de la Main,*" in which he accuses d' Arpentigny of not having noticed the prevalence of mixed hands and of disregarding the importance of their significations. *Vide* also *op. cit.,* p. 521.

attached to each of the forms represented. Without these hands, that is to say, without the mixed intelligence peculiar to them, society,—deprived of its lights and shades, and without moral *alkalis* to effect the combination of its *acids*, and to amalgamate and modify them,—would advance only by struggles and leaps.

¶ 613.
Wars between
exclusive races.

If the laws of war were cruel even to atrocity among the earliest peoples of whom history makes mention, it is because in those days, each nation having sprung from a single tribe or from a family free of all admixture of foreign blood, differed radically from all others in temperament and in instinct. In flooding their rivals with their own blood, and in destroying their cities, they obeyed the laws of antipathy which were continually urging against one another, classes destined by their organisations to a never-ending antagonism.

¶ 614.
Inter-tribal wars
of the Arabs and
Kaffirs.

The Arab tribes, sprung from practically identical roots, are not pitiless in the wars in which they engage between themselves, but they are so in the wars which they wage against Europeans. It is the same with the Kaffirs. "The way in which these nations fight with one another," says Lichtenstein, "bears the imprint of a generosity very different to the usages adopted by other peoples. As soon as war is declared, which is always done by an ambassador bearing the tail of a lion or of a panther, the chiefs receive orders to join the king with their vassals; when the army approaches the enemy's territory, another ambassador is sent forward to warn them of the fact, and if the enemy says he is not ready, or that his forces are not yet properly assembled, the attacking force halts and waits until the others shall be ready to fight. Finally, to render ambuscades impossible,—manœuvres which would be looked upon as dishonourable,—they select as a battle-

field an open space without rocks or bushes. Then
they fight with -as much stubbornness as valour.
When one of the armies is defeated, the same generosity makes itself apparent in the conduct of the
conqueror, who does not fail to send a portion of the
spoil back to the vanquished, regard being had, as they
say, to the fact that one must not let even one's enemy
die of hunger. But this moderation," adds Lichtenstein, "only obtains between one tribe of Kaffirs and
another : for if they are at war with the Boers or the
Hottentots, they endeavour to exterminate them by
every means in use among other nations, whether
savage or civilised."⁴⁴⁶

As regards Europe, it is evident that wars have
been less cruel in proportion as by the progress of
navigation and commerce the people have become
more mixed. These ideas, thus lightly touched upon,
deserve a longer discussion and a fuller development, but this book, as may be seen, is merely an
essay.

¶ 615.
European wars.

Just as there are *absolute* truths and *absolute* beauties, there are also *relative* ones. Between Apollo
and Vulcan, between the Muses and the Cyclops—I
take these extreme symbols on account of their clearness—*Mercury*, the god of practical eloquence and
industrial arts, hovers upon his *petasus* and *talaria*,
and wields the *caduceus*. Well, it is to mixed
hands that the intelligence of mixed works of inter-
mediary ideas belongs ; of sciences which are not
really sciences, such as administration and commerce ;
of arts, which are not the outcome of poetry ; and of
the beauties and the relative realities of industry.

¶ 616.
Intermediate
ideas and talents

⁴⁴⁶ M. H. C. LICHTENSTEIN, "*Reisen im Südlichen
Africa in den Jahren* 1806-1807 (Berlin : 1811,
2 vols.). The passage referred to may be found on
pp. 277-8 of vol i. of Anne Plumtre's translation, entitled "*Travels in Southern Africa by Henry Lichtenstein*" (London : 1812, 2 vols.).

¶ 617.
Influence of
industry.

Industry raises [or lowers] everything to the dead level of commonplace.   In rendering material life lighter by the multiplication of articles of mere utility, in rendering the cultivation of the mind more easy by the multiplication of the means and instruments of study, industry *civilises*—by interest indeed —but she destroys art and science, which civilise by the agency of love, by materialising them, and by substituting for creation and for intellectual effort— imitation by mechanical process.

¶ 618.
Its definition
and advantages.

One might define industry as the magical art which draws money from everything.   The man who is born with the talent of industry, practises the arts, the sciences, eloquence, and even virtue,—with a very few exceptions—only to derive material benefits from them ; money is his aim, not glory, not perfection ; among the ancients Mercury the god of industry was also the god of liars and of thieves.

¶ 619.
Peculiar forms
of industries.

Murder, by the charge at quick-step, is one of the industries of the Swiss ; murder, by rhetoric and by the careful combination of phrases, was the industry of the Attorney-General Marchangy.[447]   For some men the priesthood itself is only an industry.   In our-

[447] Louis Antoine Marchangy, a celebrated French magistrate and writer, born at Clamécy (Nièvre) 1782, died at Paris 1826.   He was in turn an enthusiastic Buonapartist and a bigoted Royalist [!].   It is not clear to my mind whether M. d'Arpentigny alludes to his magisterial or to his literary eloquence in the above paragraph.   His great prose epic, "*La Gaule Poétique*" (Paris: 1813, 9 vols.), is one of the heaviest compositions ever published, as Larousse justly remarks, "L'emphase, les banalités pompeuses, la declamation monotone, sont les traits distinctifs de cette œuvre ;" the same author calls him "the purveyor to the political scaffold."   Under Louis XVI. he became in turn Procureur du Roi and Avocat Général à la Cour de Cassation [1822], and in these capacities became renowned for his diabolical ingenuity in twisting and turning men's phrases so as to use them against themselves.

rural districts there are persons whose sole occupation is to watch the grass growing, and to this labour,
which they lighten with frequent potations, they give
the title of an *industry.* There are professions in
which this word *industry* would be a disgrace; there
are others again which it ennobles.

Apt for many pursuits mixed hands nevertheless
often excel in none in particular; a great moral indifference is their endowment. The hand which
belongs to a particular type, on the contrary, is the
sacred shrine in which God has placed the imperish-
able germ which is destined to renew or to reveal
every art, every science hitherto ignored, or for a
long time lost sight of. Its promptings, too imperious
to be disobeyed, too significant to be mistaken, give
it the clear knowledge of *itself;* it knows what it
wants, and, like the animals which are guided by an
infallible instinct, it desires nothing that it cannot
possibly attain.

¶ 620.
Jack-of-all-
trades, master of
none.

Pascal, Descartes, Newton, Buffon, and the rest of
them who have divined so many things, must have
had hands of a pronounced and single type. From
their inspired brains sprang spontaneously sciences
already perfected; these great men, occupied by their
labours alone, all of them led a life which was studious
and more or less esoteric—for solitude is liberty.

¶ 621.
Contrast of men
of single types.

Men whose hands present the forms of a particular
type have minds which are more powerful in one
direction than versatile; men whose hands are of
mixed types have minds which are more versatile
than powerful. The conversation of the former is
instructive, that of the latter is amusing; it is for
these latter above all that a powerful education,
judiciously adapted to the development of the most
prominent faculty of their minds, is an immense
benefit.

¶ 622.
Mixed and
single types
compared.

CONTINUATION.

MIXED HANDS [*continued*].

Artistico-elementary hands, etc.

¶ **623.**
The artistico-elementary hand.

THICKER and less supple than the true artistic hand, the artistico-elementary hand, whose ungraceful outlines indicate an intelligence turned towards things which are sordid, presents nevertheless neither the extreme hardness nor the rustic expansiveness of elementary hands. Its fingers are large, without prominent joints [or with one only developed] and as if swollen up, the thumb is large and the fingers arc conic [*vide* PLATE X.].

¶ **624.**
Its prevalence in Normandy.

This hand is sufficiently numerous in Normandy to draw into the sphere of its moral action the genius of the other types sparsely distributed throughout the province. Richness, is in the present day, the only side of art which the Normans appreciate in their hearts and without restraint. They love it for its own sake, and sacrifice everything to it, even to their sensuality, to which they allow none but the cheaper sorts of pleasures ; they are always greedy, rather than avaricious.

¶ **625.**
Rusticity of Norman towns

The appearance of their towns is always somewhat rustic, and to see the costumes of nine-tenths of their inhabitants you would say that the citizen element had been expelled by an invasion of peasantry. Rouen, Saint-Lo, Falaise, and the rest of them, in the midst of the green champaign which is resplendent all around them, recall those hideous reptile carcases which the folly of the ancient Egyptians used to case in gold and porphyry.

The Normans have a code of morals, if *customs* can constitute them, but they have none if they are constructed of *principles*. They are legal, but not just; they are devout, but not pious. Although naturally brave, war is antipathetic to them, not on account of the privations which it involves [which they could bear without complaint], but on account of the little profit which it gives. Glory without money seems to them as a vapour, vain and ridiculous; it is reserved for the Bretons, a nation governed by obstinate and passionate instincts, to wage war for the maintenance of a principle; the Normans have never drawn the sword save for a material interest.

¶ 626.
Their morals.
Customs contrasted with principles.

On his return from the Holy Land, whither he had been led by a pious and chivalrous idea, the brave Robert of Normandy found his throne occupied by Henry, his younger brother. He appealed to the people, but the latter turned a deaf ear to his appeal, finding it just and good that an adventurer, capable of preferring glory to actual material advantages, should pay the penalty of a so flagrant infraction of the laws of good sense, with not only a double crown, but his liberty.

¶ 627.
Robert of Normandy.

"A king in a state of poverty," says Euripides, "is nothing." [448] Wealth is what is most revered among

¶ 628.
Euripides

[448] Our author, I presume, refers to the concluding lines of the *Phœnissœ*, where the tyrant Œdipus, deploring his fall, calls attention to the misery of his present state compared to his former magnificence.*

---

* EURIPIDES, ΦΟΙΝΙΣΣΑΙ, l. 1758, *et seq. ad fin.*:—
"'Ω πάτρας κλεινῆς πολῖται, λεύσσετ', Οἰδίπονε ὅδε,
ὃς τὰ κλείν' αἰνίγματ', ἔγνω καὶ μάγιστος ἦν ἀνήρ
ὃς μόνος Σφιγγὸς κατέσχον τῆς μιαφόνου κράτη,
νῦν ἄτιμος αὐτὸς οἰκτρὸς ἐξελαύνομαι χθονός.'

[¶ 628]

men ; between Greek and Norman there exists but the difference of a hand.

¶ **629.**
The Norman
character.

The Normans have an intelligence which is no delicate, but cunning : they calculate rather than reason ; their language, generally negative, never becomes exalted, though there are times when it becomes inflated, even to bombast [Malherbes, Brébeuf, and even Corneille, often confound emphasis with subtlety]. They are a people of clear judgment, for whom the purse represents the man ; at the same time brutal and cunning, benignant and subtle ; without *art* but full of *artifice*. Very well, then, in the same way that art is [as I have said] a means of causing the *true* to be appreciated, artifice is a means of causing the *false* to be esteemed.

¶ **630.**
Value of the
sordid mind.

Still, it is good—indispensable indeed—that a great number of men are devoted by their instincts to the cultivation of wealth for its own sake, and laying aside the pleasures of every kind of which it is or can be the source, it is by these men, insensible as they are to every happiness save only that of being rich, that the fortunes are made and remade without which science, art, and poetry, those Muses disdainful of mechanical occupations and manual labour, would languish, unhappy and discouraged, for want of physical leisure.

¶ **631.**
Socialism in
North America.

The legislators of the western states of North America, in proscribing domesticity, and by these measures condemning their fellow-citizens to ignoble and futile labours which are in Europe the natural occupations of weak understandings, have given a more unrefutable proof of their want of appreciation of the fine arts and of the higher sciences than any which have been furnished by barbarians or iconoclasts.

¶ **632.**
The aspect of
the type with its
character.

The aspect of the artistico-elementary hand proclaims its egoism and avidity ; large and short, closing more easily than it opens, it seems to have been

[¶ 632]

formed only for the purpose of grasping and holding; it is probably from it that we derive this edifying axiom, What is good to take is good to keep.[440] Inapt at the professions which are governed by sciences, it excels at negotiation; it is not industrious, but industrial.

Normandy, full as she is of manufactories, has not invented, nor even perfected, a single machine. There come from its factories nothing—with the exception of cloths—but vulgar products; in agriculture she is not, intellectually, as highly developed as the fertility of her soil.

¶ 633. Norman manufactures.

It is in Normandy, in the verdant and cunning country of apples, of the Forbidden Fruit, that we find the lower limbs of the law, the sharp attorneys, and the loud-voiced counsellors, who will bark at anything for a crown-piece.

¶ 634. Norman lawyers

Education, which ameliorates the Normans, a race born, after all, for the pleasures and transactions of society [a kind of reasonable, calm, and wise imposture entering largely into these things] corrupts the Bretons, on the contrary; the character of the Breton is irreproachable in the country; the character of the Norman shows to best advantage in cities.

¶ 635. Education among the Normans and the Bretons.

Like the Normans, the Jews are distinguished by a great commercial capacity. These people, who for so many centuries have been separated from their fellow-men by their love of the letter of the law, [a pursuit even less fruitful than that of form], and their hatred of foreign blood, are happy and flourish,

¶ 636. The Jews.

---

[440] This reminds one of the remark of the witty young Frenchman of whom Bacon speaks in his "*Instauratio Magna*," who "was wont to inveigh against the manners of old men, and would say that if their minds could be seen as their bodies are, they would appear no less deformed . . . for the bending of their fingers, as it were to catch, he would bring in rapacity and covetousness."

pre-eminently in the places where ignorance, slavery, and fanaticism concur to degrade the masses. They are unimportant wherever order and good manners reign hand in hand with liberty. When Europe was in a state of barbarism they were as they are to-day; now that it is civilised they are as they were then— so petrifying is the cultivation of the letter to the exclusion of the spirit. They no longer exist as a *people*, but they have never lost their nationality; *Jews*, wherever they are, they are citizens nowhere.[450] The greatest calamities when they do not actually touch them, are for them merely spectacles, and as they attach themselves neither to the soil nor to manners, nor to political interests, but only to the interests which are peculiarly their own, they flee before the storm to reappear with the conquerors, and proceed calmly to the increase of their fortunes amid corpses and ruins.

¶ 637.
The Jews in
Poland.

The Jews in Poland form nearly two-thirds of the population of the towns; in summer they wear a tight cassock of smooth and shiny texture, in winter a velvet cap something like a thick turban, and a furred robe, which they gird about themselves with

[450] I find a very interesting commentary to this paragraph in a work to which I have referred more than once in these pages—Colonel Tcheng-Ki-Tong's "*The Chinese painted by Themselves.*" He cites an authentic record of a colony of Jews which emigrated to China under the Han Dynasty, B.C. 200, which is in precisely the same condition now as it was then. "Here then," says he, "is an authentic tradition 2,000 years old. It is only in the Jewish people one finds such attachment to nationality. Take any people you please, at the end of four or five generations they will be completely naturalised; the Jews never. They remain the same wherever they go, attached to their religion, their nature, their customs; and this permanence of a particular race in the midst of a people numbering 400,000,000, is not an unimportant fact from the point of view of general history" [p. 191-4].

a band of red woisted, which serves them for a pocket. They let their hair and beards grow and straggle as they please, they have aquiline noses, and pale, oval faces; they have almond-shaped, black eyes, which are brilliant with cupidity. They are insinuating and polite; very thin as a rule; one might mistake them as they stand in the corners of the shops, where they remain ordinarily immobile and standing upright, for shadowy cypresses or pear trees trimmed spindle-fashion. They scatter around themselves an indefinable idea of Capernaum and Jericho, which reminds one of the impression produced upon one by the prints in ancient Bibles. They do not indulge in any corporal exercise, or any agreeable pastime, making their trade their sole occupation. Lying, so as to buy in the cheapest market; lying, so as to sell in the dearest, their inglorious lives are spent between these two lies. They also have a predilection for the professions of the factor, the broker, the agent, the stock-jobber, the publican, the banker,—in a word, all the pursuits where sharpness of mind are of more consideration than the gifts of science, profound knowledge of art, or skill of hand. They trade openly upon luxury and drunkenness, but one must do them this justice, that they never lose their gravity, whether sheltered beneath the thyrsus of Bacchus or the caduceus of Mercury. Their hands are the same as those of the Normans, only with a weaker palm, and *quasi-square* finger-tips.

Brittany contains a great number of individuals of high intelligence who, within the closed circle of material interests, would easily be overreached by a Jewish or Norman child. Humble and resigned, they ask nothing better than to be preserved from the cares of business and of figures; they do not measure happiness by the volume of a man's belongings; they

¶ 638.
The Breton intelligence.

do not worship God in the image of a crown-piece; and they do not hear resounding in their dreams the magic whinny of the tax-collector's mule. To wander over the flowery heath, to dream, lying in the high grass, to follow God in the woods and on the feet of the sun, to fill themselves with the poetry of old books and old legends, to bear with pleasure the yoke of faith, to prefer to luxury or even prosperity, not money [like the Normans, who are temperate only by reason of their avarice] but meditation and repose; it is along these blest paths that we track the white foot of the Muse, that the incense of mystical roses perfumes the air, that the halo of the guardian angel illumines the way, and that the heart seeks for, and finds, happiness. These hands are psychico-elementary.

¶ 639.
The social and intellectual state of Brittany.
But it is perhaps germane to our subject to remark here that if the Bretons have for a long time been too much despised, the reaction which has taken place in their favour, consequent upon the startling apparition of a few rare genii born among them, has led many of the writers of to-day to praise them beyond all reason. No doubt they are frank, courageous, and capable of a disinterested devotion, but the *social* man among them is too far behind the *instinctive* subject. Whilst the whole of France is progressing in enlightenment and prosperity, the sorcerer, the petty squire, and the parish priest remain amongst them the objects of their most tenacious fetich-worship; they can anticipate nothing out of the ordinary run, they can appreciate nothing which is not customary. You may see them in their dirty villages wandering about with an air which is at the same time indolent and savage, clad in shapeless garments cut from the skins of heifers and of goats. Certainly, France would march in the rear of the nations, if, persuaded by the lovers of nature in whatever form she presents her-

self, she were to place them at her head instead of trailing them in tow behind her.

The Vendeans are a people of a limited but clear intelligence; opinionated rather than fanatic, they are simple without being ingenuous; they have not in their hearts the poetry which the Bretons have, nor in their minds the imagination of the Normans. Nor do their costumes present the striking singularity, or the quasi-Oriental elegance which we encounter here and there among these two nations; robust rather than active, without being lazy, they are slow, their humour is proud, irritable, and morose, only slightly sensual and limited in their desires, by reason of their want of imagination rather than by reason of their want of temperament, they manifest for their country a love which in their hearts is equal to none other.

From La Vendée there come most estimable officers, honest accountants, and incorruptible cellarers, but never men of mark. Like those wines whose flavour is not appreciable excepting upon the ground where they grew, expatriation deprives the Vendeans of all their virtue. They are very considerable as a nation, but quite inconsiderable as individuals; nature has decorated the ignorance of the Bretons with a few flowers of poetry; theirs, on the other hand, is as arid as a sand desert. They cherish the empire of custom and usage, and are only remarkable for their characters. Where they show themselves in all their startling originality is in the Bocage, a mysterious district bordered with the foliage of the oak [like the uniform of a marshal of France], where flow the fountains which have received from the Druids, their venerable godfathers, the not yet contested power of miraculous cures. There are so many high, quick-set hedges whence spring huge trees, so many cherry, pear, and apple trees border the roads and cluster round the houses, that one could not venture into this

[¶ 641]

part of the country without a guide. Almost inaccessible to artillery, and even to cavalry, war can only be waged there on foot ; it is a theatre better suited for the development of the spirit of cunning and ruse, and to the personal prowess which belong to the soldier, than to the general's combinations in advanced strategy. The peasant lives there a free life, circumspect and silent in these retreats, where all things are silent,—the air, the water all are still ; and where, were it not for the hammer of the farrier, one would hear nothing in the villages but the song of the birds. These are elementary hands with square fingers.

¶ 642.
Norman and Vendean hospitality

Vendean hospitality, much over-rated as it has been, is prescribed by custom ; in Normandy the practice of this virtue is facultative. Among the Vendeans it is an honour to the nation, among the Normans it is an honour to the individual ; in the latter case gay as a pleasure, in the former solemn as a duty.

¶ 643.
Effect of knowledge of one's own attributes.

Whence comes it that the universe constantly has its eyes turned in the direction of the ancient Greek world ? It is that the peoples of whom it was formed had, not only great instincts and great virtues, but had also a profound knowledge of those instincts and of those virtues. The Vendeans and the Bretons have also great virtues, but it is a question whether they would preserve them if they had any intelligence of them. Our own species, however, can only give us credit for those virtues which have their guarantee in the single attribute which places us above other created beings—intelligence. The more soundly does the somnambulist sleep, the more sure and certain is his step as he walks upon roofs and on the edges of precipices ; but who thinks of applauding this skill of which he is unconscious ? In the same way must we only very lightly esteem the virtues of a people plunged in the darkness of an evident intellectual somnambulism.

Conclusion.

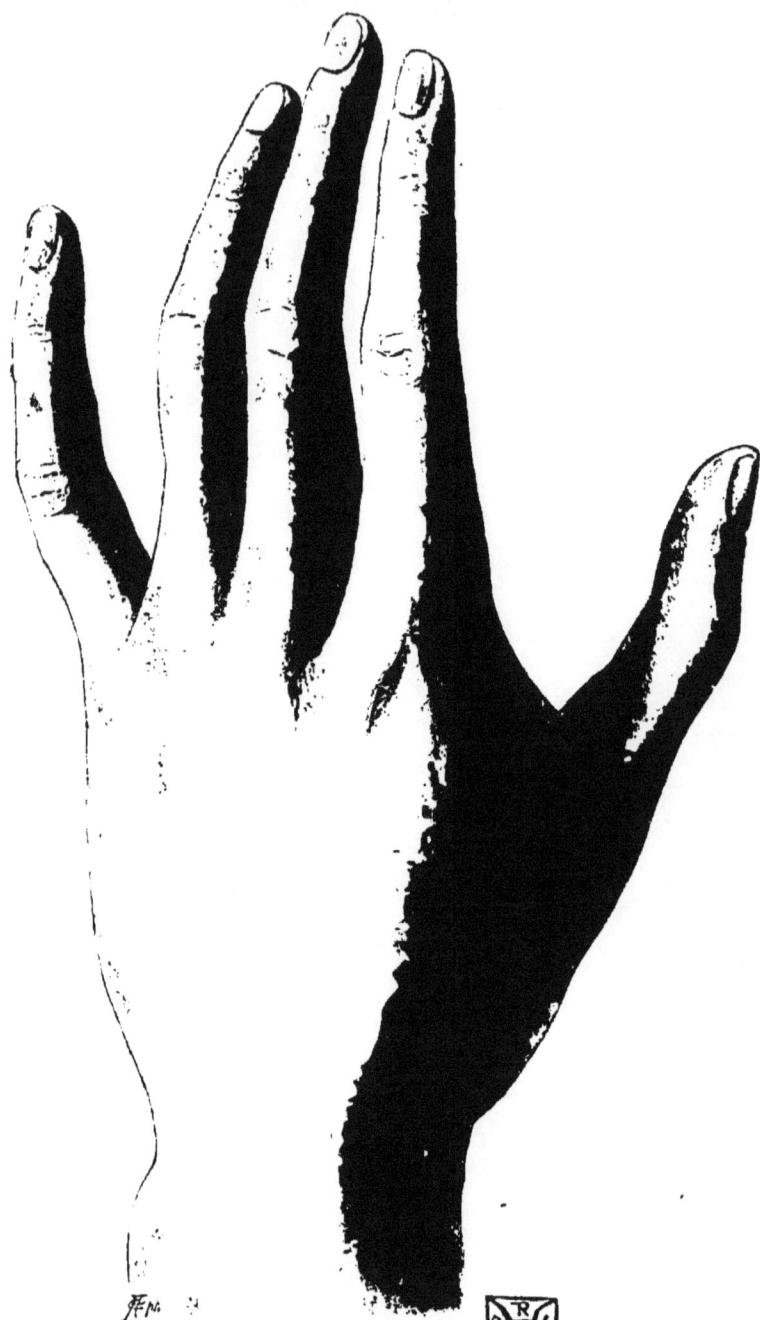

PLATE XI. —A FEMALE HAND.

[Page 401.

And Death once dead, there's no more dying then.

SUB-SECTION XXIV.

A FEW WORDS UPON THE HANDS OF WOMEN.

THE tendencies of each type are, among women, the same as they are among men, only, those which are peculiar to the spatulate and square types, are much less imperious and intense among women, by reason of the suppleness of their muscles, than they are among us [45] [*vide* PLATE XI.].

¶ 644. The types among men and women.

Out of a hundred women in France, I calculate that forty belong to the conic type, thirty to the square, and thirty to the spatulate type. These two latter types, of which the all-absorbing faculty is the mind, outweigh the former, of which the all-absorbing influence is the imagination.

¶ 645. French women's hands.

[451] "He must be," says Dr. Carus in his work, "*Ueber Grund und Bedeutung der Verschiedenen Formen der Hand in Verschiedenen Personen*" (Stuttgart : 1846), "but a superficial observer of mankind who could not at once recognise the sex from a simple inspection of the hand. The hand of woman is smaller, more delicate, and much more finely-articulated than that of man ; it has a softer palm, and joints, which are but slightly prominent," etc.

¶ 646.
Man and woman

The man creates, the woman develops; *we* have *principle*, she has *form;* we make *laws*, she makes *morals*.

¶ 647.
St. Martin.

"The man is more true than the woman," said St. Martin, "but she is better than he. The man is the *mind* of the woman, but she is the *soul* of the man."

¶ 648.
Genesis.

To compensate the woman for her weakness, says the Book of Genesis, as interpreted by Fabre-d'Olivet, God has clothed her in one of His envelopes—beauty; and straightway she becomes the incarnation of the faculty of volition on the part of the man.[452]

¶ 649.
Man and woman contrasted.

Then again, we value things with our brains, they with their hearts; we are the more sensual, whilst they are the more sensitive; their instincts deceive them less often than our reasoning powers; we have the faculty of reflection and know what can be learnt, they have intuition and know what can be divined.

¶ 650.
Women in Europe and Asia.

Europe, where they are free, and which they fill with brilliancy and movement, owes them these three beautiful things,—good morals, liberty, and opulence ; whereas Asia, where they live in a state of slavery, crouches metaphorically in a state of inertia, and loses itself in misery, despotism, and the lowest forms of passion. Light, truth, and liberty are one and the same thing.

[452] This reference is to a most fascinating work, unfortunately comparatively unknown to the general reader, "*La Langue Heberaïque restituée et le Veritable Sens des Mots Hébreux rétabli et prouvé,*" by Fabre-d'Olivet (Paris: 1815-6). Part II. consists of a portion of Genesis in Hebrew, English, and French, and the passage in question occurs in Gen. ii., 21 and 23 [p. 315], which read :—" Alors Shôah, l'Être des êtres, laissa tomber un sommeil profond et sympathique sur cet homme universel, qui s'endormit soudain : et rompant l'unité *d'une de ses enveloppes extérieures* il prit l'un d'elles et révetit de forme et de *beauté corporelle sa faiblesse originelle;*"—v. 23, —" Et il l'appela Aïsha, *faculté volitive efficiente* ou cause du principe volitif intellectuelle, Aïsha, dont elle avait été tirée en substance."

Few women have knotty [*i.e.*, jointed] fingers; synonymously few women are gifted with the talent of combination. In the matter of intellectual labours, they generally choose those which require more tact than science, more quickness of conception than strength, more imagination than judgment. It would be otherwise if they had prominent-jointed fingers; then they would be less impressionable and less given to yield to the inspirations of fantasy, and like as the intoxicating qualities of wine are neutralised by the addition of water, so would theirs be by reason.

¶ 651.
Jointed fingers among women.

It is convenient—setting aside the form of the exterior phalanges—to range women under two principal categories: those with large thumbs, and those with small. The former, more intelligent than sensitive, extol history; the latter, more sensitive than intelligent, are captivated by romance. For pleasure and consideration for others, recommend me to a large-thumbed woman; love, under her clear-sighted guidance, attains its end without scandal, her passion, which she follows without consulting her head, has more root in her senses than in her heart. Leave her alone and trust to her skill, at the right moment she will come to the assistance of your timidity, not because she has much sympathy with your torments, but in the interest of her own pleasures. Besides complete security, her many graces of mind will add to the joy of winning her.

¶ 652.
Large and small-thumbed women.

Women with a little thumb are not endowed with so high a degree of sagacity. To love is the whole of their science, but the charm attached to this powerful faculty is such, that there is no delight equal to it.

¶ 653.
Small-thumbed women.

The cares of maternity being extremely difficult and complicated, their practice requires an instinct more intelligent than that which is revealed by elementary hands; these hands, therefore, are extremely rare among women. Women exercise an almost

¶ 654.
Absence of the elementary hand among women.

[¶ 654]

absolute empire among populations in which the elementary hand is in a majority among the men [as, for instance, in Lower Brittany and La Vendée]; for there is no type which does not dominate over the elementary, morally speaking.

¶ 655.
Elementary-handed nations.
The peasants of these countries marry willingly, and very commonly, with women who are older than themselves. The same heaviness of spirit which renders them insensible to the charms of youth and beauty, delivers them over, helpless, to the superior intelligence of the woman who has reached maturity.

¶ 656.
Helen.
The Greeks of the heroic age were not more particular; Helen was nearly forty, when, on her return from Argolis, flying before Orestes who wanted to destroy her, " she went haphazard, tracing here and there her footsteps, brilliant with the splendour of her golden sandals." [453] She must have been still beautiful for men to occupy themselves thus concerning her shoes!

¶ 657.
The Caroline and Mariana Islands.
In the Caroline and Mariana Islands the political power belonged, until the conquest of these archipelagi by the Spaniards, to the women, who, contrary to the men, who have very large hands, have very small ones. [454]

[453] *Vide* the exquisitely touching account which Euripides puts into the mouth of Phryx of the terror of Helen and the agonised fury of Orestes.*
[454] Arago's accounts of these islands are full of indications of this consideration in which women are held. Almost the greatest crime a man can commit *against Heaven* is to beat his wife [J. ARAGO, " *Souvenirs d'un Aveugle. Voyage autour du Monde*" (Paris: 1839), vol. iii., p. 26]; and in another place he says:—

* EURIPIDES, ΟΡΕΣΤΗΣ :—

" 'Α δ' ἴαχεν, ἴαχεν ὤμοι μοι·
λευκὸν δ' ἐμβαλοῦσα πῆχυν τέρνους
κτύπησε κᾶτα μάλεον πλαγάν·
φυγᾷ δὲ ποδὶ τὸ χρυσεοσάνδαλον
ἴχνος ἔφερεν, ἔφερεν," etc., etc.

Born for the dance, love, and festivals, the women of Otahiti have hands which are small and conical, but fleshy and thick.

¶ 658.
The Tahitians.

English women have, as a rule, the finger-tips delicately square; they are contented with love as they find it in the married state, and devote themselves even to *manual labour*.

¶ 659.
English women.

The institution of the Harém being immemorial in Asia, I conclude that the women of these countries have delicate hands with little thumbs. They devote themselves even to *death*.

¶ 660.
Asiatic women.

Charlotte Corday, Sophie de Condorcet, and Lucile Desmoulins had very fine fingers.

¶ 661.
Illustrations.

The legislators of the East Indies are not, like ours, pre-occupied solely by the real needs of women and her duties, but also concerning her caprices and the fancies inherent to her nature. "Brahma," says Manu, "has endowed woman with love of rest, and of ornament, with passion, with fury, with evil instincts, and with perversity.[455] He desires that her name shall be easy to pronounce, soft, distinct, agreeable, and propitiously sounding ; that it shall end in long vowels, and shall resemble the words used in a benediction.[456] She must be con-

¶ 662.
Manū on women.

"What is quite clear, and what has been said before us by Spanish historians, is, that the women of those days had on all occasions the pre-eminence over the men, that they presided at all public deliberations, and that the code of laws had been drawn up for them alone. The Spanish rule, crushing with all its despotism this archipelago so brilliant and so variegated, has not been able to abolish this custom, which to my mind is most rational, established, as one might say, in the primitive manners."—*Op. cit.*, vol. ii., ch. 21, p. 370. *Vide* also the official report of this same journey, entitled "*Voyage autour du Monde, entrepris par Ordre du Roi, par M. C. de Freycinet*" (Paris : 1839), vol. ii., p. 2.

[455] "*The Ordinances of Manu*" [*vide* note ·³¹, p. 202], lect. ix., 19.

[456] *Ibidem*, lect. ii., 33.

[¶ 662]

tinually in a good humour; she must have the graceful gait of the swan or of the young elephant; she must keep her body thin by living solely upon flowers, roots, or fruits.[457] She must be brilliantly attired, regard being had to the fact that, when a young woman is brilliant in her ornaments, her family shines by her reflected light, whilst, if she is not resplendent, her family are not honoured by her." [458]

¶ 663.
Penelope.

Mercury, said the Greeks, overcame the virtue of Penelope under the form of a goat : what must have been their ideas of women if they spoke thus of the most chaste among them.[459]

¶ 664.
The Chinese.

The Chinese are more just towards them, and in their eyes the death of the mother of a family is not regarded as so slight an evil as the death of the father; at least so one would infer from the text of the Chou King,[460] who does not recommend widowers less urgently than widows to the care of the mandarins.[461]

[457] *Ibidem*, lect. v., 150 and 157.
[458] *Ibidem*, lect. iii., 55-62.
[459] It was of this connection that the god Pan, the son of Penelope, is said to have been born, according to Lucian Hyginus, and other authors. Homer, who does all in his power to exalt Penelope to the position of a model of prudence and chastity, declares that this occurred before her marriage with Ulysses. Mercury, we are told, assumed the form of a beautiful white goat when Penelope was tending the flocks of her father Icarius on Mount Taygetus. Later authors, however, discard Penelope's claims to virtue, and adduce a much more confused parentage to the birth of Pan, with a more esoteric signification to his name.
[460] "*The Shoo King, or the Historical Classic of China*" translated by W. H. Medhurst (Shanghae : 1846).
[461] "The Chinese woman is usually imagined as a pitiful being, scarcely able to walk, and imprisoned in her household among the servants and concubines of her husband. This is another flight of imagination, to be cut short, however much it may hurt the feelings

In France, women of the spatulate hand and small thumb are distinguishable by a great fund of affectionate freedom, by an imperious desire of action and of movement, and by their intelligence of real life. Theirs, among the higher classes of society, is the proud and ancestral grace of such women as Clorinde and Bradamante,[462] and of the belted patricians; theirs, as of Diana and the magnanimous Hippolyte,[463] are the swift horses and snowy hounds; theirs, among the middle classes, are these households full of noisy, laughing children, whose hands are ever active and whose voices are never still, where the Persian cat lives at peace with the spaniel

¶ 665.
Spatulate and small-thumbed women.

of veracious travellers" [*vide* note[16], p. 149]. . . . "We consider the depths of science a useless burden to women; not that we insult them by supposing they are inferior to us in ability to study art and science, but because it would be leading them out of their true path. Woman has no need to perfect herself; she is born perfect; and science would teach her neither grace nor sweetness. . . . Family life is the education which forms the Chinese woman, and she only aspires to be learned in the art of governing her family."—"*The Chinese painted by Themselves*," *op. cit.*, p. 45, etc.

[462] Clorinde and Bradamante, the ideal amazons of French and Italian literature. The first, a fair Saracen, the beloved of Tancred, a heroine of Tasso's "*Gerusalemme Liberata*," who clad herself in armour and fought among the Saracens, and in this disguise was killed by her unconscious lover in single combat. The second the sister of Renaud de Montauban, one of the heroines of Ariosto's "*Orlando Furioso*," who in a similar manner distinguished herself among the Paladins. Both these names are constantly used as synonyms for beautiful and brave women; as, for instance, by Théophile Gautier in "*Mademoiselle de Maupin*" (Paris: 1869):—"On concevra que ce n'est pas trop d'un volume pour chanter les aventures de la *diva*, Madeleine de Maupin, de cette belle *Bradamante*," etc.

[463] A queen of the Amazons, given in marriage to Theseus by Hercules. She was the mother of Hippolytus, mentioned in note[266], p. 222.

and the tame dove. Theirs, in the farmstead, is the
passionate love of horses, of the white-coated heifer,
and of the other domestic animals, and the occupation of transactions with the neighbours, and long
nights of hard work. Theirs, finally, in the granary
and in the barn, are the resources of an inde-
fatigable physical activity, calm resignation under
strokes of ill-luck, and some of the robust peculiarities of the women of the, Don.[464]

¶ 666.
Madame Roland.
Madame Roland had fine large hands with spatulate
finger-tips. With a head filled with practical ideas
and a soul strongly inclined towards the ideal, she
understood the beauty of passion, though she preferred to it that of self-sacrifice. At the same time
stoic and passionate, positive and enthusiastic, tender
and austere, she loved three things with an intense
devotion : her country, her liberty, and her duty.
Careful always to think well, to speak well, and act
well, she relieved her mind after the study of theoretical mechanics by reading Plutarch and Rousseau.
Gifted with the kind of beauty peculiar to active women
she combined in herself an elegant carriage, a beautiful complexion, magnificent hair, and a splendidly
developed figure. Her mouth, which was rather
large, shone with freedom and serenity of mind. Her
looks were soft and frank, her manner was open, calm,
and resolute ; born brave and strong like most women
of her type, she was never untrue to herself, whether
in poverty, in splendour, or on the scaffold.[465]

[464] M. d'Arpentigny has on a previous page given a
short sketch of the wives of the Cossacks of the Don,
with particulars of their *sang froid* and laborious and
housewifely occupations. I have omitted the passage,
in common with a good many others in this sub-section,
as being unnecessary as illustrations, and offensive to
English taste as information.
[465] Marie Jeanne, the wife of Jean Marie Roland de la
Platière, known to history as "Madame Roland," was

Order, arrangement, symmetry, and punctuality reign without tyranny in the homes which are governed by these calm managers with square fingers and a small thumb.

¶ 667.
Square and small-thumbed women.

But what do I see! children silent and gloomy, servants trembling and sulky ; who is it, then, who keeps them in this state of restraint and worry. It is the peevish voice and vigilant watchfulness of petticoat government, represented most surely by a large thumb.

¶ 668.
Large-thumbed women.

Do you lay siege to the heart of a beautiful woman whose fingers are square ? Speak the language of common sense and of solidity of mind, and do not confound *singularity* with *distinction;* remember that she has less imagination than mind, and that her mind is more just than original. Amongst her axioms are these : —Silence is strength, and mystery is an ornament. Do not forget that she has the social instinct strongly

¶ 669.
Square-handed women.

one of the most famous of the famous women of the revolution. Born in 1754, she became at an early age remarkable for the power and extent of her intellect. Her husband, twenty years her senior, was returned to the Convention as *député* for Lyons, and became one of the leaders of the Girondins. His wife was arrested on her husband's flight in 1793, and was guillotined in the November of that year. Hers was the celebrated phrase, " O Liberty ! what crimes are committed in thy name ! " She is known to posterity by *"La Correspondance de Madame Roland avec les Demoiselles Cannet"* (Paris : 1841), and " *Lettres Autographes de Madame Roland, addressées à Bancal des Issarts"* (Paris : 1835). " This woman, combining with the graces of a Frenchwoman," says Thiers, "the heroism of a Roman matron, had to suffer every species of misfortune. She loved and reverenced her husband as a father. She experienced for one of the proscribed Girondists a vehement passion, which she had always repressed, . . . she considered the cause of liberty, to which she was enthusiastically attached, and for which she had made such great sacrifices, as for ever lost " [" *The National Convention,"* chap. xv.]. This bears out M. d'Arpentigny's opinion of this heroine.

[¶ 669]

developed, and that she combines with respect for what is regulated by good taste, a great love of influence and of command.   Her mind is as far removed from rarity as from vulgarity.[466]

¶ 670.
Illustrations.

The square type as far as women are concerned is perfectly represented by the prudish, clever, ambitious, and witty Madame de Maintenon. With the exception of Clementina, all the heroines of Richardson—creatures, all of them, more intelligent than sensitive, who, like our Madame de Sevigné, were more sprightly than tender-hearted, belong to this type.

¶ 671.
Nunneries.

Religious institutions governed by rigidly-severe and narrow rules, where nothing is left to the discretion, recruit nearly all their adherents from among the subjects of the square type.

¶ 672.
Small, rosy hands.

See these little soft, supple hands, almost fleshless, but rosy, and with little developed joints ;[467] they love brilliant phrases, which like lightning cast a sudden bright flash of wit around them ; they live with their minds alone.   The love whose fetters they bear was born in a boudoir ; it invented the madrigal, the amorous epigram, and never shows itself save in powder and ruffles.

¶ 673.
Artistic-handed women.

With women whose hands are hard, whose fingers are conic, whose thumbs are small, tint your language with glowing colours, excusing, justifying, applauding

---

[466] The whole of this chapter is almost as good a "lesson in love" as the speeches of Truewit in the fourth act of Ben Jonson's *Silent Woman*.

[467] *Othello.*   "Give me your hand.   This hand   moist, my lady.
Desdemona.   It yet has felt no age, nor known no sorrow.
Othello.   This argues fruitfulness, and liberal heart :—
Hot, hot and moist : this hand of yours requires
A sequester from liberty, fasting, and prayer,
Much castigation, exercise devout ;
For here's a young and sweating devil here
That commonly rebels. 'Tis a good hand,
A frank one.
Desdemona.   You may indeed say so ;
For 'twas that hand that gave away my heart."
OTHELLO, Act ii., Sc. 4.

peccadilloes of the more tender description. They love all that is brilliant, and rhetoric has more empire over their minds than logic. They are governed by three things : indolence, fancy, and sensuality. The sparrows of Cupid nestle in their dimples, and they have in their hearts the prayer which the Corinthians raised every morning to Venus :—"O goddess! grant that to-day I may do nothing that is displeasing, and that I may say nothing that is disagreeable ; " for to please is their highest need, and they like to be loved and admired as much as they like to be esteemed.

Such were without doubt the hands of the beautiful and triumphant Amazons, who composed the "flying" squadron of Catherine de Medicis.[63]

¶ 674. The Amazons of Catherine de Medicis.

Fingers which are delicate, smooth, and pointed in a woman's hand, when they are supported by a palm which is narrow and elastic without softness, indicate tastes in which the heart and soul have more voice than the mind or senses,—a charming mixture of exaltation and of indolence, a secret distaste for the realities of life, and for recognised duties, more piety than devotion. These characters, which are at the same time calm and radiant, expend their sovereign influence upon inspiration and grace. Good sense, which of all kinds of faculty is the most prolific, but not the most exalted, pleases them far less than true genius. It is for the purpose of exposing themselves

¶ 675. Psychic-handed women.

[63] After the "Peace of St. Ambrose " [12th March, 1563], Catherine de Medicis gave a series of the most magnificent *fêtes*, at which the honours were performed by the band of one hundred and fifty young ladies of the highest families, known as " les filles de la reine," of whom she made such telling use in the struggles which immediately preceded this epoch. "Elle voulait," says Michelet ["*Hist. de Fr.*" (Paris: 1855), vol. xi., p. 279],"travailler la noblesse, l'amuser, la séduire. Son principal moyen, s'il faut le dire, c'étaient ' les filles de la reine,' cent cinquante nobles demoiselles, ce galant monastère qu'elle menait et étalait partout."

[¶ 675]

to the heavenly rays of a pure love, that they have been sown, like spotless lilies, upon the bright plains of the day.

¶ 676.
Georges Sand.
I have in my mind a writer whose mind is carried onward by her heart, and whose ideas are always intermingled with her sentiments; she has lyricism and observation, measure and spontaneity; expansive and passionate, she has been able to interest all hearts in the throbbing of her own. She has shown herself upon the mountain tops, and the earth has sparkled with rays of light, and towards her have risen minds elevated by love and by a great ideal; the intoxication of distracted hearts, the calm of hearts that have become appeased, one understands them all as one reads, and one feels better after having read; above all religions [by reason of an idea of God which is superior to those which they have evolved], she has beauty for a worship, and liberty for a code of morals; still simple in her life, she is happy only when among simple people. What shall we regard as happy if not this master-mind with the resplendent brow and magnificent presence so dear to her surroundings, so dear to all whom the sibyl has endowed with the golden wand, and the fairy, with the magic ring which gives universal knowledge, and to whom these two sources of our best pleasures, labour and admiration, are so easy of practice and acquisition. The hand of Madame Georges Sand—for it is of her that I speak—realises all that I have just said, but with developed joints which modify it sufficiently appreciably.

¶ 677.
Sentiments of women and their effects.
The delicate sentiments which education alone can give to the greater number of us men, women possess naturally. They spring up in their tender souls, like the fine grass upon a light soil; women have an innate knowledge of things appertaining to the heart, but the perfect intelligence of the real and positive

[¶ 677]

world is wanting in them. It is less to their physical weakness than to the nature of the ideas attached to their organisation that they owe the fact that they see us reigning over them as masters. In vain should we have the strength to subdue horses, to exercise the more laborious trades, and to brave the elements of sea and sky, if our hearts, like theirs, greedy of emotions, and always ready to flee to something new, were to vacillate at the least breeze like the foliage of the aspen ; if this were so the empire which we hold would speedily slip from our grasp.

*   *   *   *   *   *

If these notes, all incomplete though they be, shall help you, O reader, to escape the rocks which lie hidden beneath the deceitful waves of the River of Life, you will glorify the professor who has laid them before you.

¶ 678.
Conclusion.

THE END.

"Et est completus per Eduardum Heron-Allen, Die XXIV. Januarii, MDCCCLXXXVI."

Appendices.

Imperia

Chez un sculpteur, moulée en plâtre,
J'ai vu l'autre jour une main
D'Aspasie ou de Cléopatre,
Pur fragment d'un chef-d'oeuvre humain ;

Sous le baiser neigeux saisie,
Comme une lis par l'aube argenté,
Comme une blanche poésie
S'epanouissait sa beauté.

A-t-elle joué dans les boucles
Des cheveux lustrés de don Juan ?
Ou sur le caftan d'escarboucles
Peigné la barbe du sultan ?

Et tenu, courtisane ou reine,
Entre ses doigts si bien sculptés
Le sceptre de la souveraine,
Ou le sceptre des voluptés ?

Impériales fantaisies,
Amour des somptuosités ;
Voluptueuses frénésies,
Rêves d'impossibilités.

On voit tout cela dans les lignes
De cette paume, libre blanc,
Où Vénus a tracé des signes,
Que l'amour ne lit qu'en tremblant !

Théophile Gautier.

APPENDIX A.

ON page 181 of the 15th edition of the late Adrien
Desbarrolles' work "*Les Mystères de la Main*," a
· description of the hand of M. d'Arpentigny occurs. I
reproduce it in this place, as I think it can hardly fail
to be of interest to my readers. It runs as follows :—

We give here a description of the hand of M. d'Arpentigny, drawn up by means of his system : we will
explain his tastes and aptitudes by applying the pre-
cepts of his own science to its inventor.

We might have carried it a step farther by consulting cheiromancy, but everything must come in its
proper order.

Our only chance of clearness in so abstract a science
lies in keeping even step with it, going from point to
point ; and giving a condensation thereof after havirg
studied separately every branch of the art.

The hand of M. d'Arpentigny is, in the first place,
remarkable for its rare beauty : its long and pointed
fingers give it an extreme elegance, and thanks to a
phalanx of logic [in the thumb], and a joint of philosophy [in the fingers], he is gifted with all the useful
qualities of his type. We need not call attention to
the inspirations of the professor, the discovery of his
system of cheirognomy affords proof enough of their
existence. Drawn by his pointed fingers towards a
love of form, he encourages a love of the beautiful in

art, poetry, and works of imagination ; his taste is keen and delicate, but drawn by his attraction for all that pleases the eye and the ear, he sometimes attains to research. Though continually held in check by his great logic, which gives him a love of truth and simplicity, the nature of his pointed fingers regains from time to time the upper hand. He speaks well, writes charmingly and wittily, his style is never heavy, and is sometimes even characterised by brilliant inspirations which are sadly out of harmony with the material century in which we live.

He pays but little attention to the circumstance that he is noble, he is simple, but at the same time he moves in the highest society, of which he has the easy manners. His whole personality is fraught with a natural aristocracy, and he has a horror of people who are vulgar. His conversation is charming, and always very instructive, sprinkled here and there with brilliant, though quietly expressed epigrams.

His pointed fingers would lead him towards religion, but his joint of philosophy renders him essentially a sceptic; he has aspirations with which he struggles continually and savagely, one would say that he reproaches himself for secret enthusiasms of which he will not seek the causes.

With fingers merely pointed he would have had only inspirations of his system ; vague and fugitive, he would certainly not have made use of them ; the philosophic joint, however, which leads him to the research of causes, has explained to him what his imagination merely hinted at, and logic has come to encourage him and to make his convictions profound.

Notwithstanding his pointed fingers, his modesty is charming, and he seems almost astonished when people congratulate him on a great discovery.

But the philosophic joint, useful though it undoubtedly is, has also some grave inconveniences. It

renders a man independent, and the love of independence which it inspires, not at all appreciated in a
military career, prevented him from rising to the
grade for which his superior intelligence fitted him.

His fingers, smooth by reason of the absence of the
joint of material order, in giving him, to a marked
degree, all the qualities of the artist, naturally have not
recommended to him the arrangement and œconomy
of which they have so wholesome a horror. But
being large at the bases, they give him a taste for
sensual pleasures ; they have by this means made life
as bearable as possible for him, causing him to stoop
and gather one by one, without too particular a choice,
all the flowers which are to be found on the road of
life. To this the softness of his hands has added the
charms of an intelligent laziness.

M. d'Arpentigny appreciates the charms of indolence,
and thence perhaps it is that comes his indifference
for success in the world, for the great reputation
which ought to have accrued to him ; thence comes
also his distaste for the discussions, the controversies,
and the academic struggles which fall to the lot of
every inventor.

His road lay athwart the brilliant sunshine, he has
preferred to walk in the shade ; and without the
rather large upper phalanx to his thumb, which gives
him a certain obstinacy, probably he would have left
his system in the same shadow, as much by reason of
his horror of worry and intrigue, as on account of
his disdain of his fellow-men.

M. d'Arpentigny was endowed with all the qualities
of the inventor,— the pointed fingers which receive inspirations from on High ; causality, the great sceptic
which discusses and examines them ; and the logic
which finally adopts them, calmly deciding what there
is of truth in the intuitions of his pointed fingers, and
in the doubts of his inherent causality. His long

fingers, by the love of minutiæ which they give him, have led him to his studies, making him pursue his system with care, even to its minutest details.

But what is a good quality in discovering a system, may become a fault, a defect in expounding and teaching it.   M. d'Arpentigny, being without the order and arrangement of square fingers, and without the material order found in a development of the second joint of the fingers, has allowed himself to wander away amid the charms of description of citation and of science.   Carried away by his philosophic instinct, he has discovered at every step subjects for admirable reflections, highly interesting to the reader, and doubtless equally so to himself, for he often loses sight of the point whence he started, returning to it regretfully as a thing too positive, only to lose himself once more amid the mazes of his high imagination.

His pointed thumb also,—a very rare form,—which augments the power of his intuitions, is long enough to give him a certain amount of strength of resistance, but not enough to make him triumph over the philosophic indifference, by which he allows himself in other respects to be dominated very willingly.   This alone prevented our inventor from becoming the high priest of a sect ; he forged for himself out of the science a sparkling ring, but it never occurred to him to make thereof a crown.   With a logic which interferes seriously with the promptings of his will, with a philosophic joint which strips of their embroidered vestments all the splendours of the world, he came naturally to the conclusion that the science was too noble, too grand, and too proud to become a mere crutch for his ambition.

It will reach posterity clad in all the greater glory.

[*From the French of*]

ADRIEN DESBARROLLES.

APPENDIX B.

BIBLIOGRAPHIA CHEIROSOPHICA.

THE following Bibliography cannot, of course, in any way aim at completeness; it pretends to be no more than a transcript of the catalogue of my cheirosophical library, to which I have added the titles of a few works to which I have had on various occasions to refer. I have adopted the alphabetical in preference to the chronological arrangement, as the latter necessitates a separate index for purposes of reference; and also it is difficult, when that plan is adopted, to gain any idea of the collective works of a particular writer. I have drawn up this list of books in odd moments for my own use; I publish it now in the hope that it may prove as useful to other students of cheirosophy as it has proved to me. I shall, of course, be very grateful to any reader who will call my attention to any works at present omitted from this catalogue, in order that by completing it, I may enhance its value to the cheirosophist in subsequent editions of my work.

ED. HERON-ALLEN.

St. John's,
Putney Hill, London, S.W.
May 7th, 1886.

1. Achillinus, *A.*, and Cocles, *B.* (in one vol.)—"Alexander Achillinus Bononiensis de Chyromantiæ Principiis et Physionomiæ," 12 leaves; *and* "Bartholomæi Coclitis Chyromantiæ ac Physiognomiæ Anastasis cum Approbatione Magistri Alexandri d'Achillinis." (Bononiæ: 1503.) Fol., 2nd Edition (Bononiæ: 1523.) Fol., 178 leaves. *Vide* No. 36.

2. Achillinus, *Alexander.*—"De Intelligentiis, de Orbibus, de Universalibus, de Elementis, de Principiis Chyromantic et Physionomie," etc. (Venice: 1508.) Fol.

3. Albertus Magnus.—"Geheime Chiromant. Belustigungen, Kunst aus der Hand wahrzusagen." (Leipsic: 1807.)

4. Andrieu, *J.*—"Chiromancie. Etudes sur la Main, le Crâne, la Face." (Paris: 1860, 1875, and 1882.) 12mo.

5. Andrieu, *Jules.*—"Chiromanzia. Fisiologia sulla Mano, sul Cranio, e sul Volto." (Milan: 1880.) 12mo.

6. Anianus.—"Compotus cum Commento. Liber qui Compotus inscribitur una cum figuris et manibus necessariis tam in suis locis qui in fine libri positus incipit feliciter." (Rome: 1493.) 4to, 42 leaves.
   Other Editions: [Paris: 1500?] 4to, 40 leaves; [Basle: 1500?] 4to, 39 leaves.

7. Anianus.—"Compotus Manualis Magistri Aniani Metricus cum Commento, et Algorismus." (Argu: 1488.) 4to.
   Other Editions: (Rothomagi) [1502?] 8vo; (Paris: 1519) 4to.

8. Anonymous.—"La Cognoissance de la Bonne et Mauvaise Fortune, tirée de la Main." (Rouen: N.D.)

9. Anonymous.—"Wahrsagekunst aus den Linien der Hand. . . . . . Nach einer alten Zigeunerhandschrift bearbeitet." (N.D. or PL.)

10. Anonymous.—"Opus Pulcherrimum Chiromanticum multis additionibus noviter impressum." (Venice: 1499.) 4to, 36 leaves.

11. Anonymous.—Die Kunst der Chiromantzey usz Besehung der Hend; Physiognomcy usz Andblik des Menschens," etc. (Strasburg: 1523.) Fol.

12. Anonymous.—"La Science Curieuse, ou Traité de la Chyromance." (1667.) 4to.

13. Anonymous.—"La Chiromantie Universelle réprésentée en Plusieurs Centaines de Figures, contenue en lxxxviii Tableaux: avec leur Explication generale et particulière, et une Instruction exacte de la Methode pour s'en pouvoir servir." (Paris: 1682.) 4to.

14. ANONYMOUS.—"Wegweiser, Ganz neuer und accurater, Chiromantischer." (Hannover: 1707.) 8vo.

15. ANONYMOUS.—"Die Chiromantie, nach Astronomischen Lehrsätzen Lehrende, nebst der Geomantie," etc. (Frankfort: 1742.) 8vo.

16. ANONYMOUS.—"Die Chiromantie der Alten, oder die Kunst aus den Liniamenten der Hand wahrzusagen," etc. (Cologne: 1752.) 8vo.

17. ANONYMOUS.—"Schauplatz, Neueröffneter, geheimer philosophischer Wissenschaften: Chiromantia, Metoposcopia," etc. (Regensburg: 1770.) 8vo.

18. ANONYMOUS.—"The Hand Phrenologically Considered: being a Glimpse at the Relation of the Mind with the Organisation of the Body." (London: 1848.) 8vo.

19. ANONYMOUS.—"Les Petits Mystères de la Destinée. La Chiromancie, ou la Science de la Main; la Physiognomie, ou la Science du Corps de l'Homme," etc. (Paris: 1861.) 12mo.

20. ANONYMOUS.—"Dick's Mysteries of the Hand; or, Palmistry Made Easy, etc., etc., based upon the Works of Desbarrolles, D'Arpentigny, and Para d'Hermes." Translated, etc., by A. G. and N. G. (New York: 1884.) 12mo. *Vide* Nos. 24 and 54.

21. ANTIOCHUS TIBERTUS.—"Ad Illustrem Principem Octavianum Ubaldinum Merchatelli Comitem a Tyberti Epistola." (Bononiæ: 1494.)

22. ANTIOCHUS TIBERTUS.—"Antiochi Tiberti de Cheiromantià Libri III. denuo recogniti. Ejus idem argumenti de Cheiromantià," etc. (Moguntiæ: 1541.)

23. ARISTOTLE.—"Chyromantia Aristotelis cum Figuris." (Ulmæ: 1490.) 22 leaves.

24. ARPENTIGNY, *Casimir Stanislas d'.*—"La Chirognomonie; ou l'Art de reconnaître les Tendences de l'Intelligence d'après les Formes de la Main." (Paris: 1843.) 8vo.

25. ARPENTIGNY, *C. S. d'.*—"Die Chirognomie, oder Anleitung die Richtungen des Geistes aus den Formen der Hand zu erkennen." Bearbeitet von Schraishuon. (Stuttgart: 1846.) 8vo.

26. ARPENTIGNY, *C. S. d'.*—"La Science de la Main, ou l'Art," etc. (Paris: 1865.) Third Edition, 8vo.

27. BAUGHAN, *R.*—"The Handbook of Palmistry." (London: N.D.) 8vo.

28. BAUGHAN, *R.*—"Chirognomancy; or, Indications of Temperament and Aptitudes manifested by the Form and

Texture of the Thumb and Fingers." (London: 1884.) 8vo. *Vide* Nos. 24 and 54.

29. BEAMISH, *Richard.*—"The Psychonomy of the Hand; or, the Hand an Index of Mental Development, according to MM. D'Arpentigny and Desbarrolles." (London: 1865.) Second Edition, 4to.

30. BELL, *Sir Charles.*—"The Hand: its Mechanism and Vital Endowments as evincing Design and illustrating the Power, Wisdom, and Goodness of God."—*Bridgewater Treatise.* (London: 1839.) 8vo.

    Other Editions: [London: 1852; 1860; 1865; 1874, etc.]

31. BELOT, *Jean.*—"Les Œuvres de M. Jean Belot, Curé de Milmonts, Professeur aux Sciences Divines et Celestes, contenant la Chiromence, Physionomie," etc. (Rouen: 1640; Lyon: 1654) 8vo. Second Edition (Rouen: 1669.) Third Edition (Liége: 1704.)

32. BULWER, *John.*—"Chirologia; or, the Naturall Language of the Hand, composed of the Speaking Motions and Discoursing Gestures thereof," etc. (London: 1644.) 12mo.

33. CAMPBELL, *Robert Allen.*—"Philosophic Chiromancy: Mysteries of the Hand revealed and explained," etc. (St. Louis: 1879.) 12mo.

34. CARUS, *C. G.*—"Ueber Grund und Bedeutung der Verschiedenen Formen der Hand in Verschiedenen Personen." (Stuttgart: 1846.) 8vo.

35. COCLES, *B.*—*Vide* No. 1.

36. COCLES, *Bartholomæus.*—"Expositione del Tricasso sopra il Cocle." (Venice: 1525.) 8vo.

37. COCLES, *B.*—"Tricassi Cerasarensis Mantuani supra Chyromantiam Coclei Dillucidatiqnes Præclarissimæ," etc. (Venice: 1525.) 8vo. *Vide* No. 119.

38. COCLES, *B.*—"Bartolomæi Coclitis Bononiensis, Naturalis Philosophiæ ac Medicinæ Doctoris, Physionomiæ et Chiromantiæ Compendium." (Argentorati: 1533.) 8vo.

    Other Editions: (Ditto: 1534; 1554; 1555.) 12mo.

39. COCLES, *B.*—"Physionomiæ et Chyromantiæ Compendium." (Argentorati: 1536.)

40. COCLES, *B.*—"Le Compendion et Brief Enseignement de Physiognomie et Chiromacie de Berthelemy Cocles. Monstrant par le regard du Visage, signe de la Face et Lignes de la Main les Mœurs et Complexions des Gentz; selon les Figures par le Livre de painctes." (Paris: 1550.) 12mo, not paged, pp. 240. *Siy. O. v.*

41. Cocles, *B.*—"Ein Kurtzer Bericht der gantzen Phisionomey unnd (*sic*) Ciromancy gezogen aus . . . B. Cocliti von Bononia," etc. [(Strasburg:) 1537.] 8vo.

42. Cocles, *B.*—"Enseignemens de Physionomie et Chiromancie," etc. (Paris: 1638.) 8vo.

43. Cocles, *B.*—"La Physiognomie Naturelle et la Chiromance de B. Cocles." (Rouen: 1698.) 12mo.

44. Corvus, *Andreas.**—"Excellentissimi et Singularis Viri in Chiromantia exercitatissimi Magistri Andrea Corvi Mirandulensis." (Venice: 1500.) 8vo.

45. Corvus, *A.*—"L'Art de Chyromance de Maistre Andrieu Corum . . . . Translatee de Latin en Français par Jehan de Verdellay." (Paris: 1510.) 8vo.
    Reprinted *sub tit.* "Les Indiscretions de la Main." (Paris: 1878.)

46. Corvus, *A.*—"Excellentissimi A. Corvi Mirandulensis opus . . . de Chiromantiæ Facultate Destinatum." (Venice: 1513.) 8vo.

47. Corvus, *A.*—"Opera Nova de Maestro A. Corvo da Carpi, habitata a la Mirandola, trattata de la Chiromantiæ," etc. (Marzania: 1519.) 8vo.

48. Corvus, *A.*—"Excellente Chiromancie monstrant par les Lignes de la Main les Mœurs et Complexions des Gens." (Lyon: 1611.) 12mo.

49. Craig, *A. R.*—"The Book of the Hand; or, the Science of Modern Palmistry, chiefly according to the Systems of D'Arpentigny and Desbarrolles." (London: 1867.) 8vo. *Vide* Nos. 24 and 54.

50. Craig, *A. R.*—"Your Luck's in your Hand; or, the Science of Modern Palmistry." (London and New York: N.D. [1884.]) Third Edition, 8vo.

51. Cringle, *Tom* [pseudonym of *William* Walker].—"The Hand and Physiognomy of the Human Form. (Melbourne: 1868.) 8vo.

52. Cureau de la Chambre, *M.*—"Discours sur les Principes de la Chiromancie." (Paris: 1653.) 8vo.

53. Cureau de la Chambre, *Martin.*—"A Discourse on the Principles of Chiromancy." Englished by a Person of Quality. (London: 1658.) 8vo.

54. Desbarrolles, *Adrien.*—"Chiromancie Nouvelle. Les

---

* Some doubt exists whether or no "Andreas Corvus" was a pseudonym of Bartholomæus Cocles.

Mystères de la Main révélés et expliqués."   (Paris: 1859.)
First Edition, 8vo.
  Fifteenth Edition (Paris: Dentu: N.D.)
55. DESBARROLLES, *A.*—"Almanach de la Main pour 1869, ou
la Divination raisonnée et mise à la portée de tous."
(Paris: 1868.) 16mo.
56. DESBARROLLES, *A.*—"Mystères de la Main. Révélations
Complètes, suite et fin." (Paris: N.D, [1879.]) Large 8vo.
57. DU MOULIN, *Antoine.*—"Chiromance et Physiognomie par
le regard des Membres de l'Homme." (1556; 1638; 1662.)
58. ENGEL.—"Die Entwickelung der Menschlichen Hand."
"Berichte der Kaiserlichen Akademie der Wissenschaften
zu Wien, Mathemat.-naturwissenschaftl. Klasse, B. xx.,
p. 261.
59. FABRICIUS, *Johann Albert.*—"Gedanken von der Erkennt-
niss der Gemüther aus den Temperamenten der Chiromantie und Physiognomie." (Jena: 1735.)
  FINELLA.—*Vide* PHINELLA.
60. FLISCO, *Count M. de.*—"De Fato, per Physiognomiam et
Chyromantiam," etc. (N.D. [1666.]) 4to.
61. FRITH, *Henry,* and HERON-ALLEN, *Edward.*—"Chiromancy; or, the Science of Palmistry," etc. (London:
1883.)
62. GOCLENIO, *Rudolphus* [junior].—"Aphorismorum Chiro-
manticorum tractatus compendiosus ex ipsius artis funda-
mentis desumptus," etc. (Sicho: 1597.) 8vo.
63. GOCLENIO, *R.*—"Uranoscopiæ, Chiroscopiæ, Metoposcopiæ,
etc., contemplatio," etc. Editio nova. (Frankfort: 1608.)
16mo.
64. GOCLENIO, *R.*—"Uranoscopiorum, Cheiroscopiorum, et
Metoposcopiorum, hoc est tractatus," etc. (Frankfort:
1618.) 8vo.
65. GOCLENIO, *R.*—"Physiognomica et Chiromantica Specialia
nunc primum in lucem emissa. Accesserunt . . . obser-
vationes chiromanticæ cum speciali judicio." (Marpurgi
Cattorum: 1621.) 8vo. (Halle: 1652.) 8vo.
66. GOCLENIO, *R.*—"Trattato di Chiromantia." (Amsterdam:
1641.) 8vo.
67. GOCLENIO, *R.*—"Physiognomica et Chyromantica Specialia
ante annos in lucem emissa. Nunc denuo recognita."
(Hamburg: 1661.) 8vo.
68. GOCLENIO, *R.*—" . . . besondere Physiognomische und
Chiromantische Anmerkungen," etc. (Hamburg: 1692.)
8vo.

69. HARTLIEB, *Johann.*—"Die Kunst Ciromantia, 1448." (Augsbourg: 1745.) Fol.

70. HASIUS, *Joannes.*—"Prefatio Laudatoria in Artem Chiromanticam: in laudem Joannis Hasii Memmingensis artis Jurium et Medicinarum Doctoris, Chyromantiæ Principis." (1519.) Sm. 4to, pp. 60.

71. HEBRA, *H.*—"Untersuchung über den Nagel." (Vienna: 1880.) 8vo.

72. HERON-ALLEN, *Edward.*—"Codex Chiromantiæ."—*Odd Volumes Opusculum, No. VII.* (London: 1883.) 12mo.

73. HERON-ALLEN, *Edward.*—"A Manual of Cheirosophy: being a Complete Practical Handbook to the Twin Sciences of Cheirognomy and Cheiromancy," etc. (London: 1885.) Sq. 8vo.

74. HERON-ALLEN, *Edward.*—"The Science of the Hand; or, the Art of Recognising the Tendencies of the Human Mind by the Observation of the Formations of the Hand." Translated from the French of . . D'Arpentigny. With an Introduction, Appendices, and a Commentary on the Text. (London: 1886.) Sq. 8vo.

75. HÖPING, *Johann Abraham Adolph.*—"Chiromantia Harmonica, das ist Übereinstimmung der Chiromantia," etc. (Jena: 1681.) 8vo.

76. HÖPING, *J. A. J.*—"Institutiones Chiromanticæ." Mit Fleiss verfertiget durch J. A. J. Höping. 2 vols., 12mo.

77. HUMPHREY, *George M.*—"On the Human Foot and Human Hand." (Cambridge: 1861.) 8vo.

78. INDAGINE, *Joannes ab.*—"Introductiones Apotelesmaticæ Elegantes in Chyromantiam, Physiognomiam, Astrologiam Naturalem, Complexiones Hominum, Naturas Planetarum," etc., etc. [(Strasburg:) 1522.] Fol. (in two parts).
Other Editions: (Frankfort: [1522]) 12mo; (Argentorati: 1531 and 1541) fol.; (Paris: 1543 and 1547) 8vo; (Ursellis: 1603) 8vo; (Augusta Trebocorum: 1663) 8vo.

79. INDAGINE, *J.*—"Die Kunst der Chiromantzey usz Besehung der Hend, Physiognomey usz Anblick des Menschens," etc. (Strasburg: 1523.) Fol.

80. IDAGINE, *J.*—"Chiromantia, Physiognomia, Periaxiomata de Faciebus Signorum," etc. (Argentorati: 1534.) Fol.

81. INDAGINE, *J.*—"Feltbuch der Wund Artzney sumpt vilen Instrumenten der Chirurgen uss dem Abucasi contrafayt. Chiromantia J. Indagine," etc. (Strasburg: 1540.) Fol.

82. INDAGINE, *J.*—"Chiromence et Physiognomie par le regard

des Membres de l'Homme." Le tout mis en François par
A. du Moulin. (Lion: 1556.) 8vo.
Other Editions: (Rouen: 1638) 12mo; (Paris: 1662)
12mo.

83. INDAGINE, *J.*—"The Book of Palmistry and Physiognomy:
being Brief Introductions both Natural, Pleasaunt, and
Delectable unto the Art of Chiromancy, or Manual Divina-
tion and Physiognomy," etc., etc. Written in Latine by
John Indagine, *Priest*, and translated into English by
Fabian Withers. (London: 1651.) 8vo.
Sixth Edition, 1666; seventh Edition, 1676.

84. INGEBER, *Johann.*—"Chiromantia, Metoposcopia, et Physiognomia Practica; oder Kurtze Anweisung, wie man auss
denen vier Hauptlinien in der Hand . . . urtheilen
kann," etc. (Frankfort A/M : 1724.) 8vo.

85. JOB, *J. G.*—"Anleitung zu den Curieusen Wissenschaften,
nemlich der Physiognomia, Chiromantia," etc. (Franckfort A/M: 1721.)

86. KOLLMANN, *Arthur.*—"Der Tast Apparat der Hand der
Menschlichen Rassen und der Affen in seiner Entwickelung und Gliederung." (Hamburg u. Leipzig: 1883.) 8vo.

87. LUTZ, *L. H.*—"Cheirosophia concentrata; d.i., Eine Kurtze
Unterweisung vermittelst deren jeden Menschens ganzten
Lebens Beschaffenheit in seinen Händen vor Augen könne
gestellet werden." (Nürnberg: 1672.)

88. MARTIN, *C.*—"Methode de Chirogymaste, ou Gymnase des
Doigts." (Paris: 1843.) 8vo.

89. MAYER, *Philip.*—"La Chiromancie Medicinale. Accom-
pagnée d'un traité de la Physionomie, et d'un autre des
Marques qui paroissent sur les Ongles des Doigts."
Traduit de l'Allemand par Ph. H. Trenchoes de
Wezhausen. (La Haye: 1665.)

90. MAYER, *Ph.*—"Chiromantia et Physiognomia Medica, mit
einen Anhange von den Zeichen auf die Nageln der
inger," etc. (Dresden: 1712.)

91. MOREAU, *Adèle.*—"L'Avenir Dévoilé. Chiromancie Nouvelle. Etude des Deux Mains." (Paris: 1869.) 8vo.

92. NAURATH, *Lud. Ern. de.*—"De Manuum Morphologiâ et
Physiologiâ." (Berolini: 1833.) 8vo.

93. PERUCHIO, *Le Sieur de.*—"La Chiromance, la Physionomie
et le Geomance," etc. (Paris: 1656.) Sm. 4to. Second
Edition (Paris: 1663.)

94. PEUSCHEL, *C. A.*—"Abhandlung der Physionomie, Metoposcopie und Chiromantie." (Leipzig: 1769.) 8vo.

95. PHILOSOPHI [*Begin*] "Ex Divina Philosophorum Academia nature vires ad extra Chyromantico Diligentissime Collectum." [*End*] "Ex Divina Philosophorum Academia Collecta Chyromantica scientia naturalis ad Dei laudem finit." ["Per Magistrum Erhardum Ratodolt de Augusta Venetiis."] [1480?] 4to, 25 leaves, without title, pagination, or catchwords.

Other Editions : [1484] 26 leaves ; [1490?] 32 leaves ; - [Venice : 1493] 24 leaves ; [Oppenheim : 1499] 32 leaves.

96. Translation : "Incomentia l'Arte Divina de la Chyromantia recolta da la Schola de Philosophi." (Venice : [1480?]) 28 leaves.

97. PHINELLA or FINELLA, Philippo.—"De Quatuor Signis quæ apparent in Unguibus Manuum." Auctore P. P. (Naples: 1649.) 12mo.

98. POMPEIUS, Nicolaus.—" Figuræ Chiromanticæ ad systema Nicolai Pompeii." (Hamburg : 1682.) 8vo.

99. POMPEIUS, N. —"Præcepta Chiromantica . . . prælecta olim ab ipso jam vero recognita." (Hamburg : 1682.) 8vo.

100. PRÆTORIUS, Johann.—"Cheiroscopia et Metoposcopia." (Jena : 1659.) 4to.

101. PRÆTORIUS, J.—"Ludicrum Chiromanticum Prætorii, seu Thesaurus Chiromantiæ," etc. (Jena : 1661.) Sm. 4to, pp. 340.

102. PRÆTORIUS, J.—"Philologemata Abstrusa de Pollice ; in quibus Singularia Animadversa vom Diebes-Daume, et Manu," etc. (Leipzig : 1677.) 4to.

103. PRÆTORIUS, J.—"Collegium Curiosum . . . oder ein sehr Nützliches Werck darinnen curieus . . . abgehandelt wird, was zur Physiognomie, Chiromantie, etc., gehöret." (Frankfort A/M : 1704.) 8vo.

104. REQUENO, Vincenzo.—"Scoperta della Chironomia, dell' Arte di Gestire con le mani." (Parma : 1797.) 8vo.

105. RONPHILE [pseudonym of Rampalle].—"La Chyromantie Naturelle de Ronphile." (Lyon : 1653) 8vo ; (Paris : 1671) 8vo.

106. RONPHILE.—"Die in der Natur best-gegründetc. . . . Chyromantie oder Hand-Wahrsagung." (Nürnberg : 1695.) 8vo.

107. ROTHMANN, Joannes.—"Chiromantiæ Theorica Practica. Concordantia Genethliaca Vestustis Novitate addita." Autore Joanne Rothmanno Med. et Philos. (Erphordiæ : 1595.) Sm. 4to.

108. ROTHMANN, *J.*—Κειρομαντια, or the Art of Divining by the Lines and Signatures engraven in the Hand of Man by the Hand of Nature," etc. Written originally in Latine by Io. Rothmanne, Div. Phisique, and now faithfully englished by Geo. Wharton, Esq. (London: 1652.) 8vo. *Vide* No. 125.

109. SANDERS, *Richard.*—"Physiognomie and Chiromancie, Metoposcopie, etc., handled; with their natural predictive significations," etc. (London: 1653.) Fol.

110. SANDERS, *R.*—"Palmistry, the Secrets thereof disclosed. . . . With some choice Observations of Physiognomy and the Moles of the Body," etc. (London: 1664.) 12mo.

111. SCHALITZ, *Christian.*—"Die von Aberglauben, Vanitäten und Teuscherei gereinigte Chiromantia und Physiognomia." (Leipzic: 1703.) 2nd edition. (Franckfort: 1729.) 8vo.

112. SCHEELER, *Karl von.*—"Abimelech, der wunderbahre Prophet, oder die Chiromantie und Physiogmonik womit man sich und anderen aus den Linicamenten wahrzusagen im Stande ist." (Reutler : N.D.)

113. SCHOTT, *Gaspar.*—"Thaumaturgus Physicus, sive Magiæ Universalis Naturæ et Artis Pars IV. et ultima." (Herbipoli: 1659.) 4to.

114. SOHN, *Fr.*—"Kunst aus der Handhöhle, den Fingern und den Nägeln wahrzusagen, oder die Chiromantie der Alten." (Berlin : 1856.) Second edition.

115. SPADONI, *N.*—"Studio di Curiosità, nel quale tratta die Fisionomia, Chiromantia, e Metoposcopia." (Venice: 1675.) 16mo.

116. SPADONI, *N.*—"Studium Curiosum, darinnen von der Physiognomia, Chiromantia, und Metoposcopia . . . gehandelt wird." *Vide* No. 115.

117. T——, *J. G. D.*—"Hochstfürtreflichstes Chiromantisch-und-Physiognomisches Klee-Blat, bestehend aus drey herrlichen Tractaten," etc. (Nürnberg : 1695.) 8vo.

118. TAISNIER, *Joannes.*—"Opus Mathematicum Octo Libros complectens innumeris propemodum figuris idealibus manuum et physiognomiæ aliisque adoratum quorum sex priores libri absolutissimæ Cheiromantiæ Theoricam Praxim, Doctrinam, Artem, et experientiam verissimam continent, etc., etc. Authore Joanne Taisnierio Hannonio Mathematico expertissimo." (Coloniæ Agrippinæ : 1562.) Large 8vo.

119. TRICASSO, *J.*—"Tricasso da Cesari, Mant. : Epitoma

chiromantico allo illustre e magnifico Signor Conte Joan Baptista di affaitati Cremonese," etc. (N.D.) 12mo.

120. TRICASSO, *J.*—"Chyromantia de Tricasso da Cesari Mantuano al Magnifico et Veneto Patritio Dominico di Aloisio Georgio novamente revista e con somma diligentia corretta e stampata." (1534.) Sm. 8vo.

121. TRICASSO, *J.*—"Epitoma Chyromantico di Patritio Tricasso da Cerasari Mantovano." (Venice: 1538.) 8vo.

122. TRICASSO, *J.*—"Tricassi Cerasariensis Mantuani enarratio pulcherrima principiorum Chyromantiæ ex qua facillime patere possunt omnes significationes quorumcunq: signo-rum Chyromanticorum. Ejusdem Tricassi Mantuani opus Chyromanticum absolutissimum nunc primum in lucem editum. Item Chyromantiæ incerti authoris Opera Bal-duini Ronflei Gandavensis in lucem edita cum ejusdem in Chyromanticen brevi Isagoge." (Noribergæ: 1560) 12mo sm. 4to, not paged, pp. 312, *sig.* M. M. ii.

123. TRICASSO, *J.*—"La Chiromence de Patrice Tricasse des Ceresars Mantouan de la derniere reveue et correction de l'Autheur et nagueres fidelement traduicte de l'Italien en langage Français." (Paris, 1552 and 1561.) 12mo.

Vide Nos. 36, 37.

VERDELLAY, *Jehan de* [*pseudonym* of *Andreas* CORVUS].

124. WARREN, *Claude.*—"The Life-size Outlines of the Hands of Twenty-two Celebrated Persons." (London: 1882.) Fol.

125. WHARTON, *Sir George.*—"The Works of that most excellent Philosopher . . . including a translation of Rothmann's 'Chiromancy,'" etc. (London: 1683.) 8vo.

Vide No. 108.

Index.

Lacenaire.

Pour contraste, la main coupée
De Lacenaire l'assassin,
Dans des baumes puissants trempée
Posait auprès, sur un coussin.

Momifiée et toute jaune,
Comme la main d'un Pharaon,
Elle allonge ses doigts de faune
Crispés par la tentation,

Tous les vices avec leurs griffes
Ont, dans les plis de cette peau,
Tracé d'affreux hiéroglyphes,
Lus couramment par le bourreau.

En même temps molle et féroce,
Sa forme a pour l'observateur
Je ne sais quelle grâce atroce,—
La grâce du gladiateur.

Criminelle aristocratie !
Par la varlope ou le marteau
Sa pulpe n'est pas endurcie,
Car son outil fut un couteau !

Théophile Gautier

INDEX.

اكـنون كـه دم ز عـمـر°حـروم نشد

كـم بـود ز اسـرار كـه مفـهـوم نشد

چون نيك همي بنگرم از روي خرد

عمّرم بگذشت و هيچ معلوم نشد